Y0-DWK-658

THE FUTURE OF SMALL TELESCOPES IN THE NEW MILLENNIUM

ASTROPHYSICS AND SPACE SCIENCE LIBRARY

VOLUME 289

THE FUTURE OF SMALL TELESCOPES IN THE NEW MILLENNIUM

Volume III – Science in the Shadows of Giants

Edited by

TERRY D. OSWALT
Florida Institute of Technology,
Melbourne, Florida, U.S.A.

KLUWER ACADEMIC PUBLISHERS
DORDRECHT / BOSTON / LONDON

A C.I.P. Catalogue record for this book is available from the Library of Congress.

ISBN 1-4020-0950-X (HB)
ISBN 1-4020-0951-8 (Indivisible set)

Published by Kluwer Academic Publishers,
P.O. Box 17, 3300 AA Dordrecht, The Netherlands.

Sold and distributed in North, Central and South America
by Kluwer Academic Publishers,
101 Philip Drive, Norwell, MA 02061, U.S.A.

In all other countries, sold and distributed
by Kluwer Academic Publishers,
P.O. Box 322, 3300 AH Dordrecht, The Netherlands.

Printed on acid-free paper

Printed in the Netherlands.

Contents

Preface

As I first scanned the table of contents of Volume III, I felt a certain sense of disappointment. Where were the chapters on my favorite small telescope research topics and triumphs?

A moment's reflection showed that there were several answers to that question: answers that go to the heart of the purpose of these three volumes. First, this third volume would have had to be many times its current size. This impractical size would only have been dictated by what I personally had been expecting to see. Individual readers will have their own pet topics that they would have expected which, in most cases, would have dictated a similar size but with different chapters.

Why was my expectation so huge? Because it is difficult to think of a topic in modern astronomy to which small telescopes have not made a significant contribution. This contribution, as shown by several of the authors in Volume I, has been disproportionately large compared to the cost of small telescopes and, especially, to the budget provided them in recommendations of the 2001 Astronomy and Astrophysics Survey Committee. So, to spare most of the potential readers of this volume that initial sense of disappointment that I felt would have required one, and most probably, more than one, chapter on every subtopic of astronomy. Considering the many new topics that have been added to astronomy in the last 40 years, satisfying this requirement would probably take a series two to three times the size of Gerard Kuiper's 13 volume Stars and Stellar Systems and The Solar System!

Second, many of the chapters were not of the type that I expected. While there are several chapters that do address past achievements of small telescopes, with discussions of the need to continue or expand such research in the future, most of the chapters are addressing new areas of research for small telescopes. This is a higher standard which, regrettably, we have come not to expect from many books composed of quickly written snapshots of current research between refereed papers by the authors. It is not that these chapters have ignored the rich history of valuable contributions of small telescopes to their topics but that each lays out specific future roles with excellent scientific goals. These are goals that are achievable with modern small telescopes and scientifically exciting enough to be competitive with research suggested for any size telescope.

Third, the volume covers a much wider range of topics than I expected. The selection, is excellent: from our solar system, through nearby stars, extrasolar planets, Galactic clusters, external galaxies, to cosmological-scale research of gravitational lensing, gamma ray bursts and the importance of small telescopes to cosmological research. Not surprisingly, all of the topics I had lamented missing in the first glance at the contents have scales within this range.

Where am I to go for my missing papers on research done with small telescopes? The answer is found in the first volume of this series. As Helmut Abt points out in his chapter in Volume I, over three-quarters of papers published in major refereed journals based on ground-based telescope observations are based on work done at what are now considered to be small telescopes. Unfortunately, these journal articles, focused on reporting their research results, usually will not provide us the focus on the use of small telescopes in the new millennium so ably done in the chapters of this volume.

Wm. Bruce Weaver
Monterey Institute for Research in Astronomy
Marina, California USA
November 2002

Chapter 1

Reflections on Past and Future Photometric Calibration Situations and the Role of Small Telescopes

Arlo U. Landolt
Louisiana State University
Baton Rouge, Louisiana USA

Abstract: The original implementation of the most often used of the various photometric systems will be reviewed. The role of small telescopes, herein defined as those whose diameters are less than 2.5-m, in current and future efforts at enhancing photometric standards will be recounted, and future needs driven both by science and technology will be explored. The community must focus on and invest in these cost-effective activities.

Keywords: photometric calibration, standard stars

1. INTRODUCTION

The entire idea of photometric calibration is to enable astronomers to inter-compare their results and to allow them to compare their data with models. An attempt will be made to re-visit the past, and to forecast future demands that will be placed upon standardization. The theme of the discussion will be that it is a vital function that has been done successfully by small telescopes in the past, and which can be done via their use in the future, at a considerable cost savings to the astronomical community.

As the community lays out its research needs, it must focus and invest wisely with a breadth of supporting instrumentation that will ensure the fulfillment of data acquisition needs from all aspects. Hopefully, this chapter will convince readers of the absolute need for thorough calibration efforts, the basis of which can most efficiently and economically be supplied by small telescopes. Abt (1980, 2003) has assembled data and has made

T.D. Oswalt (ed.), The Future of Small Telescopes in the New Millennium, Vol. III, 1-9.

arguments concerning the cost effectiveness of small telescopes. There is no evidence that such scientific cost effectiveness has diminished.

Prior to encountering more completely referenced examples a little later in this paper, one should recall that virtually all optical and infrared photometric standard star systems were developed with telescopes of 0.4 to 1.5-m in diameter (Johnson and Morgan 1953; Johnson 1963; Crawford and Mander 1966; McClure and van den Bergh 1968; Crawford and Barnes 1970; Landolt 1973, 1983, 1992; Golay 1974; Canterna 1976; Cousins 1976; McClure 1976; Straizys 1977; Graham 1982; and Geisler (1990, 1996). A 0.9-m telescope was used to obtain the rotation curves of galaxies, which led to the requirement of the existence of missing mass. The Lyman-α forest, as well as the first gravitational lens (Walsh, Carswell and Weymann 1979), was discovered with a 2.1-m telescope. The first extra-terrestrial planets were found with 1.0 and 2.0-m class telescopes (Mayor and Queloz 1995). Schmidt telescopes, always 'small' in the current lexicon, have been used in a variety of surveys which inevitably uncover exotic celestial objects. The first variable white dwarf star was discovered at a 0.9-m telescope at the Kitt Peak National Observatory (KPNO; Landolt 1968), and the first pulsar was discovered optically with the University of Arizona's Steward Observatory 0.9-m telescope (Cocke, Disney and Taylor 1969). Uranus' ring system was found using a small telescope (Elliott, et al. 1977; Millis, et al 1977). In fact, in the sense of advancing science of its time, Galileo's discoveries of the four brightest moons of Jupiter and the spots on the sun were accomplished with a small telescope.

2. THE MAKING OF PHOTOMETRIC SYSTEMS OVER THE YEARS

Astronomical history is appealing because it can provide the thread of reasoning underlying the need for, use of, and development of observational techniques. Insofar as astronomical photometry is concerned, an excellent summary as seen at the middle of the past century has been provided by Weaver (1946a, b, c, d, e, f). A more recent fine review of two centuries of astronomical photometry has been written by Hearnshaw (1996). More specialized references may be found in review articles by Weaver (1962) and Whitford (1962). Additional summaries, including references, may be found in the volume edited by Philip (1979) and in an article by Drilling and Landolt (1999).

The foundations of astronomical photometry, and the standard stars which define them, were essentially all derived from data obtained with telescopes of modest aperture, i.e., 0.4 to 1.5-m diameter. A partial listing of the most used photometric systems includes, in no particular order: UBV [Johnson and Morgan (1953), Johnson (1963), Landolt (1973)]; UBV(RI)c [Cousins (1976), Graham (1982), Landolt (1983, 1992)]; uvby H-beta [Stromgren (1966), Crawford and Mander (1966), Crawford and Barnes (1970)]; Geneva system [Golay (1974)]; DDO [McClure and van den Bergh (1968), McClure (1976)]; Washington [Canterna (1976), Geisler (1990, 1996)]; and, Vilnius [Straizys (1977, 1992)]. Comments concerning many of these photometric systems have been made in Philip (1979). All these series of photometric standard stars were obtained with small telescopes of their time. Relatively few of the data acquired through use of the newest instrumentation and largest telescopes at any given moment could have been interpreted with any kind of completeness without these photometric standards.

Photometry is a technique necessary in a myriad of ways for the interpretation, understanding and modeling of various celestial objects or events. There occur projected against the celestial sphere a wide range of phenomena for which observers and experimenters need intensity and color information. Therefore, there is a long-term need for accurate photometric standard stars, those with known intensities and colors at a variety of wavelengths. Use of such standard stars permits the combination and inter-comparison of the brightness and color measurements of different celestial phenomena made by different investigators. For example, the interplay of theory and observation requires accurately calibrated magnitudes and color indices of celestial objects to further the understanding of stellar evolution, ages and distances of star clusters, studies of variable stars, and to define the distance scale through our Milky Way Galaxy, the local group of galaxies, and onward to the edge of the Universe. In fact, stars throughout the Hertzsprung-Russell (H-R) diagram need the most accurate possible magnitudes and color indices to effectively solve just mentioned research investigations. On the one hand, it well may be that astronomers will be unable to, or even need to, interpret their data to the one-percent level. However, it certainly should not, and does not ever need to be the case, given appropriate financial support and telescope time, that photometric calibration is the dominant source of error. The extension of photometric sequences to fainter magnitude limits means that the limits of interpretation will be scientific, and not those of calibration.

Why is calibration work important to astronomy? The examples that follow are illustrative, near the author's understanding, and in no way cite all the beautiful investigations undertaken across the spectrum. For instance, one needs accurate calibrations, if colors and color gradients from surface photometry of galaxies is the goal. The best possible calibrations are important in the theory observation connection for stellar evolution, ages and distances of star clusters, for variable star investigations, and so on. They also are indispensable in establishing the distance scale through our Milky Way Galaxy, out through the Local Group, and to the edge of the Universe. Hence, the following examples, such as Andersen, Clausen and Gimenez (1993), and their work involving the fundamental dimensions of individual stars via their studies of eclipsing binary systems, are cited to indicate the diversity and breadth of the astronomical and astrophysical investigations which have made use of standard stars, the point being that standardization is necessary for intensity and color studies of objects as close to Earth as our solar system and as distant as the furthermost reaches of the Universe.

Murayama et al. (2000) reported the discovery of, and calibrated the characteristics of a very low surface brightness galaxy. Erben et al. (2000) discussed an apparent mass-detection of a matter concentration projected near the cluster of galaxies designated Abell 1942. Harris, Pritchet, and McClure (1995) studied globular cluster systems in cD galaxies within rich Abell clusters. Neely, Sarajedini, and Martins (2000) used photometry to describe the characteristics of the galactic globular cluster NGC 6144. Carretta et al. (2000) reviewed results on distances and absolute ages of galactic globular clusters. Aguerri et al. (2000) obtained accurate UBVRI surface photometry to study different morphological structures present in disk galaxies. Saviane, Held and Bertelli (2000) reported on the stellar populations of the Fornax dwarf spheroidal galaxy. Kuhn et al. (1995) investigated the spectral energy distribution of the $z = 3$ quasar HS 1946+7658. Shanley et al. (1995) imaged the luminous "supersoft" x-ray source nova GQ Muscae 1983. Callanan et al. (1995) did a detailed study of the millisecond binary pulsar PSR 1957+20. Postman and Lauer (1995) investigated the use of the brightest cluster galaxies as standard candles. Cappellaro et al. (1995) obtained extensive photometric observations of the type II supernova SN 1990K. Germany et al. (2000) discussed data for SN 1997cy, an object with extraordinary observational properties, and whose location corresponds to a short duration gamma ray burst. Platais et al. (2000) obtained the absolute proper motion and photometry of Nova Velorum 1999 = V382 Vel, as part of an effort to get magnitudes and proper motions for novae which in turn will lead, via the secular parallax method, to absolute magnitudes.

Cieslinski et al. (2000) found the possible rapid irregular variable star V729 Sagittarii to be a periodic cataclysmic variable. Dolphin (2000) studied the star formation rate histories of two fields in the Large Magellanic Cloud. Delgado and Alfaro (2000) searched for pre-main sequence stars in the young galactic cluster NGC 6910. Bassino, Waldhausen and Martinez (2000) used CCD photometry to describe the remains of an old open cluster in the region of NGC 6994. Pozzo et al. (2000) reported on the serendipitous discovery of a group of low-mass pre-main-sequence stars in the region of Gamma Velorum. Quillen et al. (1995) made an estimate of the gas inflow rate along the bar in NGC 7479. Weaver, A'Hearn et al. (1995) imaged comet Shoemaker-Levy-9 with the Hubble Space Telescope. Benetti et al. (1995) confirmed a microlensing event toward the galactic bulge. Hainaut et al. (1995) made post-perihelion observations of Comet P/Halley. Freedman et al. (1994) discovered Cepheids and derived a new distance to M81 as part of a Hubble Space Telescope key project. A bit farther back in time, Lasker et al. (1988) standardized the guide stars to be used for the Hubble Space Telescope. Grauer (1984) used a two-star photometer in the discovery of rapid optical variations in the quasi-stellar radio source 4C 29.45. The Amateur Sky Survey (TASS) [Richmond et al. 2000] operates a telescope in a drift scan mode that monitors a band around the sky centered on the celestial equator. Their data go down to about 13th magnitude. Pinfield et al. (2000) searched a six-square degree field centered on the Pleiades, for low-mass stars and brown dwarfs. French (1987) calibrated CCD observations of asteroids and their light curves. Pilachowski et al. (1989) monitored the sky brightness at Kitt Peak National Observatory.

High quality photometric data are a necessity for theoreticians, too, for such data enhance comparisons between theory and observation by furnishing a diagnostic of the accuracy of, for example, stellar evolution models (Green, Demarque and King 1987). Van den Berg and Bell (1985) calculated stellar evolutionary sequences and isochrones which may be used in conjunction with photometric studies of globular star clusters. Green (1988) discussed an empirical calibration of UBVRI photometry with such models. Krawchuk, Dawson and De Robertis (2000) used spectral energy distributions from model atmospheres to synthesize colors for low-mass stars, and compared those colors to observed colors to elucidate fundamental parameters of low-mass stars.

3. FUTURE NEEDS

One cannot say too often that photometry done at small telescopes complements that done at larger telescopes. More important, photometry as practiced at small telescopes is essential to calibrate, tie down, the data obtained at the large telescopes. One only need peruse the citation indices for the publications which define the different photometric systems to affirm this statement. Robotic telescopes have a complementary role to play in certain kinds of observations (See chapters in Filippenko 1992; Henry and

Eaton 1995), situations where one does not have to pay quite so close attention to local environmental conditions while the data are being taken.

The attendees at the Eleventh Cambridge Workshop on Cool Stars, Stellar Systems and the Sun (Pasquini 2001) strongly felt the need for the continued availability of small telescopes. The gathering made several points, and this author paraphrases: Small telescopes can do both preparatory and follow-up observations for large telescope projects like target identification, astrometry, photometry and low-resolution spectroscopy. The point was made that small telescopes should be used whenever large telescopes absolutely were not required. This latter point confronts the high operational costs associated with the large telescopes. They, too, emphasized the need for libraries of standard stars, including photometric, spectrophotometric and spectroscopic standards. Finally, the assembled astronomers noted that the smaller telescopes again had strong advantages in availability, unless of course they were closed, and in support costs, when it came to projects such as planetary searches, variability studies, long-term series observations ranging from minutes to years, and so on.

Small telescopes historically readily have been available to pursue long-term dedicated projects, i.e., at least prior to the bloodletting of the past few years. Small telescopes supply the only efficient approach to long-term fundamental programs. The large installations can ferret out new kinds of objects, but sufficient observing time never is available to do the follow-up. Certainly long-term time-series projects almost always are out of the question at the big telescopes.

Small telescopes also have a vital role to play in support of space missions. Again, it is the situation where discoveries are made in space, but the necessary long follow-up observations must be, and can be, made from the surface of the Earth. It is a matter of *economics*.

The continued availability of a suite of small telescopes promises complete studies of investigations wherein one needs a long time series of data. In the situation where standard star sequences are being established, many measures are needed over a multi-year time span to ensure constancy of the standard stars. And, at every wavelength, the improvement and enhancement of the standard star production process is never ending since telescopes continue to increase in size, and equipment continues to become ever more sensitive. As telescopes grow in capability, new kinds of objects are discovered, thereby demanding additional or new calibrators.

For years, the cry was that photometric sequences had to be done in a compact area of the sky since the CCDs were small. Now exactly the opposite problem already has confronted the community. Perhaps the most critical need in the near future concerns the many wide-field surveys of the sky now in the planning stages. All will need new standard stars, the establishment of which will take years, initially to identify the candidate objects to be made into standard stars, and then the years necessary to follow observationally the proposed new standards to ensure that they are constant in every way, and hence worthy of being called photometric (or other) standards.

The situation only now is being realized, much later in the chronology of availability of mosiaced CCDs, then might have happened. There are many fields of standard stars, but they cover only small regions on the sky, and in most photometric systems, too bright to be of much use with modern instrumentation. Currently, the observer has to image the standard star field once with each CCD chip of the mosaic, and then image the next CCD chip, and the next, and so on. When the standard star field has been imaged by each chip of the mosaic through the first filter, one then turns attention to the second filter. Since the readout time is on the order of two or so minutes for each chip, and perhaps there are eight chips, and two or more filters, the time needed to calibrate the mosaic is considerable. Extinction measures also need to be made. One needs standard star sequences of a range in color which cover a sufficiently large area on the sky so that one could image all chips of the mosaic with one telescope setting. Such sequences do not exist. Such photometric sequences can be defined. However, the effort will be a lengthy one because the candidate stars must be identified, and then measured over a long enough time period to ensure constancy in intensity. The project will be even more difficult because the small telescopes and associated equipment at acceptable sites essentially have disappeared from the observational scene.

A similar situation pertains for small and intermediate sized telescopes and their equipment combinations. Here improved technology will require that current standard stars be re-observed to an ever-higher precision. The product will be a cornucopia of surprises among the 250,000 brightest stars, most of which can no longer be observed at national observatories since it has become the style to close small telescopes.

4. ACKNOWLEDGEMENTS

The author acknowledges with thanks comments from G. C. Clayton, J. A. Graham, D. James, and B. B. Snavely on an initial draft of this chapter.

5. REFERENCES

Abt, H. A. 1980, PASP, 92, 249.
Abt, H. A. 2003, volume 1 of this Kluwer series.
Aguerri, J.A.L., et al. 2000, AJ, 119, 1638.
Andersen, J., Clausen, J. V., and Gimenez, A. 1993, A&A, 277, 439.
Bassino, L.P., Waldhausen, S., and Martinez, R.E. 2000, A&A, 355, 138.
Benetti, S., Pasquini, L., and West, R.M. 1995, A&A, 294, L37.
Callanan, P.J., van Paradijs,J., and Rengelink, R. 1995, ApJ 439, 928.
Canterna, R. 1976, AJ, 81, 228. [Washington system].
Cappellaro, E., et al. 1995 A&A, 293, 723.
Carretta, E., et al. 2000, ApJ, 533, 215.
Cieslinski, D. et al. 2000, PASP, 112, 349.
Cocke, W.J., Disney, M.J., and Taylor, D.J. 1969, Nature, 221, 525.
Cousins, A. W. J. 1976, Mem. R. Astron. Soc. 81, 25. [modified and added R,I].
Crawford, D. L. and Mander, J. V. 1966, AJ, 71, 114 [H-beta].
Crawford, D. L. and Barnes, J. V. 1970, AJ, 75, 978 [uvby].
Delgado, A.J., and Alfaro, E.J. 2000, AJ, 119, 1848.
Dolphin, A.E. 2000, MNRAS, 313, 281.
Drilling, J. S. and Landolt, A. U. 1999 in Allen'a Astrophysical Quantities, 4th Ed., edited by A. N. Cox (New York: Springer-Verlag), p. 386.
Elliott, J. L., Dunham, T., and Mink, D. 1977, Nature, 267, 328.
Erben, Th., et al. 2000, A&A, 355, 23.
Filippenko, A. V. 1992, Editor of ASP Conf. Series, Vol. 34..
Freedman, W. L. [plus 14 others] 1994, ApJ, 427, 628.
French, L. M. 1987, Icarus 72, 329.
Geisler, D. 1990, PASP, 102, 344. [Washington system].
Geisler, D. 1996, AJ, 111, 480 [Washington system].
Germany, L.M. et al., 2000, ApJ, 533, 320.
Golay, M. 1974, Introduction to Astronomical Photometry (Dordecht: D. Reidel Pub. Co.) [Geneva system], p. 141.
Graham, J. A. 1982, PASP, 94, 244.
Grauer, A. D. 1984, ApJ. 277, 77.
Green, E.M. 1988, in Calibration of Stellar Ages, edited by A.G.D. Philip (Davis, Schenectady), p. 81.
Green, E.M, Demarque, P., and King, C.R. 1987, Revised Yale Isochrones and Luminosity Functions (Yale University Observatory, New Haven).
Hainaut, O., et al. 1995, A&A, 293, 941.
Harris, W.E., Pritchet, C.J., and McClure, R.D. 1995, ApJ, 441, 120.
Hearnshaw, J. B. 1996, Measurement of Starlight: Two Centuries of Astronomical Photometry (Cambridge: Cambridge University Press)[QB 815 H43 1996].
Henry, G. W. and Eaton, J. A. 1995, Editors of ASP Conf. Series, Vol. 79.
Johnson, H. L. 1963, in Basic Astronomical Data, edited by K. Aa. Strand (Chicago: University of Chicago Press) [UBV].
Johnson, H. L. and Morgan, W. W. 1953, ApJ, 117, 313. [UBV].
Krawchuk, C.A.P., Dawson, P.C., and De Robertis, M.M. 2000, AJ, 119, 1956
Kuhn, O., et al. 1995, ApJ, 438, 643.
Landolt. A.U. 1968, ApJ, 153, 151.
Landolt, A. U. 1973, AJ, 78, 959 [UBV].
Landolt, A. U. 1983, AJ, 88, 439 [UBVRI].

Landolt, A. U. 1992, AJ, 104, 340 [UBVRI].
Lasker, B.M., together with 15 co-authors 1988, ApJS, 68, 1.
Mayor, M. and Queloz, D. 1995, Nature, 378, 355.
McClure, R. D. 1976, AJ, 81, 182 [DDO].
McClure, R. D. and van den Bergh, S. 1968, AJ, 73, 313. [DDO].
Millis, R. L., Wasserman, L. H., and Birch, P. V. 1977, Nature, 267, 330.
Murayama, T. et al., 2000, AJ, 119, 1691.
Neely, R.K., Sarajedini, A., and Martins, D.H. 2000, AJ, 119, 1793.
Pasquini, L. et al., 2001, in Cool Stars, Stellar Systems and the Sun: Eleventh Cambridge Workshop, edited by R.J.C. Lopez, R. Rebolo, and M.R.Z Osorio, (San Francisco: Astronomical Society of the Pacific), p. 423.
Philip, A. G. D. 1979, in Problems of Calibration of Multicolor Photometric Systems, edited by AGDP (Schnectady, NY: Dudley Observatory Reports).
Pilcahowski, C.A. et al., 1989, PASP, 101, 707.
Pinfield, D.J. et al., 2000 MNRAS, 313, 347.
Platais, I. et al., 2000, PASP, 112, 224.
Postman, M., and Lauer, T.R. 1995, ApJ, 440, 28.
Pozzo, M. et al., 2000, MNRAS, 313, L23.
Quillen, A.C. et al., 1995, ApJ, 441, 549.
Richmond, M.W. et al., 2000, PASP, 112, 97.
Saviane, I., Held, E.V., and Bertelli, G. 2000, A&A, 355, 56.
Shanley, L., et al., 1995, ApJ, 438, L95.
Straizys, V. 1977, Multicolor Stellar Photometry, Photometric Systems and Methods (Mokslos, Vilnius) [Vilnius system].
Straizys, V. 1992, Multicolor Stellar Photometry (Tucson: Pachart Publishing House) [Vilnius system].
Stromgren, B. 1966, Ann. Rev. Astron. Astrophys. 4, 433 [uvby].
van den Berg, D., and Bell, R.A. 1985, ApJS, 58, 561.
Walsh, D., Carswell, R. F., and Weymann, R. J. 1979, Nature, 279, 381.
Weaver, H.A., plus 20 co-authors, 1995, Science, 267, 1282.
Weaver, H. F. 1946a, Popular Astronomy, 54, 211.
Weaver, H. F. 1946b, Popular Astronomy, 54, 287.
Weaver, H. F. 1946c, Popular Astronomy, 54, 339.
Weaver, H. F. 1946d, Popular Astronomy, 54, 389.
Weaver, H. F. 1946e, Popular Astronomy, 54, 451.
Weaver, H. F. 1946f, Popular Astronomy, 54, 504.
Weaver, H. F. 1962, edited by S. Flugge, in Handbuch der Astrophysik (Berlin: Springer-Verlag), 51, 130.
Whitford, A. E. 1962, edited by S. Flugge, in Handbuch der Astrophysik (Berlin: Springer-Verlag), 51, 240.

Chapter 2

Science of the Inner Planets From Small Ground-Based Telescopes

Ann L. Sprague
Lunar and Planetary Laboratory
The University of Arizona
Tucson, Arizona, USA

Abstract: For Mercury, Venus, Mars, and the Moon, 4-m and smaller telescopes make profound contributions to our knowledge. In fact, most non-spacecraft discovery has been made with telescopes from 10-cm to 3-m aperture. There is much more that can be done with the same telescopes and improved spectroscopic and imaging instrumentation as it becomes available. Discovery with these telescopes continues as this chapter is being written. More available time with suitable instrumentation on 3-4-m telescopes would open up a new opportunity of discovery for the inner planet atmospheres and surfaces. Some examples of the science discoveries of the past, and current issues waiting to be explored, are given.

Key words: planetary atmospheres, spectroscopy, planetary surfaces, remote sensing of the inner planets

1. GROUND-BASED SMALL TELESCOPIC DISCOVERIES

1.1 Mercury

Today observations of Mercury are relatively simple with electronic pointing, guiding and daytime observing. In previous decades Mercury was difficult to observe because of its proximity to the Sun. The greatest triumphs were those of Pettengill Dyce & Shapiro (1967) who discovered Mercury's rotation rate to be 59 ± 5 days (while not quite correct, previously Mercury had been thought to be in synchronous rotation with the Sun like the Moon is

T.D. Oswalt (ed.), The Future of Small Telescopes in the New Millennium, Vol. III, 11-35.

with Earth), and microwave measurements of temperature extremes on the surface (Morrison, 1970). Nothing was known about Mercury's surface morphology or composition and there were only upper limits for the thin exosphere until the three flybys of Mariner 10 in 1974 and 1975. As remote sensing technologies for telescopes and instrumentation improved, so did our ability to make diagnostic measurements of the surface and exosphere with a new generation of visible and infrared imaging spectrographs as well as rapid imaging VCR recorders. A few major discoveries have been made using very small telescopes, 60-cm to 3-m in diameter. These are briefly outlined below. There is ample opportunity for making new and exciting observations of Mercury. More measurements are needed and can be made with an appropriate spectrograph from any telescope with adequate pointing and guiding control. It is hoped that this brief review of observational successes will inspire many more observations and new discoveries.

1.1.1 Infrared spectroscopy and composition of surface materials

Ground-based spectroscopic observations in the near-infrared absorption bands characteristic of FeO have found Mercury's surface low in FeO relative to the Moon.

As shown in Figure 1, the surface is generally low in crystalline Fe^{2+}, the electronic state found in FeO that is so common on the Moon and many asteroids. Laboratory spectra of two lunar soils are also shown for

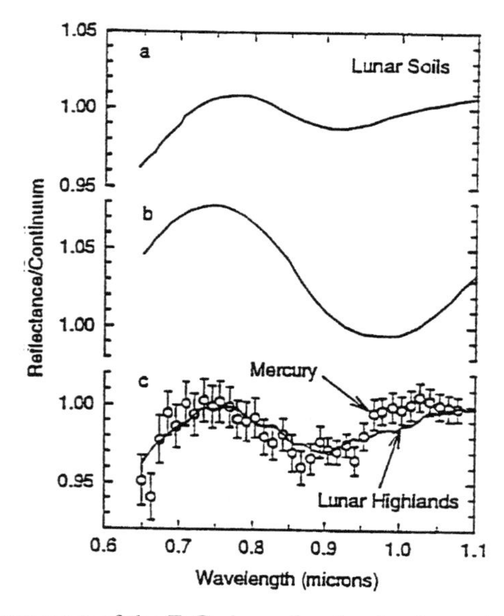

Figure 1. Spectral measurement of the FeO absorption band ~ 1–μm in Mercury's reflected continuum (McCord & Clark, 1979) compared to laboratory spectra of two lunar soils (a) noritic low-Ca orthopyroxene (b) basaltic high-Ca pyroxene.

Adapted from Vilas et al. (1988).

comparison. McCord and Clark (1979) put an upper limit of ~6% FeO at locations observed. This quantity was lowered to ~3% by Vilas *et al.* (1984) in a series of measurements covering most longitudes.

With the advent of new, high resolving power mid-infrared spectrographs it was possible to search for emission features indicative of specific mineral species and rock types. Regions on Mercury's surface exhibit spectral features similar to those of feldspar-rich basalt with Na-rich feldspars. In addition, one spectrum from ~34° longitude exhibits similarity to spectra from particulate lunar breccias composed of 90% anorthosite and 10% pyroxene. No evidence for highly iron-rich mineralogies has been found so far (Sprague *et al.* 1994).

Figure 2 shows some of these Mercury mid-infrared spectra along with laboratory spectra from a variety of feldspars and mixtures with pyroxenes for comparison. No match is perfect but comparisons are suggestive of labradorite and some (up to 25%) pyroxene, of unidentified composition. With the added constraint provided by near-IR spectroscopy, low-iron pyroxene basalts, alkali-rich plutonic rocks, and anorthosites are thought likely. These observations and interpretations are consistent with microwave

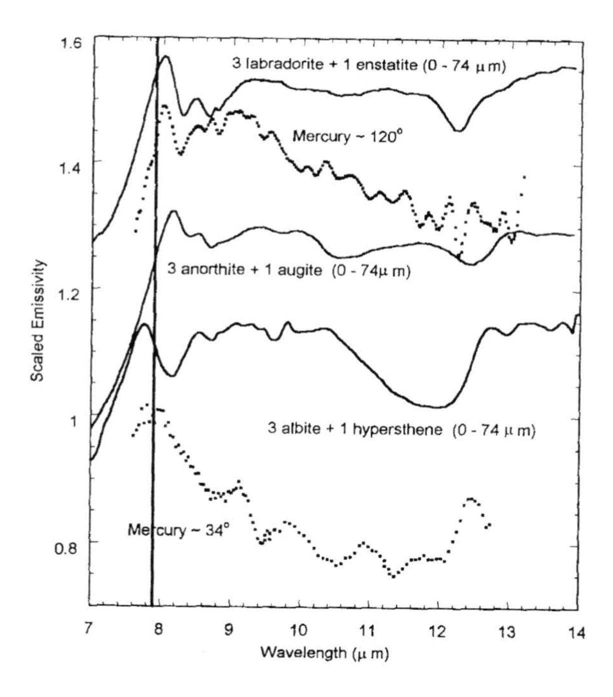

Figure 2. Telescopic mid-infrared spectra from Mercury's surface along with some laboratory spectra of terrestrial rock and mineral soils showing the similarity of Mercury's surface signature to laboratory feldspars. Data from the IRTF.

Adapted from Sprague et al. (1988a).

imaging and modeling of Mercury's regolith up to a few meters deep that show it to be of lower electrical conductivity (less iron and titanium-rich) than that of the Moon.

1.1.2 Spectroscopic study of Na, K, and Ca in the thin atmosphere

A study of scattering properties of the Moon's and Mercury's surface, lead to the exciting discovery of a thin sodium (Na) atmosphere at Mercury. Measurements were made with the McDonald 2.7-m telescope (Potter & Morgan 1985). Figure 3 shows the bright Na D1 and D2 emission lines rising from the Fraunhofer absorption in the reflected continuum from Mercury's surface.

The Doppler shift caused by Mercury's relative motion with the Sun shifts the emission lines from line center. A year later, potassium (K) in the atmosphere was also discovered with the same "small 2.7-m telescope" (Potter & Morgan 1986). Spatial variations are present in almost all observations (*cf.* Killen *et al.* 1990; Sprague *et al.* 1990).

The atmosphere is very tenuous (a few thousand kg of Na and K combined). Because the ionization time of Na and K is short and the new ions

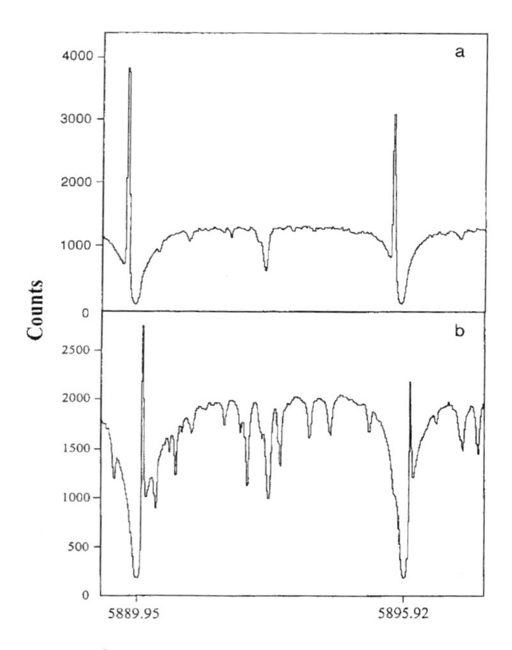

Figure 3. Na D1 (5895.92Å) and D2 (5889.95Å) emission lines at Mercury. Top spectrum shows dry conditions in Earth's atmosphere--bottom spectrum is from more humid conditions. Emission lines are Doppler shifted from Fraunhofer absorption line center. Wavelength scale is in Ångstroms. *Courtesy A. Sprague*

are swept into the interplanetary medium by electric fields in the solar wind, the neutral atoms must be continuously supplied to the atmosphere. Sources are surface materials, meteoritic material, and minor amounts from the solar wind. Atoms almost never collide with each other - they interact only with the surface. Thus the atmosphere is called a surface-bounded exosphere. It has been the object of ground-based high resolving power spectroscopic studies for over a decade. Spectral line shape and width measurements at the 3.7m Anglo-Australian Telescope and the McDonald 2.7-m (Killen *et al.* 1999) have modeled the equivalent temperature of Na atoms in the exosphere and found it to be consistent with a photo desorption release mechanism from the regolith (600 to 1000 K).

Other searches for new constituents require much larger telescopes. Calcium was discovered using the Keck 10-m and found to be extending to high altitudes off very high southern latitudes (Bida *et al.* 2000) and is mentioned for the sake of completeness despite the requisite of the large aperture for discovery. This brings the number of known species in Mercury's atmosphere to six. H, He, and O were measured by instruments on Mariner 10. Helium is outgassed from the interior, H is from the solar wind. Na, K, Ca, and O are probably from meteorites and the regolith.

1.1.3 Visible Light Spectral Imaging of the Thin Atmosphere

The exciting discovery of Na emission patches in Mercury's surface bounded atmosphere was made with the use of the McDonald 2.7-m telescope and the 1.4-m McMath-Pierce telescope on Kitt Peak (Potter & Morgan 1990). A 5x5 arcsec image slicer was used for the first images. Overlapping observations were required to cover the whole planet. The brightness of emission from Na and K are changeable on timescales from hours to years, and often the bright patches of emission are associated with surface features. Such an example is illustrated by Figure 4 which shows Na bright spots corresponding rather well with the radar bright regions along 355° longitude.

The fact that many of the bright emissions are localized in spots may indicate the surface at these locations is Na-rich. Bright emission spots of Na and K have also been observed in the region of Caloris Basin and an equatorial region of high radar reflectivity, southwest of Caloris, and over the Kuiper-Murasaki crater complex (Sprague *et al.* 1998b). However, sometimes bright emissions do not seem to be associated with any particular surface feature. In some observations there is no bright emission over Caloris basin. Observations are few but valuable. A 10x10 arcsec image slicer is now available for such observations. Also needed is a high resolving power spectrograph. It is used at the McMath Pierce Solar 1.4-m telescope on Kitt Peak, outside of Tucson, Arizona.

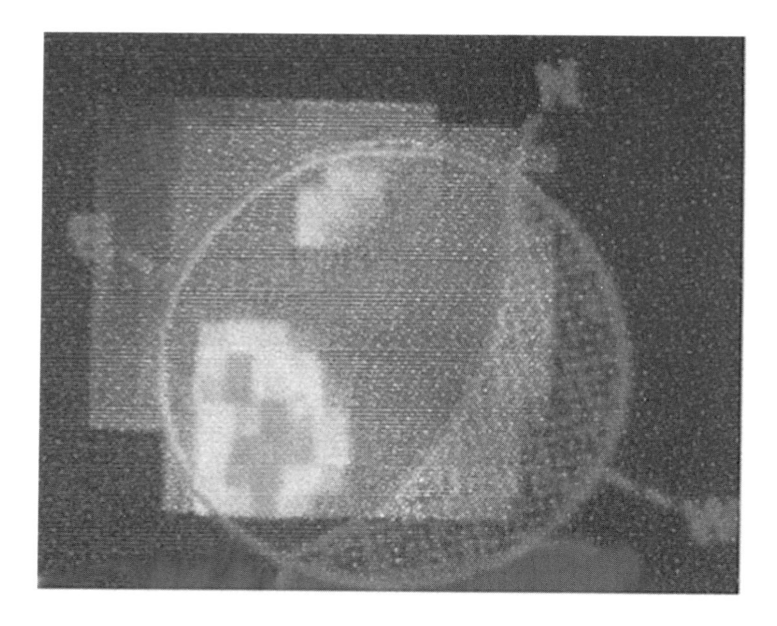

Figure 4. Bright atmospheric Na emission patches over radar-bright regions on Mercury's surface located at 355° longitude. Bright emissions could also be related to the magnetosphere in some as yet unknown way. *Adapted from Potter & Morgan (1990)*

1.1.4 Visible Light Imaging of Mercury's Surface

Recent imaging of Mercury in visible and near-infrared wavelengths has resulted in some remarkable successes. With the use of sensitive equipment and short exposure times, moments of extremely good seeing permit the capture of relatively good views of Mercury's surface. Most longitudes of Mercury have now been imaged with enough clarity to show distinct and unambiguous features on the surface. A composite of these images as cylindrical projection maps of Mercury is shown in Figure 5. The top panel is from Warell & Limaye (2001). The bottom left panel shows an image of longitudes 270 - 320° never imaged by Mariner 10 from Baumgardner *et al.* (2000) and Mendillo *et al.* (2001) and bottom right of the same panel, a cylindrical projection of entire portion imaged by Mariner 10 is also shown to complete the coverage.

The Baumgardner *et al.* (2000) observations were made at the Mount Wilson Observatory in California during a joint observing run with staff from the Boston Museum of Science. The Warell and Limaye research program uses the 0.5m Swedish Vacuum Solar Telescope (SVST) operated on the island of La Palma by the Royal Swedish Academy of Sciences in the Spanish Observatorio del Roque de los Muchachos of the Instituto de Astrofisica de Canarias. These images are the first set of optical wavelength images

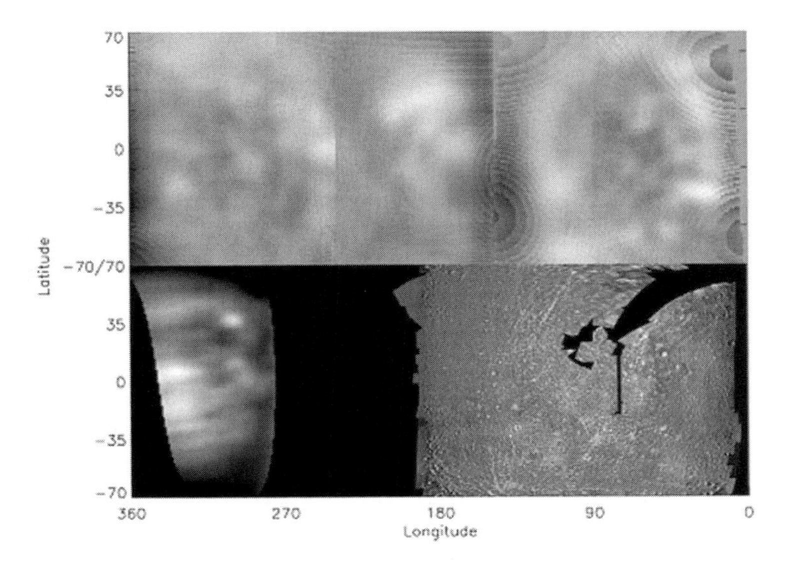

Figure 5. Telescopic imaging of all longitudes of Mercury by Warell and Limaye (2001) and Baumgardner et al.(2000) Lower right is a cylindrical coordinate map of images from Mariner 10 (courtesy of A. Tayfun Oner). *Figure from Mendillo et al. (2001)*

showing resolved surface features on Mercury for longitudes between 270 and 360°. The best spatial resolution is a region ~240-km on a side, a spatial resolution that could be improved with a larger aperture.

Overall the images give the appearances of regions of low albedo maria and higher albedo highlands with some fresh impact excavation regions; however, this may not be the correct interpretation. The brightest albedo feature seen in the bottom left is located at ~38° N, 298° longitude. Dark albedo features centered at ~15° N, 285°; 0°, 305° longitude and 26° N latitude, are located in regions showing low radar reflectivity, and thus in regions of relatively smooth terrain. This lends support to the interpretation of these regions as "lunar, maria-like" material. However, the albedo differences could be differences of surface maturity or even grain size.

1.2 VENUS

1.2.1 Atmosphere

Small telescopes are ideal for long-term visible and near-infrared observations of the planet Venus, because this planet is often the third brightest object in the sky (after the Sun and the Moon). Observations from telescopes with modest apertures (0.5 to 2-m) led to the discovery of some of the most prominent features of the Venus atmosphere, and the suspicion that the atmosphere was ~95% CO_2 (Belton *et al.* 1968) and the composition and particle size distribution of the planet-encircling cloud decks. Polarization measurements made with a 71-cm telescope were correctly interpreted by Sill (1983), Hansen & Arking (1971), and Young (1975) as 75% H_2SO_4 particles, with a modal radius of 1-μm.

The discovery of intense near-infrared (NIR) emission from the night side of the planet has provided new opportunities for studying the deep atmosphere and surface of the planet. Analysis of observations acquired by the Infrared Imaging Spectrometer (IRIS) at the 3.9-m AAT lead to the discovery that thermal radiation originating at altitudes between 35 and 50-km escapes through the clouds in "spectral windows" at wavelengths near 1.74 and 2.3-microns. Images of the Venus night side taken at these wavelengths at a variety of telescopes ranging from JPL's 60-cm Table Mountain facility, the 3-m IRTF and Las Campanas telescopes, the 3.9-m CFHT and AAT, and the KPNO 4-m telescope show high-contrast bright and dark features. Coordinated, long-term monitoring of the motions of these NIR features by these facilities indicate that they are produced by horizontal variations in the opacity of the middle ($50 < z < 57.5$-km) and lower ($47 < z < 50$-km) cloud decks (*cf.* Meadows & Crisp 1996).

The discovery of sulfur dioxide (SO_2) in Venus' atmosphere was made by Barker (1979) with the 1.7-m McDonald telescope (see Figure 6). The distribution and abundance of SO_2 below the clouds of Venus was measured by instrumentation on the 3.6-m Canada-France Hawaii telescope in 1989 and 1991 (Bezard *et al.* 1993). This is of intense interest because SO_2 gas is a tracer for volcanic activity. This important measurement can be made using high resolving power in the 2.45-μm window on the night side of Venus.

Other thermal emission windows permitting observations beneath the clouds in Venus 90 bar atmosphere have been discovered at 1.10, 1.18, 1.27, 1.31, 1.74, and 2.3μm. For example, high resolution measurements between 1.09 and 2.5μm of water vapor bands beneath the cloud tops at three different altitude ranges showed a constant water vapor mixing ratio is likely present (de Bergh *et al.* 1995). Based on these observations it is shown the atmosphere is dryer than that measured by instrumentation on Pioneer Venus and Venera mass-spectrometers and gas-chromatographs. This temporal change is unexplained.

1.2.2 Surface

Because of Venus' thick atmosphere, surface observations have been impossible to date except by radar. This does not preclude that in the future

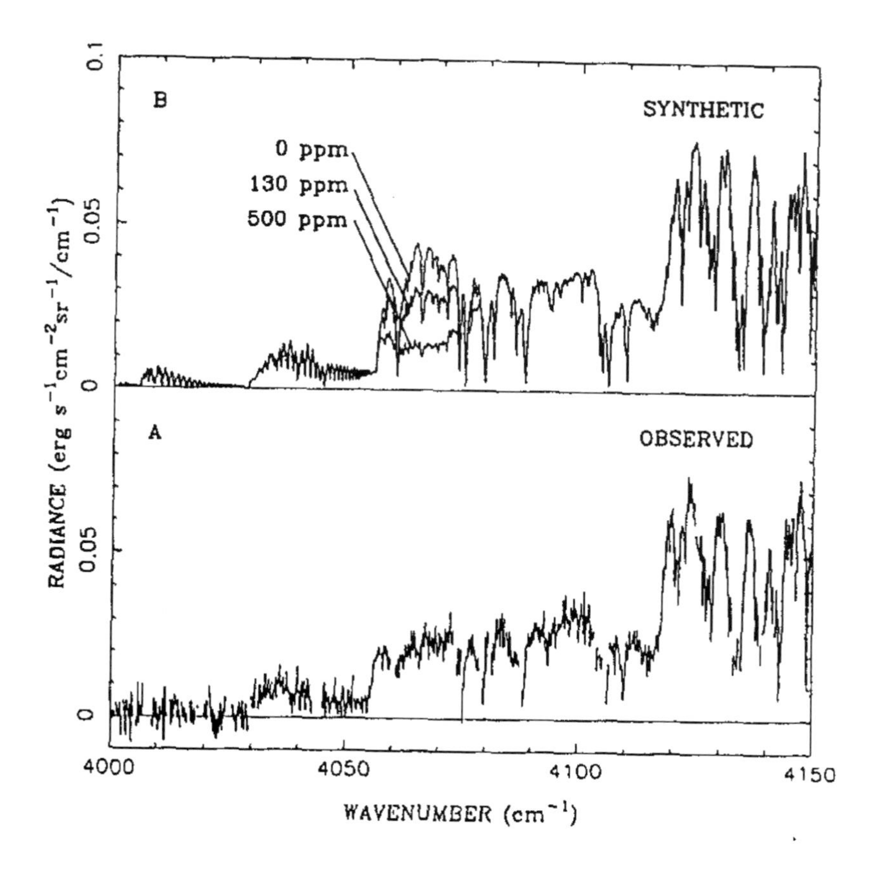

Figure 6. Synthetic (top) and small telescopic Fourier transform spectroscopic data from the Canada France Hawaii (CFH) telescope at Mauna Kea, showing SO_2 absorption in the atmosphere of Venus below the thick cloud layer. *From Bézard et al. (1993).*

some ingenious method will be discovered and 3- to 4-m telescopes used for surface studies.

1.3 MARS

1.3.1 Atmosphere

Atmospheric species at Mars as for the other terrestrial planets, can be studied through absorption spectra of sunlight reflected from the planet's surface or through emission spectra of gases excited by sunlight. Discoveries of the major constituents in Mars' atmosphere (CO_2, H_2O, O_2, and dust) have been made with small ground-based telescopes.

1.3.1.1 Water Vapor

Early observations (Owen 1966; Parkinson & Hunten 1972, 1973) roughly determined the CO_2 pressure and dust optical depth at Mars before the successful Mariner spacecraft a few years later. Also, a program has been in place at McDonald Observatory for nearly 30 years using the 2.1 and 2.7-m for the study of Mars atmospheric water vapor (Spinrad *et al.* 1963; Schorn *et al.* 1970; Tull 1971; Barker *et al.* 1970; 1976). These observations were the only systematic and seasonal measurements of atmospheric water vapor known until the Viking program in the late 1970s. The ground-based and Viking Mars Atmospheric Water Detector experiments are summarized by Jakosky & Barker (1984) and the ground-based measurements continue today at the McDonald Observatory 2.7-m.

Measurements with the 1.5m Catalina Observatory near Tucson, Arizona have also been important, and resulted in better understanding of the low latitude sink for autumn water vapor (Rizk *et al.* 1991), daily and hourly fluctuations at one geographic location (Sprague *et al.* 1996), and diurnal variations (Hunten *et al.* 2000). Such ground-based measurements are also critical for calibration of spacecraft measurements (e.g. Mars Global Surveyor) and for ground support during missions where no or few atmospheric water vapor measurements are made (e.g. Mars Pathfinder). Ground-based measurements provide a broader spatial and temporal context.

One of the instruments on NASA's 3.0-m infrared telescope (IRTF) is a high resolution ($\lambda/\Delta\lambda$ ~ 40,000) spectrometer (CSHELL) that can take spectra in the 1.0 to 5.0–μm range. This instrument has been used to measure volatiles in Mars' atmosphere. Temporal and spatial maps of atmospheric species throughout the Martian year (*cf.* Novak 2001) are made. Data in late spring and early summer have shown a shift in ozone concentration from the northern hemisphere to the southern hemisphere. One of the reasons for this shift is anti-correlation between water and ozone in the atmosphere; absorption spectra of HDO on Mars are taken within an hour or these O_2 spectra. By taking data through Mars' year, a better understanding of the yearly atmospheric dynamics can be developed.

The spatial resolution of these maps is limited by the seeing of the Earth's atmosphere (~1 arcsec). A Doppler shift greater than 12-km s^{-1} is sufficient to distinguish the spectra from Mars from the Earth's atmosphere. Using CSHELL, or the LPL échelle spectrograph used to obtain data shown in Figure 7, the slit can be positioned across the entire planet; with the slit positioned north-south on Mars' central meridian, an atmospheric map can be taken along the sub-earth longitude. By repositioning the slit to the east and to the west, longitudinal variations can also be determined. Such measurements complement those of spacecraft that are often in Sun-synchronous orbits.

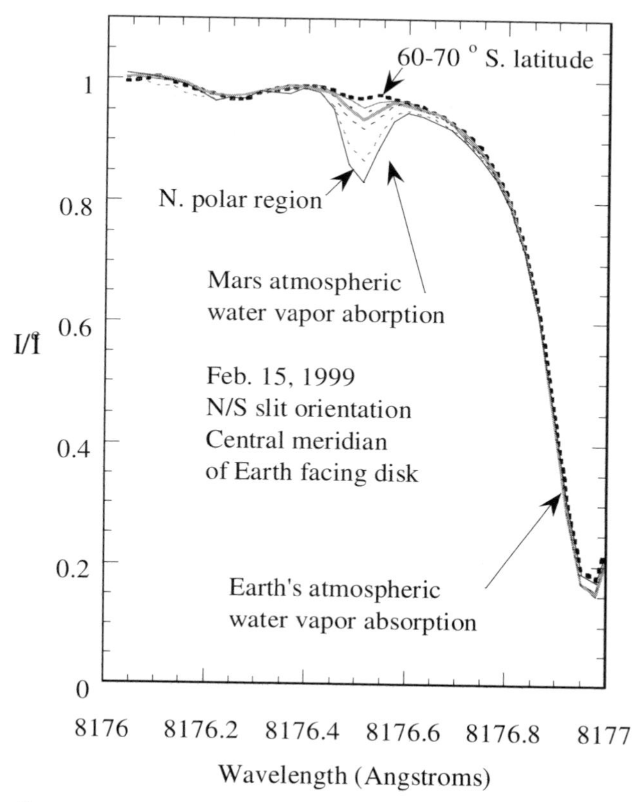

Figure 7. Spectroscopy at the 1.5-m Catalina Observatory provides latitudinal and longitudinal resolved water vapor measurements. *Adapted from Sprague et al. (2001)*

Figure 8 presents an illustration of a spectral-spatial image taken with CSHELL on NASA's IRTF and a horizontal trace of the light intensity across the spectrum below. The image is centered at 7908-cm^{-1}. Eight spectral lines from the $a^1\Delta_g$-$X^3\Sigma_g$ band of O_2 are located in this region (marked by $*$). The excited state of O_2 is produced by UV sunlight (Hartley continuum) photolyzing O_3. From measuring the intensity of these transitions, the ozone column density can be calculated. The $^1\Delta_g$ state has a relatively long lifetime (~ 1 hour); at low altitudes, the state is quenched by collisions. The emitted photons thus originate from molecules at altitudes above 20-km. From the O_2 transitions, the rotational temperature and the total band intensity can be calculated.

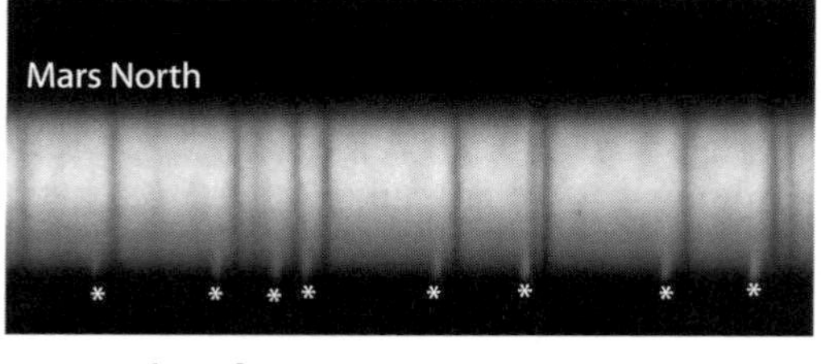

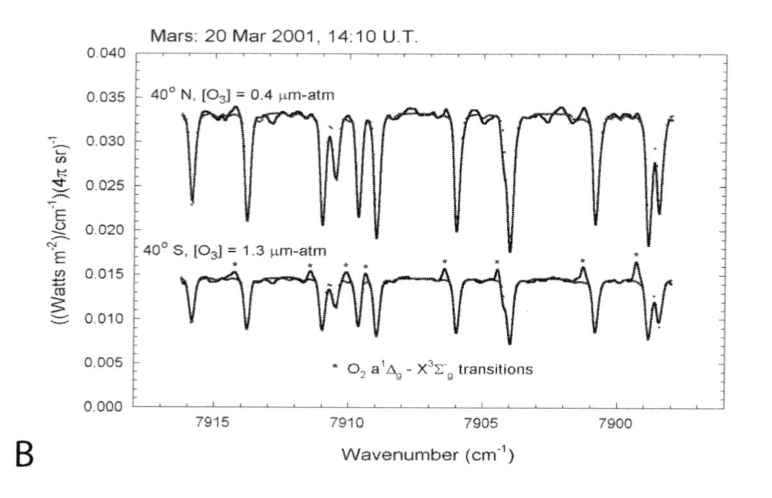

Figure 8. (A) Spectral image showing atmospheric emission at different latitudes on Mars. The emissions from Mars are Doppler shifted from the corresponding absorptions in earth's atmosphere. Intensity is greater in southern latitude. (B) Horizontal light intensity trace centered at 40°S and 40°N. Column density is much greater in the south (lower trace). Data are from Northern mid-summer (Ls = 133°). *From Novak (2001)*

1.3.1.2 Clouds and Dust

Ground based imaging is well suited for monitoring onset of dust storms, dust circulation, cloud growth, coverage and dissipation. For dust storm monitoring, small 25 to 61-cm telescopes with good imaging CCD cameras are adequate (see Figure 9).

For cloud discrimination, a telescope like the IRTF with a good near-IR spectrograph is required. A good diagnostic for H_2O ice clouds is a strong spectral curvature between 3.0 to 3.5–μm that becomes pronounced when caused by < 1–μm ice grains. CO_2 clouds are often masked by absorptions of water ice clouds (Klassen *et al.* 1999).

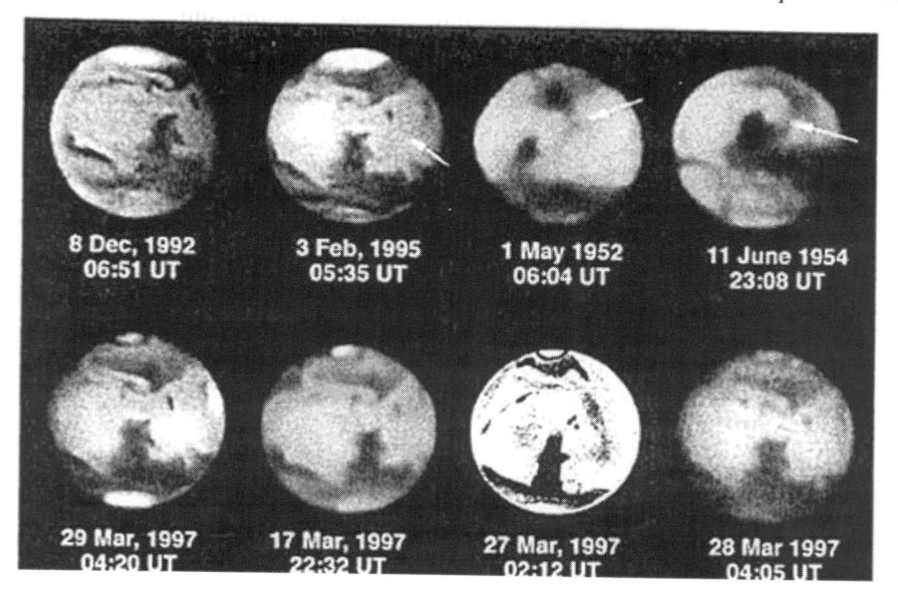

Figure 9. Syrtis Major imaged by several small telescopes to monitor dust and clouds. Top left two images obtained with 41-cm Newtonian, top right two images with 61-cm Lowell Observatory refractor. Bottom from left to right, 41-cm, 1-m, drawing, 25-cm.

Adapted from Parker et al. (1999)

1.3.2 Surface

Near infrared spectroscopy of Mars' surface has met with limited but important success in identifying the major rock and dust type, namely basalt. Differences across the surface are seen by studying the slope of the continuum and the depth and shape of the 0.87 and 1.05–μm absorption bands that are indicative of Fe^{3+} and Fe^{2+}, respectively (Singer *et al.* 1979, Bell & McCord 1989) as shown below in Figure 10.

Small telescope measurements of regions of water ice, highly desiccated metal hydrate, and CO_2 ice along the south polar cap (Bell & Crisp 1993), are also noted. In the 4.5 to 5.0–μm region, there is some hope of searching for the $2\nu_3$ overtone of the SO_4^{-2} anion present in sulfates that may lurk on Mars surface (Blaney & McCord 1995). The search is motivated by Viking analyses that showed S to be present in the Martian regolith. The search is confounded by strong CO_2 atmospheric absorption between 4.2 and 4.4–μm.

Some measurements of hydrated metals and sulfates have not been corroborated but are tantalizing and more observations are needed. It may be that the missions planned for the next decade with thoroughly map the mineral and rocks of the surface and thus remove Martian surface studies from the list of research for small telescopes. However, until that actually happens,

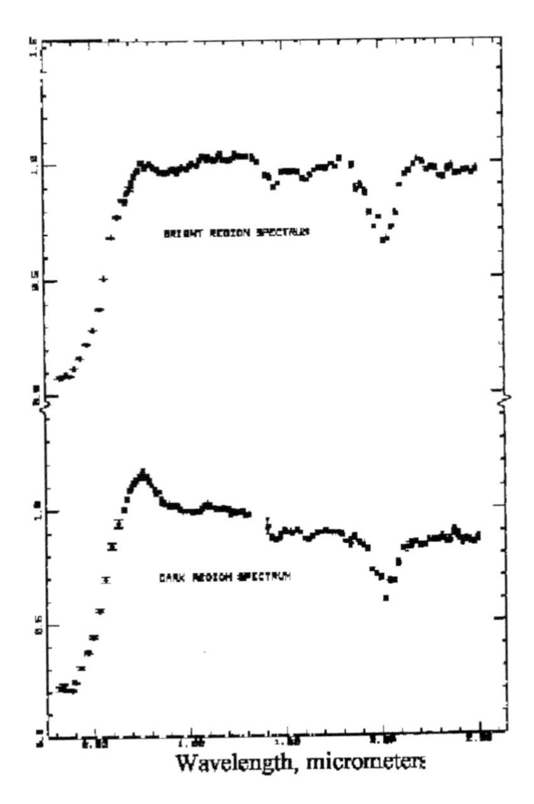

Figure 10. Mars' surface spectral observations show differences in the pyroxene content of basalts from one location to another (after Singer et al. 1979).

continued observations by ground-based telescopes in the 3-4-m class are very valuable.

1.4 MOON

1.4.1 Spectroscopy and imaging of the lunar atmosphere

The triumph of doing important planetary science with small telescopes is epitomized by the discovery of a thin Na and K atmosphere at the Moon and the subsequent spectroscopic and imaging studies that followed. Study of the atmosphere by spectrographs at McDonald Observatory 2.1-m (Potter & Morgan 1988), the 1.5-m at the Catalina Observatory, and the 18-cm Mt. Lemmon Lunar Coronograph (Sprague *et al.* 1992, 1998) showed a variable distribution of atoms apparently subject to the temperature of the lunar surface and perhaps the solar wind. High resolution observations of the lunar

Na atmosphere using the 1.8-m Astronomical Observatory of Padova (Cremonese & Verani 1997; Contarini *et al.* 1996) indicated a strong link to a meteoritic source. A good image of the Na near the surface of the Moon (Potter & Morgan 1998) showed the thermal component that had been observed by spectroscopy with the 1.5-m Catalina Observatory (Kozlowski *et al.* 1991 for K; Sprague *et al.* 1992 for Na) and in imaging over the terminator by Stern & Flynn (1995).

Figure 11 shows the Na atmosphere above and near to the lunar surface. The image shows Na emission (and thus Na atmospheric atoms) are most plentiful near the sub-solar point. This image was obtained with the 40.6-cm Sac Peak Solar Telescope in New Mexico with the use of an occulting disk.

Figure 11. Lunar Na atmosphere image obtained at the 40.9-cm Sac Peak Solar Observatory. Na is close to the surface, most plentiful near the sub-solar point. The Moon is superposed for effect.

Adapted from Potter & Morgan (1998)

Figure 12 shows the 18-cm Lunar Coronographic Telescope used on Mt. Lemmon to make systematic observations of the lunar Na and K for over a period of four years. The Mt. Lemmon coronograph was designed and fabricated specifically to measure the height distribution of Na and K atoms above the lunar limb at all phases. At the time this instrument was fabricated, little was known about the distribution of Na and K above the lunar surface or the source and release mechanisms. One big accomplishment was showing the relationship of the two-component Na and K atmosphere to the local solar zenith angle. This observation helped to clarify the source release mechanism of photon sputtering from the regolith.

An even smaller facility, the Boston University 10-cm occulting telescope have solved many of the puzzles of the origin, distribution, release mechanisms of the Na and K atmosphere by taking broad scale images when the Moon was in and out of the Earth's magnetotail (Mendillo *et al.* 1999), during lunar eclipse (Mendillo & Baumgardner 1995), and during the Leonid meteor shower (Wilson *et al.* 1999; Smith *et al.* 1999). Other observations with the 16-cm Mt. Lemmon Lunar Coronograph (Hunten *et al.* 1998) demonstrated meteoritic material as an important source of Na. Had small telescopes not been available for this work it does not seem likely that our knowledge about the Moon's Na and K atmosphere would have advanced so quickly.

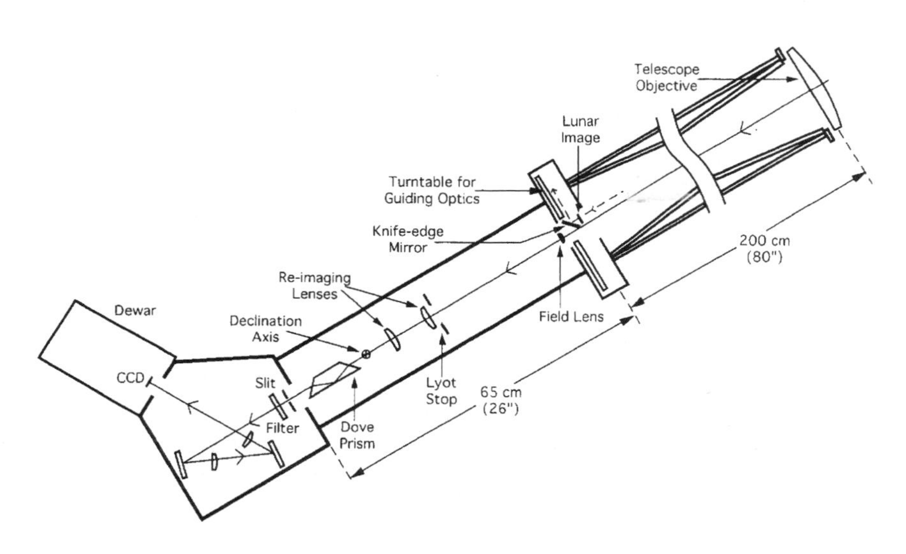

Figure 12. An 18-cm diameter coronograph designed and implemented to study the Na and K lunar atmosphere. The telescope has an occulting knife-edge. The spectrograph and cryogenically cooled CCD camera are shown at the left end of the coronograph.

Adapted from Sprague et al. (1998)

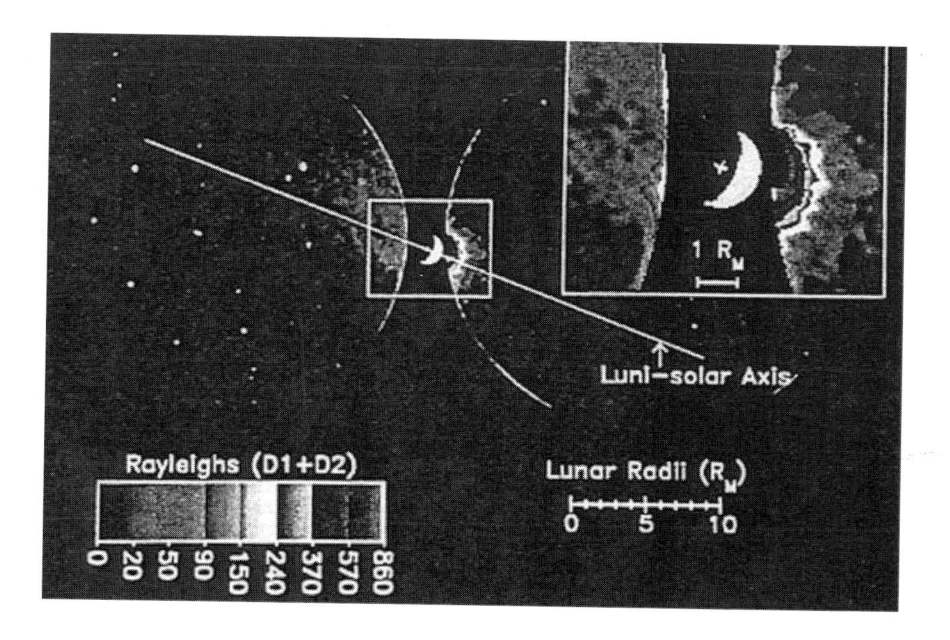

Figure 13. The Na extended lunar corona was discovered with the use of a 4-inch occulting telescope designed especially to study extended Na atmosphere of the Moon and the escaping Na atmosphere of Io. *Adapted from Mendillo et al. (1991)*

1.4.2 Spectroscopy and imaging of the lunar surface

This chapter doesn't need to remind the reader of the obvious glory of imaging the lunar surface with small telescopes. In addition, near-infrared spectroscopy remains the most effective method of mapping lunar surface composition to date (*cf.* McCord *et al.* 1981). While spacecraft multi-color imaging has revealed much about the presence of FeO and differences in soil maturity, mineral identification and mapping requires more detailed spectral analysis. Many early discoveries were made with small ground-based telescopes. Figure 14 shows near-IR spectra of the lunar surface obtained at two different locations. From these spectra, composition is deduced by modelling and comparison to laboratory lunar and terrestrial analogues. It is a very effective process. In the past, telescope time at 3 to 4-m telescopes to look at the Moon was rare and so most observations were made from the University of Hawaii 2.2-m and other smaller aperture facilities. While the Moon is large in the sky (subtends nearly 180 arc seconds) the spatial resolution is still quite large (~35-km at the equator) for reasonably good seeing of and arcsec or two.

One of the first measurements of the lunar mantle is that made in the crater Copernicus. Near-infrared telescopic reflectance spectra indicate extensive exposures of olivine. The geologic history of the region has been interpreted based on these important measurements (Pieters 1982, Pieters & Wilhelms 1984).

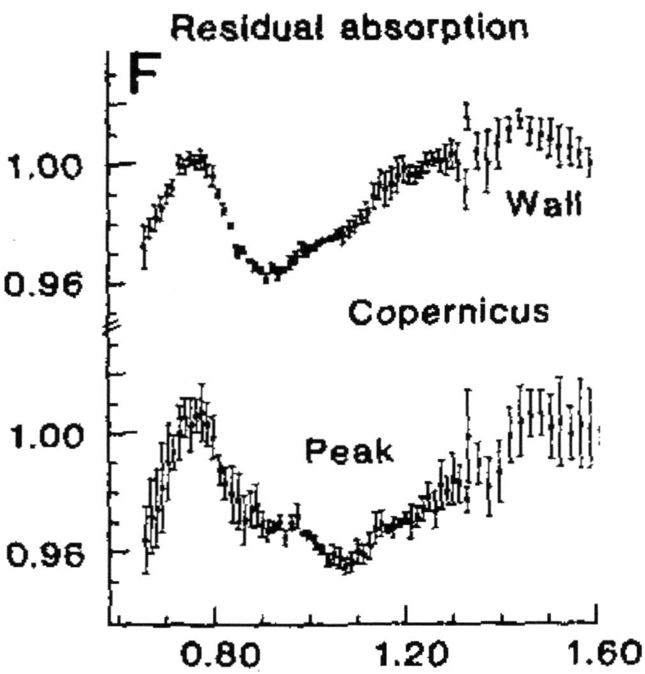

Figure 14. Most of the major mineralogical mapping of the lunar surface has been achieved with the use of telescopes like the Hawaii 88-inch on Mauna Kea. Here can be seen the distinct olivine feature of Copernicus peak and the absence of olivine at the location of the crater wall.
From Pieters & Wilhelms (1984)

2. USEFUL SMALL TELESCOPE COMPARISONS

As Table 1.1 reminds us, good spatial resolution on inner planetary surfaces and diffraction limits within the limits of atmospheric turbulence are readily available with small telescopes. The situation is improved with rapid read-out imaging systems that can "freeze the seeing of turbulent cells" and typically reach 0.5 arc second spatial resolution. For visible imaging and spectroscopy with systems near *f*/13, this means spatial scales of less than an arcsec can be achieved with telescopes as small as 0.6-m. For near-infrared work, telescope apertures of 1.5-m or greater will achieve the 0.5 arc second spatial goal. For mid-infrared spectroscopy, larger apertures are required for good spatial resolution because of the diffraction limit that doesn't drop below one arc second until the 3-m of aperture are used. Table 1 gives spatial scale in arc seconds using an assumed pixel size of 15–μm.

Table 1.

Some Handy Telescope Comparisons

Aperture	Spatial Scale (arc sec)		Diffraction Limit (arc sec)		
m	*f*/13	*f*/35	0.55–μm	3–μm	10–μm
0.61	0.40	0.15	0.19	1.02	3.41
1.55	0.16	0.06	0.07	0.40	1.34
2.24	0.11	0.04	0.05	0.21	0.93
3.05	0.08	0.03	0.04	0.21	0.68
4.01	0.06	0.02	0.03	0.16	0.52

3. FUTURE STUDIES OF THE INNER PLANETS AND THE MOON WITH SMALL TELESCOPES

3.1 Spectrographs and Imaging Systems

What has to be considered is the necessity of using sophisticated and often expensive spectrographs and imaging systems to do cutting edge science. This of course is true even on the smaller class of 60-cm to 4-m telescopes. The good news is that imaging systems are getting more and more affordable and used in combination with rapid readout and good computer software analysis tools almost any telescope becomes a valuable research tool. Many of the larger "small telescopes" have facility instruments which are maintained and sometimes operated by experienced technicians and astronomers.

3.2 Mercury

3.2.1 Atmosphere

Future scientific opportunities with small telescopes for Mercury studies are many. Much is yet to be known about the interaction of Mercury's known atmosphere with the surface and ion species with the magnetosphere. Current theories predict other species as yet undiscovered. Any discovery of a relationship of the atmosphere with the radar bright spots (*cf.* Slade *et al.* 1992, Butler *et al.* 1993) that are thought to be stored volatiles at high latitudes would be of great importance. There is still plenty to be done with 1- and 2-m telescopes to image the temporal and spatial changes in the Na and K atmospheres. The search for new species must be made with larger apertures. To map the relationship of the atmosphere with the surface geology the spatial resolution gained with rapid readout imaging systems is required. Reaching the seeing limit of rapid readout arrays and adaptive optic instrumentation is possible and would make significant inroads to our study of the atmosphere and its relationship with the Sun, surface, and magnetosphere.

3.2.2 Surface

The biggest obstacle to study of Mercury's surface is the dearth of spectroscopic instrumentation with good enough resolving power and adequate attenuation filters to permit near-IR and mid-IR spectral measurements. Also are good pointing, rapid readout, and the development of a method of adaptive optics that does not require a near-by guide star. Telescopes in the 1 to 2-m class can still do useful visible and near-infrared imaging. Mineralogic mapping of the surface in the mid-infrared offers promise. Best results are achieved with the 3 to 4-m class telescopes because the diffraction limit is smaller and spatially resolved mapping can be achieved.

Serendipitous discoveries that yield exciting new physical insights to the planet should not be discounted as that has been the history of discovery of this enigmatic planet.

3.3 Moon

3.3.1 Atmosphere

While ground-based observations by several groups have generally solved some of the major questions regarding the Na and K in the lunar atmosphere, more observations, particularly during meteor showers would be interesting. Also the search for other species with more sensitive instrumentation and greater light gathering capability of 3-4-m telescopes is an area ripe for investigation. Some questions of energy distribution of atoms and the effects of charged particle sputter remain to be investigated.

3.3.2 Surface

With the right spectrograph and attenuation filters appropriate for the bright lunar surface, much better mineralogical mapping could be done at better spatial resolution. For example, full coverage of the nearside with mapping spectrometer data with spectral resolution in the near-IR and visible of 0.3 to 2.6–μm and 10 to 20-nm respectively would be a large improvement over available data sets. Unfortunately most instruments are designed for objects with low light level and are not used on the Moon. It is within technical feasibility to obtain an image cube for the Moon at 3-km resolution at the equator and address important geological issues with mineralogy maps.

Lunar experts have tried very hard to coax mineralogy out of the Clementine UV-VIS data. Clementine has great resolution (about 200m/pixel) but only five bands. It is good for geologic context, but far from ideal for mineralogy study. Spectral reduction libraries and analysis tools are readily available, what is needed are the data.

3.4 Venus

3.4.1 Atmosphere

Much more could be done at Venus with 3 to 4-m telescopes and spectrographs capable of high resolving power measurements in the thermal radiation windows. The limitations are telescope time and funding for investigators. A program to regularly monitor for SO_2 gas would be valuable and serve as a baseline for a mission that will likely be funded in the next decade or so. The search for new components and regular measurements of abundance of known ones, including water vapor, would greatly add to our ability to model the relationship between the surface, the geologic and atmospheric history of the planet. Detailed spectral imaging investigations of the NIR radiation from spectral windows at 1.0, 1.10, 1.18–μm from the IRTF, AAT, and CFHT indicate that the thermal radiation in these spectral windows originates from the surface and lowest scale height of the atmosphere. Spectra acquired at these wavelengths have been combined with spectra from the 1.27, 1.38, 1.74, and 2.3–μm windows to dramatically improve our understanding of the abundance and distribution of water vapor, carbon monoxide, carbonyl sulfide, hydrofluoric acid, hydrochloric acid, and other important trace gases between the surface and the cloud tops.

Long-term spatially-resolved spectroscopic observations from moderate-sized telescopes would provide valuable new constraints on spatial and temporal variations in the distributions of these atmospheric constituents. Additional time-resolved near-infrared imaging observations at these NIR wavelengths with small (1 to 3-m) telescopes would yield additional information needed to track the North-South as well as the East-West motions of cloud features. This information could be combined with near UV (0.4 micron) observations of cloud-top UV feature motions to yield the first, complete, three-dimensional description of the mysterious, cloud-level super-rotation, which dominates the dynamics of the Venus cloud-level atmosphere. This information is essential to fully characterize the specific chemical and dynamical processes that maintain the clouds.

3.5 Mars

3.5.1 Atmosphere

While mission opportunities are many during this decade, not all missions carry instrumentation for water vapor measurements. Landed rovers and weather stations may have humidity sensors but these will be covering only a few locations on the surface. Continued ground-based water vapor measurements are a must for Mars. These measurements are simple to make with small telescopes and existing instrumentation (high resolving power spectrographs). The baseline for meteorology studies and climate history is crucial. Should another wet southern Spring be observed like that measured by Barker *et al.* (1970), it would be an important event. Also important are monitoring of dust storms and migration of dust during missions and mission planning. Regular imaging to monitor cloud formation and motion is also helpful for meteorological modelling and mission context.

3.5.2 Surface

Once good mineralogical maps have been made by orbital and in-situ spectrometers, it is doubtful that further work with ground-based telescopes and spectrometers for surface composition mapping would be useful.

4. REFERENCES

Barker, E.S., Schorn, R.A., Woszczyk, A., Tull, R.G., & Little, S.J. 1970. Mars: Detection of Atmospheric Water Vapor during the Southern Hemisphere Spring and Summer Season. Science 170, 1308-1310.

Barker, E.S. 1976. Martian Atmospheric Water Vapor Observations: 1972-74 Apparition. Icarus 28, 247-268.

Barker, E.S. 1979. Detection of SO_2 in the UV spectrum of Venus. Geophys. Res. Lett. 6, 117-120.

Baumgardner, J., Mendillo, M., & J. Wilson 2000. A Digital High Definition Imaging System for Spectral Studies of Extended Planetary atmospheres: 1. Initial Results in White Light Showing Features on the Hemisphere of Mercury Unimaged by Mariner 10. Astron. J. 119: 2458-2464.

Bell, J.R. III & D. Crisp 1993. Groundbased Imaging Spectroscopy of Mars in the Near-Infrared-Preliminary Results. Icarus 104, 2-19.

Bell, J.F. & T.B. McCord 1989. Mars-Near-infrared Comparative Spectroscopy During the 1986 Opposition. Icarus 79, 21-34.

Belton, M. Hunten, D.M. & R.M. Goody 1968. Quantitative Spectroscopy of Venus in the Region 8,000-11,000 Å. In "The Atmospheres of Venus and Mars," Eds. J.C. Brandt & M.B. McElroy, Gordon & Breach Science Publishers Ltd. p. 288

Bézard, B., deBergh, C. Fegley, B, Maillard, J.P., Crisp, D. Owen, T. Pollack, J. & D. Grinspoon 1993. The Abundance of Sulfur Dioxide Below the Clouds of Venus. Geophys. Res. Lett. 20, 1587-1590.

Bida, T., Killen R. & T. Morgan, 2000. Discovery of Ca in Mercury's Atmosphere. Nature 404, 159-161.

Blaney, D. & T. McCord 1995. Indications of Sulfate Minerals in Martian Soil from Earth-Based Spectroscopy. J. Geophys. Res. 100, 14,433-14,441.

Butler, B., D. Muhleman, & M.A. Slade 1993. Mercury: Full-Disk Radar Images and the Detection and Stability of Ice at the North Pole. J. Geophys. Res. 98: 15,003-15,023.

Contarini, G. C. Barbieri, G. Corrain, G. Cremonese, & R. Vio 1996. Spectroscopic Observations of the Sodium Atmosphere of the Moon, Planet. Space Sci. 44, 417-420.

Cremonese G. & S. Verani, 1997. High Resolution Observations of the Sodium Emission from the Moon, Adv. Space Res. 19, 1561-1575.

De Bergh, C., Bezard, B., Crisp, D, Maillard, J., Owen, T, Pollack, J, & D. Grinspoon. Water in the Deep Atmosphere of Venus from High-resolution spectra of the Night Side. Adv. Space Res. 15, (4)-79 – (4)88.

Hansen, J.E. & A. Arking 1971. The Identification of H_2SO_4 as the Particles in Venus' atmosphere. Science 171, 669-670.

Hunten, D.M. G. Cremonese, A.L. Sprague, R.E. Hill, S. Verani, & R. W.H. Kozlowski, 1998. The Leonid Meteor Storm and the Lunar Sodium atmosphere," Icarus 136, 298-303.

Hunten, D.M., Sprague, A.L. & L.R. Doose 2000. Correction for Dust Opacity of Martian Atmospheric Water Vapor Abundances. Icarus 147, 42-48.

Jakosky, B.M., & Barker, E.S. 1984. Comparison of Ground-Based and Viking Orbiter Measurements of Martian Water Vapor: Variability of the Seasonal Cycle. Icarus 57, 322-344.

Killen, R.M., T. H. Morgan, & A. E. Potter 1990. Spatial Distribution of Sodium Vapor in the Atmosphere of Mercury, Icarus 85, 145-167.

Killen R.M, Potter, A.E., Fitzsimmons, A. T.H. Morgan 1999. Sodium D2 Line Profiles: Clues to the Temperature Structure of Mercury's Exosphere. Planet. Space Sci. 47, 1449-1458.

Klassen, D., J. Bell III, Howell, R. Johnson, P. Golisch, W. Kaminski, C. & D. Griep 1999. Infrared Spectral Imaging of Martian Clouds and Ices, Icarus 138, 36-48.

Kozlowski, R.W.H., Sprague, A.L. & D.M. Hunten 1990. Observations of Potassium in the Tenuous Lunar Atmospohere. Geophys. Res. Lett. 17, 2252-2256.

Meadows, V.C. & D. Crisp 1996. Ground-Based Near-Infrared Observations of the Venus Nightside: The Thermal Structure and Water Abundance Near the Surface. J. Geophys. Res. 101, 4595-4622.

Mendillo, M. J. Baumgardner, & B. Flynn 1991. Imaging Observations of the Extended Sodium Atmosphere of the Moon. Geophys. Res. Lett. 18, 2097.

Mendillo M. & J. Baumgardner 1995. Constraints on the Origin of the Moon's Atmosphere from Observations During a Lunar Eclipse, Nature 377, 404-406.

Mendillo, M., J. Baumgardner, & J. Wilson 1999. Observational Test for the Solar Wind Sputtering Origin of the Moon's Extended Sodium Atmosphere, Icarus 137, 13-23.

Mendillo, M., Warell, J., Limaye, S.S., Baumgardner, J., Sprague, A.L. & J.K. Wilson 2001. Imaging the Surface of Mercury Using Ground-Based Telescopes. Planetary & Space Res. In press.

McCord, T.B., Clark, R.N., Hawke, B.R, McFadden L.A., Owensby P.D, Pieters, C.M. & J. B. Adams 1981. Moon- Near-Infrared Spectral Reflectance: A First Good Look. J. Geophys. Res. 86, 10883-10892.

McCord, T.B. & R.N. Clark (1979). The Mercury Soil: Presence of Fe^{2+}. J. Geophys. Res. 84, 7664-7668.

Morrison, D. 1970. Thermophysics of the Planet Mercury. Space Sci. Rev. 11, 271-307.

Novak, R., M. Mumma, M. DiSanti, N. Dello Russo, K. Magee-Sauer, 2001. Photochemical Mapping of Ozone and Water in the Atmosphere of Mars Near the 1997 Aphelion, submitted to Icarus.

Owen, T., 1966. The Composition and Surface Pressure of the Martian Atmosphere. Results from the 1965 Opposition. Ap. J. 146, 257-270.

Parkinson, T.D., & Hunten, D.M. 1972. Martian Dust Storm: Its Depth on 25 November 1971. Science 175, 323-327.

Parkinson, T.D., & Hunten, D.M. 1973. CO_2 Distribution on Mars. Icarus 18, 29-53.

Parker, D.C., Beish, J.D., Troiani, DM., Joyce, D.P., & C.E. Hernandez 1999. Icarus 138, 3-19.

Pettengill, G.H., R Dyce, & I. Shapiro 1967. Rotation Periods of Mercury and Mars. Astron. J. 72, 351.

Pieters, C.M. 1982. Copernicus Crater Central Peak—Lunar Mountain of Unique composition. Science 215, 59-61.

Pieters, C.M., Wilhelms, D.E. & R. Paquette. 1984. Stratigraphy at Copernicus and the Source of Olivien in the Central Peak. Lunar & Planetary Science XV, p. 643-644 abs.

Potter, A.E. & T. H. Morgan 1985. Discovery of sodium in the atmosphere of Mercury, Science 229, 651-653.

Potter, A.E. & T. H. Morgan, 1986. Potassium in the Atmosphere of Mercury. Icarus 67, 336-340.

Potter, A.E. & T.H. Morgan, 1990. Evidence for Magnetospheric Effects on the Sodium Atmosphere of Mercury. Science 248, 835-838.

Potter A. E. & T. H. Morgan, 1988. Discovery of Sodium and Potassium Vapor in the Atmosphere of the Moon. Science 241, 675-680.

Potter A. E. & T.H. Morgan, 1998. Coronagraphic observations of the lunar sodium exosphere near the lunar surface. J. Geophys. Res. 103, 8581-8586.

Rizk, B., Wells, W.K., Hunten, D.M., Stoker, C.R., Freedman, R.S., Roush, T., Pollack, J.B., & Haberle, R.M. 1991. Meridional Martian Water Abundance Profiles during the 1988-1989 Season. Icarus 90, 205 213.

Schorn, R.A., Spinrad, H., Moore, R.C., Smith, H.J., & Giver, L.P. 1967. High-Dispersion Spectroscopic Observations of Mars. II. The Water Vapor Variations. Ap. J. 147, 743-752.

Singer, R.B., Clark, R.N. McCord, T.B., Adams, J.B. & R.L. Huguenin 1979. Mars Surface Composition from Reflectance Spectroscopy. J. Geophys. Res. 84, 8415-8426.

Sill, G.T. 1983. The Clouds of Venus: Sulfuric Acid by the Lead Chamber Process. Icarus 53, 10-15.

Slade, M., B. Butler, & D.O. Muhleman 1992. Mercury Radar Imaging: Evidence for Polar Ice. Science 258, 635-640 .

Spinrad, H., Munch, G., & Kaplan, L.D., 1963, The Detection of Water Vapor on Mars. Ap. J. 137, 1319-1321.

Smith, S. M. J. K. Wilson, J. Baumgardner, & M. Mendillo 1999. Discovery of the Distant Lunar Sodium Tail and its Enhancement Following the Leonid Meteor Shower of 1998.," Geophys. Res. Lett., 26, 1649-1652.

Sprague, A. L., Kozlowski, R. W. & Hunten, D. M. 1990. Caloris Basin: An Enhanced Source for Sodium and Potassium in Mercury's Atmosphere. Science 249, 1140-1143.

Sprague, A.L., Kozlowski, R.W.H., Witteborn, F.C., Cruikshank, D.P. & D.H. Wooden 1994. Mercury: Evidence for Anorthosite and Basalt from Mid-Infrared (7.3-13.5 μm) Spectroscopy. Icarus 109: 156-167.

Sprague, A. L. & T. L. Roush 1998a. Comparison of Laboratory Emission Spectra with Mercury Telescopic Data. Icarus 133, 174-183.

Sprague, A. L., R. W. H. Kozlowski, D. M. Hunten, W. K. Wells, & F. A. Grosse, 1992. The Sodium and Potassium Atmosphere of the Moon and Its Interaction with the Surface Icarus 96, 27-42.

Sprague, A. L., D. M. Hunten, R. W. H. Kozlowski, F. A. Grosse, R. E. Hill, & R. L. Morris 1998c. Observations of Sodium in the Lunar Atmosphere during International Lunar Atmosphere Week, 1995, Icarus 131, 372-381.

Sprague, A. L., W. J. Schmitt, & R.E. Hill 1998b. Mercury: Sodium Atmospheric Enhancements, Radar Bright Spots, and Visible Surface Features. Icarus 135, 60-68.

Sprague, A.L., Hunten, D.M., Hill, R.E., Rizk, B., & Wells, W.K. 1996 Martian Water Vapor. J. Geophys. Res. 101(E10), 23,229-23,241.

Sprague, A.L., Hunten, D.M., Doose, L.A., Hill, R.A. & B. Rizk 2001. Water Vapor Abundances over Mars North High Latitude Regions: 1996-1999. Icarus, In Press.

Stern S. A., & B. C. Flynn 1995. Narrow-Field Imaging of the Lunar Sodium Exosphere, Astron. J. 109, 835-841.

Tull, R.G. 1970. High-Dispersion Spectroscopic Observations of Mars IV. The Latitude Distribution of Atmospheric Water Vapor. Icarus 13, 43-57.

Vilas, F., M. A. Leake, & W.W. Mendell 1984. The Dependence of Reflectance spectra of Mercury on Surface Terrain. Icarus 59, 60-68.

Vilas, F. 1988. Surface Composition of Mercury from Reflectance Spectrophotometry. Mercury, University of Arizona Press, Vilas, F., Chapman, C. & Matthews Eds., Tucson, Arizona.

Warrell J. & S. Limaye 2001. Properties of the Hermean Regolith: I. Global regolith Albedo Variation at 200 km Scale from Multicolor CCD Imaging. Planet. & Space Sci. In Press.

Wilson, J. , S. M. Smith, J. Baumgardner, & M. Mendillo 1999. Modeling an Enhancement of the Lunar Sodium Atmosphere and Tail during the Leonid Meteor Shower of 1998, Geophys. Res. Lett. 26, 1645-1648.

Young, A. T. 1975. The Clouds of Venus. J. Atmos. Sci. 32, 1125.

Chapter 3

Planetary Astronomy: Recent Advances and Future Discoveries With Small Aperture Telescopes

Amy A. Simon-Miller
NASA Goddard Space Flight Center
Greenbelt, Maryland USA

Nancy J. Chanover
New Mexico State University
Las Cruces, New Mexico USA

Abstract: Scientific study of the various bodies in our Solar System has been advanced in recent years with new technologies such as adaptive optics, tunable and variable filters and high-resolution infrared detectors. Small aperture telescopes (< 4-m diameter) provide many of the scientific advances and new discoveries in planetary science by allowing the temporal coverage necessary to obtain longitudinal, diurnal and seasonal sampling on a planet. Numerous studies of small bodies in our solar system, the atmospheres and surfaces of the terrestrial planets, the atmospheres of the giant planets and their large satellites, and even the atmospheres of extra-solar planets and brown dwarfs, have been completed with data from such telescopes and instruments. Although an orbiting spacecraft or in situ probe can offer intense study of a small region on a planetary body, they often lack the global view of a planet obtained by Earth-based telescopes. Modest-sized telescopes on Earth can provide self-consistent, global-scale measurements of solar system objects that are unique and complementary to spacecraft exploration. Continuing advances in infrared and filter technology, as well as adaptive optics, will enable new studies that provide answers to fundamental questions about our own Solar System and about extra-solar planets as well.

Keywords: Solar System, aperture telescopes, atmosphere

T.D. Oswalt (ed.), The Future of Small Telescopes in the New Millennium, Vol. III, 37-55.

1. INTRODUCTION

Historically, planetary science was one of the first areas of astronomy to benefit from the development of small telescopes. From the channels on Mars to Jupiter's Great Red Spot or Saturn's rings, the large bodies of the Solar System are accessible through even the smallest telescopes, and spectacular discoveries have been made in this manner. Telescopes with diameters on the order of 4-m or smaller are vital to the exploration of our Solar System. For the purposes of this discussion, all telescopes with diameters ≤ 4-m will be considered small aperture telescopes.

Planetary observations are unique in that the bodies observed can change on very short time scales. First, the rotation of the body and the corresponding longitudinal variability can be observed. Additionally, many planets experience diurnal and seasonal changes. Small Earth-based telescopes are ideal for observing these kinds of phenomena, because the instruments available typically have a large enough field of view to encompass the full disk of a planet, even Jupiter. For comparison, the Cassini spacecraft's Imaging Science Subsystem Narrow-Angle Camera (ISS NAC) has a field of view that is 1238 arcseconds on a side. During the recent distant flyby of Jupiter (10 million km closest approach), the ISS could not image the planet in a single NAC frame, while the spatial resolution of the images was still only 60 km/pixel (about twice as good as a Hubble Space Telescope Planetary Camera image). The Galileo spacecraft Solid State Imager achieved even higher spatial resolution, up to 15-km, but has never returned more than postage stamp views of the planet during its 6+ years in orbit about Jupiter. Typical "narrow" fields of view from Earth-based instruments, for example, the near infrared imager on the Apache Point Observatory's 3.5-m telescope, are on the order of 60 arcseconds on a side. This is large enough to obtain full disk observations of Jupiter, even at Earth opposition, when the planet reaches more than 40 arcseconds in extent.

Another benefit to small Earth-based telescope observations is that they are frequently less over-subscribed than the bigger telescopes or the Hubble Space Telescope (HST), where the usual oversubscription rate is 6:1, and only about 3 to 5% of the approved proposals are Solar System targets. This is partially due to the large amount of time required for planetary observations. Individual planetary image exposures can be very short, as planets are usually bright and long integration is not required for high signal to noise. However, although an individual observation of a planet can be short, a multi-color imaging sequence that achieves full longitudinal coverage can take many concurrent nights (or orbits) of telescope time. Further, due to the intrinsic variability of these objects, it is frequently desirable to achieve this full coverage several times with varying time spans in between data sets. Other observing modes, such as spacecraft mission support imaging, monitoring programs, faint object surveys and high-resolution spectroscopy are also accomplished with small telescopes because of the large time commitments required. These types of programs will not be discussed in much detail here, but are an intrinsic part of Solar System studies. Finally, ground-based telescopes can more quickly observe time-critical transient phenomena, such as the onset of a dust storm on Mars, the merger or outbreak of storms on the giant planets, or the break-up of a comet (Sanchez-Lavega et al. 2001; Weaver et al. 2001, for example). While spacecraft might obtain better spatial resolution and have no risk of bad weather, they require large lead-time because of the reprogramming required to change pre-set sequences and can enter into safing modes at inopportune moments.

In this chapter, we will outline some new technological advances that have improved the planetary science that can be undertaken with small telescopes. We also summarize some recent science results from data obtained with these instruments and small telescopes. We then discuss some of the fundamental unanswered questions of planetary science, as they are now being formulated in the planetary community's first Decadal Study. Finally, we speculate on the next decade of Solar System studies and how they can be accomplished with small aperture telescopes.

2. ADVANCES IN PLANETARY SCIENCE

While all telescope technology is continually changing, perhaps the biggest advances in telescopic observations have come from the major improvements in ground-based infrared technology and detectors. It has long been recognized that spectra or images of a planet in different filters can reveal differences in the composition of the surface or atmosphere, and that the proper absorption filters can reveal information about different heights in an atmosphere. As an example, Figure 1 shows data obtained by the Cassini Imaging Science Subsystem (ISS) on October 8, 2000 as the spacecraft began a distant flyby of Jupiter. The left image, obtained in a violet filter at 445-nm, shows the strongly contrasted, banded appearance typically associated with Jupiter. The middle image, obtained in the ultraviolet at 255-nm, and the right image, taken with a methane absorption band filter at 889-nm, are more sensitive to higher hazes and clouds in the atmosphere. These images are displayed with a spatial resolution of about 930 km/pixel, equivalent to 0.25 arcsecond seeing at Earth.

Figure 1. Images of Jupiter, taken through the Cassini ISS Narrow Angle Camera from a distance of 77.6 million km. The Great Red Spot is prominent in all three images.
Credit: NASA/JPL/University of Arizona

When the spectral range is expanded into the near infrared, other atmospheric layers can also be probed. Depending on the wavelengths chosen, data can sample very high in the atmosphere or very deep, in some cases showing the thermal emission from the hot interior of a Jovian planet. Figure 2 shows the comparison of a HST image at 0.95 microns with an Infrared Telescope Facility 5-micron image of Jupiter. Areas that appear bright at 5-microns are generally less cloudy, allowing the thermal emission

Figure 2. Images of Jupiter taken on 5 October 1996 using HST WFPC2 (left) and NASA IRTF's NSFCAM (right). North is up on both images. The HST image was made by mapping several 953-nm images to create a full-disk map centered on the mean central meridian longitude of the IRTF image (35.9° System III). The map was then reprojected onto the plane of the sky at 0.15 arsec/pix. The IRTF image was taken in a narrow-band filter centered at 4.8 microns; it represents the co-addition of four images then deconvolved with a weighted deconvolution procedure. *Credit: NASA/G. Orton/ R. Beebe*

to escape from the interior. For terrestrial planets, atmospheric constituents such as water and carbon dioxide can also be observed in the near infrared, allowing the detection of thin clouds otherwise not visible.

Surface composition, such as that of Mars or asteroids, can also be studied in the near infrared. Specific minerals, for example, pyroxenes, have spectral signatures in this wavelength region that can be used to distinguish between asteroid classes and surface mineralogy. Longitudinal variability in mineralogy and particle grain size on Mars can also be mapped, as well as the seasonal changes occurring from dust storms or polar cap sublimation. Thus, the near infrared provides more information about the mineralogy of a surface or structure of an atmosphere than visible wavelength data alone.

For these types of specific planetary studies it is important to have high-spectral and spatial resolution data. A high spatial resolution imaging spectrometer with full longitude coverage can determine small local variations in composition or structure. To obtain both good spectral and spatial coverage, new instruments have been developed that take advantage of adaptive optics technology, as well as tunable filters that allow spectral image cubes to be obtained nearly simultaneously.

2.1 Sample Adaptive Optics Systems

Many major telescopes are now being fitted with Adaptive Optics (AO) systems, particularly in the infrared, that flex and tilt the primary or secondary mirror to compensate for atmospheric motion that is detrimental to seeing. Bright planetary targets offer a unique challenge to AO systems because frequently the guide star, which provides the reference wavefront that will be corrected, must be close to the target. Planets are also extended objects and are often larger than the isoplanatic patch being corrected. Despite these problems, the improvement in seeing in many cases can be significant. Here we describe several AO systems in use by planetary scientists, though the list is not all-inclusive.

The Palomar Adaptive Optics (PALAO) system on the Palomar 200-inch Hale Telescope uses visible light from a natural guide star to correct the near-infrared light from a nearby science target over the wavelength range of 1 - 2.5 microns with high-order deformable optics (Troy et al. 2000). Cornell University built an infrared science camera, PHARO, specifically for this system (Hayward et al. 2000), although other instruments can also be used with it. A standard measure of adaptive optical image quality is the Strehl ratio, which is a measure of wavefront error compared to theoretical correction. In good seeing, with a bright guide object, the system can achieve a Strehl ratio of 70% in the K band, 20% or 30% is quite common. The image scale of the PHARO camera, .025 or .04 arcsec/pixel, subsamples the diffraction limit of the telescope (0.05-0.09 arcsec/pixel) at these wavelengths. Preliminary data have been acquired of Titan, Uranus, Neptune and Gliese150C, a low-mass companion to Gliese150A (Dekany et al. 2000).

Figure 3 shows the improvement in atmospheric seeing of Neptune images acquired at 1.2 microns with the use of the AO system. The seeing is improved from ~1 arcsec to 0.2 arcsec and previously fuzzy patches become distinct cloud features. This improvement allows distinct cloud features to be tracked to determine drift rates and evolution of clouds on Neptune (see Roddier et al. 1997, 1998). Similar improvements for Uranus data, from ~1 arcsec to 0.25 arcsec seeing, have been achieved.

Another example of an AO system currently in use is the tip-tilt mechanism for the secondary mirror on NASA's 3.0-m Infrared Telescope Facility (IRTF). This correction can be used with the 1 - 5.5 micron infrared camera (NSFCAM) or the 0.8 - 5.5 micron spectrograph/imager (SpeX). The system routinely achieves image quality of about 0.7 arcsec at K-band wavelengths with a Strehl ratio in the range 5% to 10%, although the goal is to reach the diffraction limit of the telescope (0.2 - 0.4 arcsec). About 50% of IRTF time is devoted to Solar System studies; a portion of that is considered support observations for NASA spacecraft missions (e.g., Jupiter monitoring for the Galileo mission, Mars data to support the Mars Global Surveyor mission and Saturn monitoring for the upcoming Cassini mission). Much of the IRTF's Solar System time, though, is devoted to the pursuit of new and unique scientific discoveries for which a spacecraft is not required.

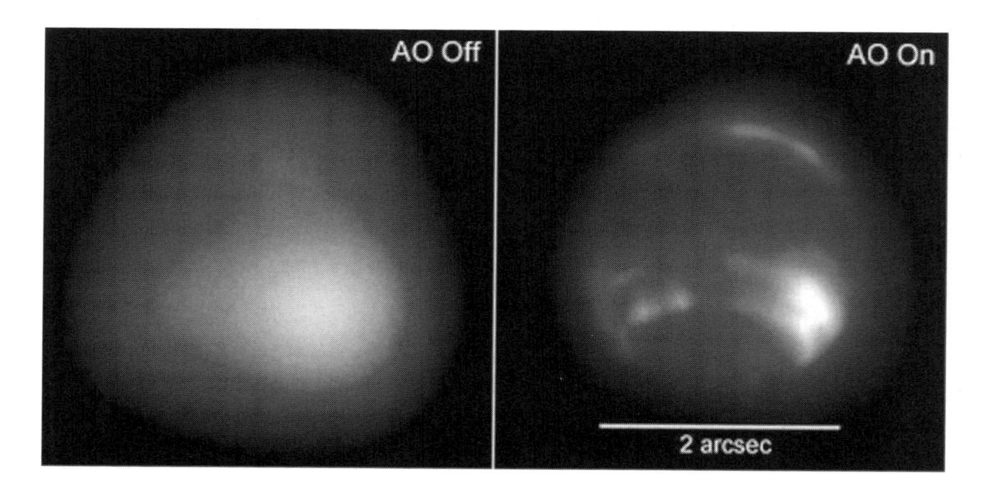

Figure 3. Near-IR images of Neptune taken with the Palomar AO system and PHARO on August 29, 1999. The AO loop was open for the left image, showing the planet's normal appearance in ~1" seeing. With the AO loop closed (right) the angular resolution is about 0.2 arcsec. *Credit: Cornell/JPL/Caltech*

A facility recently made available to the astronomical community through the National Science Foundation is the United States Air Force Research Laboratory's 3.67-m Advanced Electro-Optical System (AEOS) telescope. Part of the Maui Space Surveillance System, this telescope was designed for satellite tracking; it has high slew rates and a deformable mirror AO system that operates between 0.7 and 1.1 microns for near diffraction-limited image quality. Measured Strehl ratios range from 34% to 90%, depending on wavelength. As civilian observations had not been possible until 2000, there are relatively few planetary images or scientific publications available. An example of the outstanding performance of the system during its first open-use cycle is shown in Figure 4, a comparison of Ganymede images from AEOS, HST and Galileo. By degrading the Galileo image to the same resolution as that of AEOS, one can discern details 270-km across on Ganymede, which corresponds to a resolution of 0.09 arcsec, while HST can resolve details 350-km (0.11 arcsec) across.

2.2 Instrument Advances

In addition to the adaptive optics systems routinely operating on public and private telescopes, new instruments and techniques are being developed to allow a greater range of observations in the visible and near-infrared spectrum. These instruments and filters are designed to produce narrow-band imaging of an object at a variety of wavelengths. The advantage of such a narrow filter is in its wavelength discrimination capability. One can isolate molecular absorption bands and image a body in those bands with little or no contribution from the body's continuum signal. This provides better discrimination of phenomena such as the vertical location of clouds in giant planet atmospheres, the location and identification of chromophores in Jupiter's atmosphere, the longitudinal variation of hydrated minerals on Mars, and the variation of haze thickness with depth in Titan's atmosphere. A broad comparison of the instruments to be discussed is shown in Table 1, but detailed information is found in the instruments' respective documentation.

The first type of filter design is the Circular Variable Filter (CVF). As the name implies, a CVF is a ring-shaped, multi-layer, thin-film interference filter mounted on a filter wheel. The thickness of each layer of thin film is a function of the position on the ring. Thus, each element of the ring acts as a narrow bandpass filter, where the central wavelength and band pass are functions of the polar angle on the ring. Therefore, CVFs allow for discrete, selectable, narrow-band imaging over a range of wavelengths. CVFs are now available on many infrared cameras including the IRTF NSFCAM and Infrared Space Observatory ISOCAM. The three CVFs on NSFCAM and the three on ISOCAM span the wavelength ranges of 1.51 - 5.50 microns (with effective widths of ~1.5 - 2%) and 2.5 - 17 microns (with effective widths of ~2 - 5%), respectively. In the case of the ISOCAM, wavelength steps as small as 0.02 microns are possible at the shorter wavelengths.

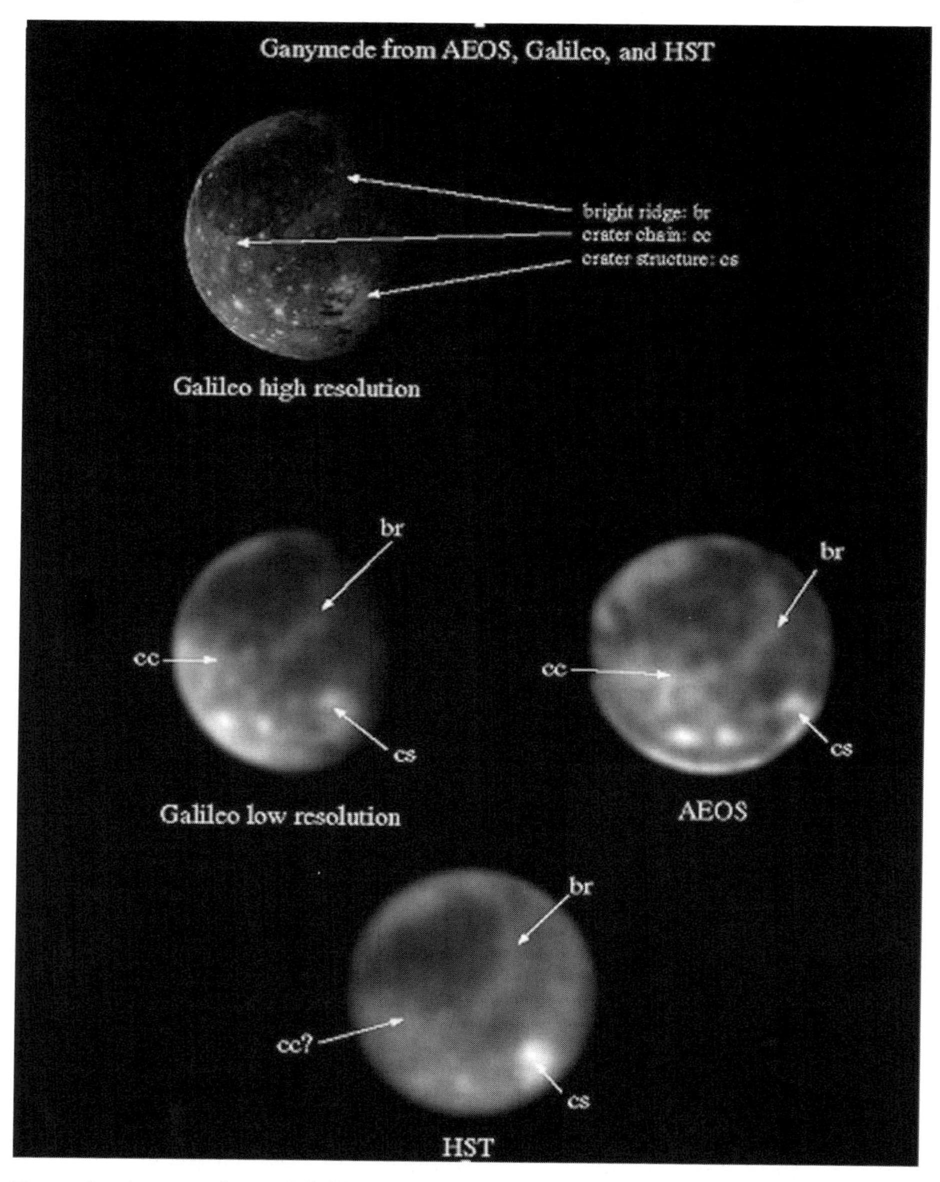

Figure 4. A comparison of Galileo spacecraft, HST, and AEOS 3.67-m adaptive optics images of Ganymede. All images were obtained at a wavelength of 0.8 microns. The upper image shows a full disk image from Galileo, with features as small as 27-km across resolved. *Credit: M. Brown/California Institute of Technology*

Table 1.
Instrument Spectral Characteristics

System	Wavelength (microns)	R = (λ/Δλ)	Notes
NSFCAM	1.51-2.52	~90	CVF#1
	2.70-4.25	~110	CVF#2
	4.45-5.50	~60	CVF#3
ISOCAM	2.27-5.12	41	CVF –SW
	4.96-9.58	37-43	CVF1-LW
	9.00-16.52	35-51	CVF2-LW
AOTF	0.5-1.05	210-530	AImS (visible camera)
	1.6-3.4	390*	*at 2.5 microns (NIR camera)
	2.5-5.2	>150	MWIR camera (test phase)
IRIS	1.48-1.78	300	H-band filter
	2.03-2.36	300	K-band filter
SpeX	0.8-2.5	2000	
	2.4-5.5	2500	

Another type of tunable device is an acousto-optic tunable filter, or AOTF. Several AOTF cameras have been developed by D. Glenar et al. (NASA/GSFC), under NASA's Planetary Instrument Definition and Development Program and Mars Instrument Development Program. General discussions of the spectral and imaging properties of AOTFs are given elsewhere (Gass and Sambles 1991; Glenar et al. 1994), but here we describe the basic principles of their operation. As shown in Figure 5, broadband randomly polarized light is incident on the AOTF and is separated into ordinary and extraordinary polarized components. When RF acoustic waves are coupled into the crystal via a transducer, the refractive index is spatially modulated, producing a phase grating that diffracts one specific wavelength of the incident light. This light is symmetrically deflected into orthogonal polarized beams on exit from the crystal, labeled o' and e' in Figure 5. All other wavelengths pass through the crystal along the incident ray direction. The spectral bandpass function for the instrument is a sinc^2 function, with a full-width at half-maximum that is typically an order of magnitude narrower than standard "narrow-band" filters.

Although TeO_2 is a usable acousto-optic material from 0.4 to 5.2 microns, the useful tuning range of a single device is usually limited to one octave by the RF transducer matching circuit. Optical efficiency within this range is

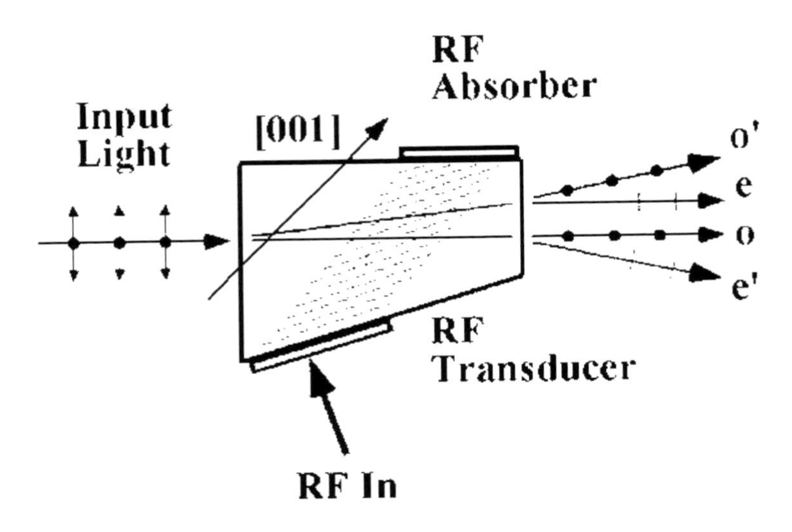

Figure 5. Ray propagation through an AOTF crystal is illustrated. Tuning the RF frequency changes the wavelength of the diffracted light (o', e').

defined by the single polarization efficiency, the fraction of input light in one polarization that is diffracted into the o' or e' output light at the passband peak. This efficiency must be divided by two when comparing AOTFs to other spectral-imaging techniques, since the randomly polarized input light is split into separately measured linearly-polarized output beams.

The Goddard AOTF instruments previously have been used at the Starfire Optical Range 1.5-m telescope at Kirtland Air Force Base, the Apache Point Observatory 3.5-m telescope, and the Mount Wilson Observatory 2.5-m telescope, to observe Jupiter, Saturn, Venus, Mars, and Titan (Glenar et al. 1997; Chanover et al. 1998, 1999, 2000; Hillman et al. 1999). Bright solar system objects are ideal targets for the AOTF instruments since ≤ 25% of the signal is imaged at a given time. While the polarimetry capabilities of the AOTF cameras have not yet been exploited, they can also be used for these kinds of studies.

As a final example, imaging spectrographs have been used extensively for the past ten years on many small aperture telescopes, such as the 3.9-m Anglo-Australian Telescope (AAT) with its InfraRed Imager Spectrographs (IRIS and IRIS2) and the IRTF's SpeX instrument. These instruments can use a normal filter with a cross-disperser (grism) to obtain many wavelengths over the filter range. The AAT IRIS instrument can directly image between 0.9 and 2.5 microns with standard infrared broadband filters. The H and K filters can also be cross-dispersed with a grism, while doing a drift scan perpendicular to the slit, to obtain spatially resolved image cubes with up to 128 narrow-band images over the filter. The new IRIS2, which came on line in 2001, is expected to have a wider field of view and better spectral resolution with more filters available in this mode.

The IRTF SpeX spectrograph uses a two-material dual-prism, instead of a grism, to maintain equal spacing of the cross-dispersed orders. It can acquire images from the slit viewer (before the prism optics) or slit spectra in the 0.8 - 2.5 micron or 2.0 - 5.5 micron range. It is possible to build spectra into near-simultaneous image cubes by slewing perpendicular to the slit, as is done with the AAT IRIS. Its throughput is 15 to 25%, again useful for bright solar system objects. Figure 6 shows an example of Mars band depth maps that can be generated by using a spectrometer to build up "spectral image cubes." They show the spatial variation of a water ice absorption feature, indicating the diurnal evolution of an aphelion cloud (D. Glenar et al. 2000, personal communication). These maps were created using data from the Kitt Peak National Observatory's Cryogenic Spectrograph (CRSP) on the 2.0m telescope. Similar data were recently acquired at the IRTF using SpeX, with higher spectral resolution.

2.3 Recent Science Results

This section summarizes some science results obtained in the past year or so with small aperture telescopes, some taking advantage of the technology just discussed. Of course, not every significant project in planetary sciences can be covered within the scope of this chapter. Instead, a sampling of projects that span the range of solar system objects have been chosen for review. We briefly outline the importance of the activity and which small aperture telescopes were key in obtaining the required data.

Starting closest to home, we begin with the Near Earth Object (NEO) searches. The Spacewatch program was instituted in 1980 to search for NEOs and identify other hazards to the Earth. A 0.9-m telescope at Steward Observatory was dedicated to the project in 1982. In 1997, a 1.8-m telescope was added, with first light achieved in 2000. This program has catalogued hundreds of asteroids and minor planets, including NEOs, helping us to understand the population of bodies in the inner Solar System and their orbital dynamics (see Bottke et al. 2000, for example). In 2000, five comets, a large Trans-Neptunian Object (20000 Varuna) and multiple asteroids were discovered with the 0.9-m telescope. Many future upgrades are planned, including large format CCD mosaics and better systematic NEO searches.

A similar program is now underway with the Lincoln Near Earth Asteroid Research (LINEAR) project. In this program, MIT's Lincoln Laboratory is using a United States Air Force 1m telescope near Socorro, NM to scan the sky at high slew rates. The telescopes were developed as part of a space surveillance system for the Air Force. This program has been in operation

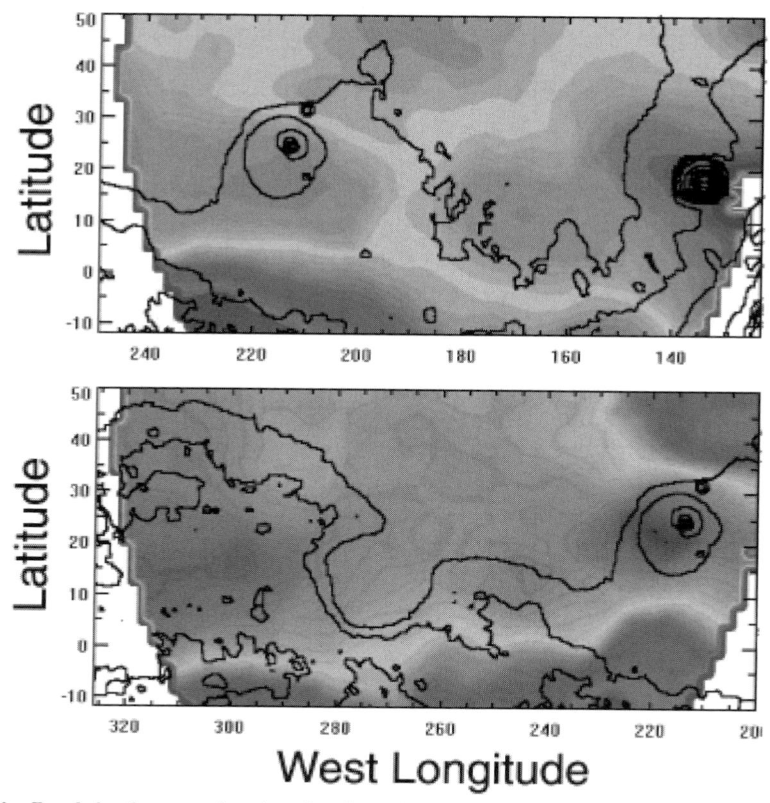

Figure 6. Band depth maps showing the diurnal evolution of a Mars H_2O cloud southwest of Elysium Mons (214° W. Longitude) during Martian Northern summer (L_s =130). Local times at this longitude are 10:10 LST (top) and 15:20 LST (bottom). Intensity represents the strength of the broad 3-micron H_2O ice absorption. Credit: D. Glenar

since 1996, with a large format CCD in place since 1997. The program has now detected millions of objects, with ~60,000 new discoveries and ~400 new NEOs detected. Probably the most exciting of these discoveries was Comet C/1999 S4 LINEAR, which in July 2000 was observed in violent outburst and disintegration (Weaver et al. 2001).

Another type of small body search that has drawn attention in recent years is the search for new faint satellites around the giant planets and asteroids. Studies undertaken with the Palomar 200-inch Hale and the 3.5-m Canada-France-Hawaii Telescopes (CFHT) have searched wide areas around the giant planets to attempt to find new satellites, particularly those in distant orbits within the planets' sphere of influence. These searches have yielded many tentative discoveries of new objects. Follow-up searches on telescopes with diameters of 0.6 to 4.0-m are used to recover the object and refine the orbital trajectory. Once an object's orbit is confirmed, it can be approved for naming by the International Astronomical Union. Since 1997, five new irregular moons have been discovered around Uranus (Gladman et al. 2000a). As the detection technique has improved, smaller telescopes have been used for discoveries; in 2000, three potential new moons of Saturn were discovered with a 2.2-m telescope (Gladman et al. 2000b, 2000c). Searches around all four giant planets will continue, with more discoveries likely as fainter magnitudes become attainable and automated detection techniques improve.

The first ground-based detection of a satellite of an asteroid was made in 1998 (Merline et al. 1999) using the 3.5-m CFHT and its adaptive optics system. Satellites of asteroids were hypothesized for many years before the first one was detected unambiguously by the Galileo spacecraft, when it flew by Ida in 1993. Ground-based detections of asteroid satellites are difficult because until recently, detectors have had neither the dynamic range to detect faint companions next to relatively bright objects, nor the angular resolution to separate out these nearby companions. Their discovery and subsequent ground-based detection are significant as they provide new insight into the collision history of asteroids, which can yield families of similar asteroids. This in turn bears on the bombardment history of the solar system as a whole. In addition, the asteroid satellites can be used to determine physical properties of asteroids; their bulk masses and densities can be determined using the orbital characteristics of the satellites and knowledge of the asteroids' sizes. A systematic survey for additional moons of asteroids using the CFHT is still underway.

Switching from small bodies to giant ones, we turn our attention to Jupiter. In 1994, many of the world's telescopes were poised to watch the impacts of the 21 fragments of Comet P/Shoemaker-Levy 9 into Jupiter's Southern Hemisphere. The impacts occurred on the dark side of the planet, beyond the morning limb. The only direct view of the event was obtained by the Galileo spacecraft while still quite distant from the planet, but Earth-based instruments operating at visible wavelengths were able to detect the massive debris fields generated by the impacts as they rotated into view. Infrared instruments, on the other hand, were able to record the massive fireballs generated as the atmosphere responded to the impact heating. One such project used the AAT IRIS instrument to generated image cubes with 128 wavelengths at H and K-band wavelengths. This yielded light curves of the impacts that showed the impact fireball and in a few cases, the incoming bolides and dust trail as well (Meadows et al. 2001). By comparing the data for different impactors, information about impact velocity and fragment size can be retrieved, while placing limits on atmospheric constituents and pressure. This is just one example of how ground-based telescopes can be used to study a time-critical or unique event. In this case, the impacts were predicted more than a year ahead of time, but the large number of significant observations generated by this event could never have been completed with just one telescope or instrument.

The next large body is Saturn, with its magnificent ring system. The visible atmosphere of Saturn tends to be rather bland, with subtle latitudinal structure. Unlike Jupiter, there are no large long-lived vortices and storms are rare. The infrared view of Saturn is quite different, however. With infrared filters, latitudinal structure can be enhanced and the rings can be much more easily removed. In one study, images were obtained by Yanamandra-Fisher and colleagues with the IRTF NSFCAM using the CVF centered at 5.2 microns, a band that probes thermal structure. The data showed an axisymmetric structure to the atmosphere and pronounced cold spots. These spots had no visible-light counterparts, although some were in dynamically active regions. This suggests that the 5.2 micron band may be a useful wavelength region for probing deeper levels of Saturn's atmosphere than can be seen in the visible (Yanamandra-Fisher et al. 2001). Another recent Saturn study used Palomar H and K-band images and spectra (using a grism with the filters). These wavelengths probe the upper troposphere and lower stratosphere hazes. Hemispheric asymmetry was seen, likely associated with seasonal variations in temperature (Stam et al. 2001). Thus, infrared data of Saturn allow us to probe many levels of the atmosphere to understand the dynamics and structure that are not apparent at visible wavelengths.

Before leaving the Saturnian system, the largest satellite in the Solar System, Titan, also presents an interesting target for infrared instruments. With its thick hydrocarbon hazes, the surface of this satellite is only visible in specific atmospheric windows and discrete cloud features are rare and difficult to track. A priori knowledge of Titan's winds is essential to the success of the Cassini-Huygens Probe Mission, slated to arrive in the Saturnian system in 2004. Using the Goddard InfraRed Heterodyne Spectrograph (IRHS) on the NASA IRTF, Kostiuk and colleagues have obtained observations of stratospheric ethane emission lines near 11.8 microns. Doppler shifts generated by zonal winds shift the frequency position of these lines, allowing the first direct measurement of Titan's global circulation (Kostiuk et al. 2001). Other near-infrared spectroscopic observations made with the 3.8-m United Kingdom InfraRed Telescope (UKIRT) Cooled Grating Spectrometer (CGS4) yielded interesting results about the weather on Titan. Spectra taken between 1.2 and 3.0 microns were used to determine that with the exception of rare, large weather systems, Titan experiences short-lived daily clouds with little surface coverage (Griffith et al. 2000). This has important implications for the existence of a methane cycle on Titan, analogous to the Earth's water cycle of clouds, rain, and oceans.

For the last Jovian planet study, we chose an example of synergistic observations made simultaneously by a spacecraft and a ground-based telescope. Neptune and its largest moon, Triton, were studied in detail by

Sromovsky et al. (2001a,b,c). HST was used to image Neptune and Triton at CCD wavelengths (near-UV to VNIR) in 1996, while simultaneous NSFCAM observations were made at the IRTF, covering the near-IR 1-2.5 micron range. The powerful combination of observations over this wide spectral range yielded important information about Neptune's variable cloud structure and intense zonal winds, and provided new insight into the thermal evolution of Triton. Neptune's lower cloud, thought to be composed of H_2S, was found to be extremely dark beyond 0.7 microns, and the vertical structure of the bright cloud features is now better constrained using the combination of CCD and NIR data. Neptune's circulation was found to be stable, with measured wind speeds matching earlier results from Voyager, with the exception of winds measured near the Great Dark Spot. Despite recent suggestions of seasonal albedo changes on Triton that would be due to "global warming" (Elliot et al. 1998), no evidence of this was found in the combined HST/IRTF data sets (Sromovsky et al. 2001b). This may imply that the distribution of volatile ices on Triton's surface has not changed very much since earlier ground-based and Voyager measurements.

Finally, planetary science extends out to the last planet, Pluto, and to the edge of the Sun's influence with the Kuiper Belt and Oort cloud. These regions are believed to be reservoirs of the short- and long-period comets, respectively, and contain objects that have likely undergone very little alteration since their formation from the protoplanetary disk. Therefore, they serve as windows to the early solar system, and observation and characterization of Kuiper Belt Objects (KBOs) and comets can provide us with clues to the dynamical and chemical evolution of our solar system. There are roughly 400 known KBOs to date, the majority of which have been discovered with modest-sized telescopes. Despite the fairly large sample size, the basic physical characteristics (*e.g.*, size and albedo) of most KBO's are not well known due to their relative faintness and their large distance from Earth. However, many modest-sized telescopes (e.g., the CFHT, the University of Hawaii 2.2-m, the 3.6-m Telescopio Nazionale Galileo and the Complejo Astronomico El Leoncito 2.1-m telescope) continue to play a role in the characterization of the brighter objects (Trujillo et al. 2001, Jewitt et al. 2001, Gil-Hutton and Licandro 2001). Kuiper Belt Objects are now known to be divided into three dynamical classes with orbits of varying semi-major axes and eccentricities, and one object was found to have a higher albedo than the "standard" adopted value (Jewitt et al. 2001), making it brighter than other observed short-period comet nuclei but darker than Pluto. Even small telescopes ($\leq$ 1-m) contribute significantly to KBO research by conducting wide field surveys for bright objects in the Kuiper Belt (Sheppard et al. 2001).

Now we move outside of our Solar System to briefly discuss extrasolar planet and brown dwarf studies, because these objects are intimately linked with the study of giant planets. Many of the dozens of known extrasolar planets are large bodies with masses many times that of Jupiter (at least 20 have masses comparable to that of Jupiter) (Marcy et al. 2000). These objects have put into question the theories of solar system formation and planetary accretion. An increased understanding of Jovian atmospheric evolution and composition will be important in comparing these new planets. Many of the initial extrasolar planet surveys used high-resolution spectra obtained with the Lick Observatory 3.0- and 0.6-m telescopes (Marcy and Butler 1996, for example). New southern hemisphere surveys have used the AAT 3.9-m telescope (Tinney et al. 2000).

Besides searching for low mass planets, other researchers seek to find brown dwarfs, the class of objects spanning the mass gap between Jupiter and the lowest mass stars. Incapable of achieving hydrogen fusion, these 'failed' stars, also classified as L or T dwarfs, can only burn deuterium and may be cool enough to form molecules and possibly condensate clouds (Lodders 1999, Marley et al. 1999). These objects have been detected in abundance by many sky surveys, including the Two Micron All Sky Survey (2MASS) with two 1.3-m telescopes and the Sloan Digital Sky Survey (SDSS) with a 2.5-m telescope (Kirkpatrick et al. 2000, Tsvetanov et al. 2000, for example). Follow-up observations of these objects are being made with the Cornell Massachusetts Slit Spectrograph (CorMASS) on the Palomar 1.5-m telescope (Burgasser et al. 2000) and the UKIRT 3.8-m telescope. Both photometric and spectroscopic follow-up observations of newly discovered brown dwarfs are essential for establishing their spectral types, which yields further information about their effective temperatures and molecular opacity sources. Like the giant planets, many molecules can be detected in the atmospheres of brown dwarfs using modest-sized telescopes, including water, methane and carbon monoxide (Noll et al. 2000, Saumon et al. 2000). This provides increased linkage between brown dwarfs and giant planets, emphasizing the importance of understanding our own solar system bodies as we begin to explore other planetary/stellar systems.

3. THE NEXT DECADE OF PLANETARY SCIENCES

There are many questions left unanswered in planetary science: 1.) How do solar systems form and do they all follow the same process? 2.) Is there or has there ever been water on Mars? 3.) Was there ever life, of even the simplest form, on Mars or Europa? 4.) What colors the clouds of Jupiter? 5.) How do new extrasolar planets compare with the Jovian planets and terrestrial planets in our solar system? 6.) Are we in danger of another mass extinction by an asteroid impact on Earth? plus many more.

What progress can be made toward answering these fundamental questions in the next decade? The planetary community is in the process of its first Decadal Study, designed to examine these fundamental questions and prioritize attempts to answer them. Small telescopes and new technology play an important role in achieving these goals.

As discussed in the Recent Science Results section (2.3), small ground-based telescopes are now being used to complete an inventory of objects in our solar system, both near the Earth and at the very edges of the Sun's influence. They are also being used to catalog extrasolar planets and low mass stars. Currently, adaptive optics technology works best in the infrared and with reasonably bright guide objects close to the science target. Improvements in the wavelength coverage, stability and correction capability of adaptive optics systems are expected over the next decade, particularly as classified military technology becomes publicly available. With advances in adaptive optics technology, small telescopes can begin to approach the diffraction limit, enabling observations of smaller and fainter bodies and better utilizing the outstanding spectral capability of the new instrumentation.

The newest technology advances in interferometry can also improve the spatial resolution available with small telescopes. Several projects are currently in the planning stages, with a few already under construction. These include Georgia State's Center for High Angular Resolution Astronomy (CHARA), which has built an array of six 1m telescopes on Mt. Wilson for optical interferometry. In a proof-of-concept test, first fringes were detected in 1999 and the site dedicated in November 2000. The project is expected to be fully operational by the end of 2001. With a 350-m largest baseline, it should achieve spectacular 200 microarcsecond resolution.

Mt. Wilson Observatory also houses the U. C. Berkeley Infrared Spatial Interferometer (ISI) and the Naval Research Laboratory (NRL) Mark III Optical Interferometer. The ISI system is a two-element interferometer with 2.0-m telescopes operating at 11 microns. The NRL Mark III is a prototype for a larger facility and operates with variable baselines between 3- and 31-m. A final example involving the NRL and the U. S. Naval Observatory is the Navy Prototype Optical Interferometer (NPOI) at Lowell Observatory. Up to six telescopes can be operated on the three 38-m arms of the interferometer, yielding baselines from 7- to 64-m.

While the majority of the science already performed at these interferometers has focused on astrometry and stellar astrophysics, many new planetary studies could be accommodated with the now-proven technology. Possible observations include the search for and direct imaging of extrasolar planets, detailed views of distant Solar System bodies like Pluto and Kuiper Belt Objects, and even the detection of individual "sunspots" or condensate clouds on nearby brown dwarfs.

With retrofits for new technology, nearly any small telescope can be transformed into a state-of-the-art facility. Does this leave any place for older systems without any planned upgrades? For planetary astronomy the answer is a clear and resounding "yes." As mentioned in the Introduction, planetary science is full of time-critical and transient events, often with little advance warning. The rapid response and flexible scheduling possible with ground-based telescopes is essential in studying these phenomena. Additionally, the time-intensive monitoring required to observe temporal and longitudinal variations often drives the need to apply to less oversubscribed telescopes. Without the ≤ 4-m class telescope, it is unlikely that many of the unsolved mysteries of the Solar System will be unraveled over the next decade.

4. REFERENCES

Bottke, W., R. Jedicke, A. Morbidelli, J. Petit and B. Gladman (2000). Science 288, 2190-2194.

Burgasser, A., et al. (2000). AJ 120, 1100-1105.

Chanover, N., D. Glenar and J. Hillman (1998). JGR 103, 31335-31348.

Chanover, N., D. Glenar and J. Hillman (1999). In Catching the Perfect Wave: Adaptive Optics and Interferometry in the 21st Century (S. Restaino, W. Junor and N. Duric, eds.), ASP Conf. Ser. 174, 75-79.

Chanover, N. and D. Glenar (2000). BAAS 32, 1025.

Dekany, R., et al. (2000). In Adaptive Optical Systems Technology (P. Wizinowich, ed), Proc. SPIE 4007, 811-815.

Elliot, J., et al. (1998). Nature 393, 765-767.

Gass, P. A. and J. R. Sambles (1991). Opt. Lett. 16, 429.

Gil-Hutton, R. and J. Licandro (2001). Icarus 152, 246-250.

Gladman, B., J. Kavelaars, M. Holman, J. Petit, H. Scholl, P. Nicholson and J. Burns (2000a). Icarus 147, 320-324.

Gladman, B., J. Kavelaars, J. Petit, H. Scholl, M. Holman, B. Marsden, P. Nicholson and J. Burns (2000b). IAU Circ. 7512, 1.

Gladman, B., J. Kavelaars, J.-M. Petit and P. Nicholson (2000c). IAU Circ., 7539, 1.

Glenar, D., J. Hillman, B. Saif and J. Bergstralh (1994). Appl. Opt. 33, 7412-7424.

Glenar, D., J. Hillman, M. LeLouarn, R. Fugate and J. Drummond (1997). PASP 109, 326-337.

Glenar, D., G. Bjoraker, J. Pearl and D. Blaney (2000). BAAS 32, 1118.

Griffith, C., J. Hall and T. Geballe (2000). Science 290, 509-513.

Hayward, T., B. Brandl, B. Pirger, C. Blacken, G. Gull, J. Schoenwald and J. Houck (2000). PASP 113, 105-118.

Hillman, J., D. Glenar, F. Espenak, N. Chanover, J. Murphy, L. Young and W. Blass (1999). BAAS 31, 1146.

Jewitt, D., H. Aussel and A. Evans (2001). Nature 411, 446-447.

Kirkpatrick, J., et al. (2000). AJ 120, 447-472.

Kostiuk T., K. Fast, T. Livengood, T. Hewagama, J. Goldstein, F. Espenak and D. Buhl (2001). GRL 28, 2361-2364.

Lodders, K (1999). ApJ 519, 793-801.

Marcy, G. and P. Butler (1996). ApJL 464, L147.
Marcy, G., et al. (2000). In Bioastronomy 99: A New Era in the Search for Life in the Universe. (G. Lemarchand and K. Meech, eds.), ASP Conf. Ser. 213, 85.
Marley, M., C. Gelino, D. Stephens, J. Lunine and R. Freedman (1999). ApJ 513, 879-893.
Meadows, V., D. Crisp, J. Barnes, G. Orton and J. Spencer (2001). Icarus 152, 366-383 .
Merline, W., et al. (1999). Nature 401, 565-569
Noll, K, T. Geballe, S. Leggett and M. Marley (2000). ApJL 541, L75-L78.
Roddier, F., C. Roddier, A. Brahic, C. Dumas, J. Graves, M. Northcott and T. Owen (1997). Planet. Space Sci. 45, 1031-1036.
Roddier, F., C. Roddier, J. Graves, M. Northcott and T. Owen (1998). Icarus 136, 168-172.
Sanchez-Lavega, A., et al. (2001). Icarus 149, 491-495.
Saumon, D., et al. (2000). ApJ 541, 374-389.
Sheppard, S., D. Jewitt, C. Trujillo, M. Brown and M. Ashley (2001). AJ 120, 2687-2694.
Sromovsky, L., P. Fry, K. Baines, S. Limaye, G. Orton and T. Dowling (2001a). Icarus 149, 416-434.
Sromovsky, L., P. Fry, K. Baines and T. Dowling (2001b). Icarus 149, 435-458.
Sromovsky, L., P. Fry, T. Dowling, K. Baines and S. Limaye (2001c). Icarus 149, 459-488.
Stam, D., D. Banfield, P. Gierasch, P. Nicholson and K. Matthews (2001). Icarus 152, 407-422.
Tinney, C., P. Butler, G. Marcy, H. Jones, A. Penny, S. Vogt, K. Apps and G. Henry (2001). ApJ 551, 507-511.
Troy, M., et al. (2000). In Adaptive Optical Systems Technology (P. Wizinowich, ed), Proc. SPIE 4007, 31-40.
Trujillo, C. A., D. C. Jewitt and J. X. Luu (2001). AJ 122, 457-473.
Tsvetanov, Z., et al. (2000). ApJL 531, L61-L65.
Weaver, H, et al. (2001). Science 292, 1329-1334.
Yanamandra-Fisher, P., G. Orton, B. Fisher and A. Sanchez-Lavega (2001). Icarus 150, 189-193.

Chapter 4

Science With Very Small Telescopes (< 2.4 meters): The NASA Deep Impact Mission's Small Telescope Science Program

S. A. McLaughlin, L. A. McFadden
University of Maryland
Maryland USA

G. Emerson
Ball Aerospace and Technologies Corporation
Colorado USA

Abstract: Deep Impact, a NASA Discovery mission, established the Small Telescope Science Program (STSP) to provide baseline information about the target comet and to complement scientific data acquisition by larger telescopes. The STSP is a network of advanced amateur, student, and professional astronomers who use very small telescopes (< 2.4-m) equipped with charged-coupled devices (CCDs) to make scientifically meaningful observations of comet 9P/Tempel 1. We describe the STSP and its goals, present preliminary photometric results based on broadband-R data acquired by STSP participants, identify other research projects enlisting STSP participants, and recommend guidelines for establishing similar, small-telescope observing programs.

Key words: comets, dust coma, broadband photometry, professional-amateur collaborations

1. INTRODUCTION

Deep Impact is a NASA Discovery mission for solar system exploration. The fundamental scientific goal of the mission is to explore the interior of a cometary nucleus. To achieve this goal, the Deep Impact flyby spacecraft will release a 370-kg auto-guided impactor towards comet 9P/Tempel 1 during July 2005. Twenty-four hours after its release, the impactor will hit

T.D. Oswalt (ed.), The Future of Small Telescopes in the New Millennium, Vol. III, 57-75.

the nucleus. Mission scientists predict the impactor will excavate a >28-meter-deep and >100-meter-wide crater. The flyby spacecraft will observe the cratering process and ejecta. Ground-based telescopes will observe the event and continue to monitor the comet for several months, looking for deviations from measurements made before the impact.

We enhance the success of the mission by improving our knowledge of the physical nature of the comet. For example, the size distribution and production rate of dust is critical to the design of the flyby spacecraft, the impactor, and the onboard instruments. However, we have few observations of this Jupiter-class comet since its discovery in 1867: Tempel 1 was lost between 1869 and 1972 because of significant gravitational perturbations by Jupiter. Therefore, Deep Impact astronomers, coordinated by Karen Meech at the University of Hawaii, began intermittent observations of Tempel 1 in 1998 using large telescopes. The observations continue through 2002. To fill the large temporal gaps in these data, Deep Impact implemented the STSP in early 2000 to engage very-small-telescope observers from around the world in this scientific endeavor.

2. DEVELOPMENT AND IMPLEMENTATION OF THE STSP

2.1 Why Very Small Telescopes?

Sky and Telescope's executive editor, Kelly Beatty, estimates there are 100 to 200 advanced amateur astronomers in the United States and many more scattered around the world (private communication). These observers typically use small, fast, wide-field telescopes equipped with commercial thermo-electrically-cooled CCD cameras. Many of these observers have broadband photometric UBVRI filters. The wide-field formats image the entire coma of Tempel 1, facilitating dust-environment studies. Wide formats capture field stars in the same frame as the comet, simplifying calibrations through use of differential photometry on the comet and field stars. Finally, a worldwide network of very small telescopes has the potential of providing good temporal coverage of the comet.

2.2 Potential Scientific Contributions

During the design phase of the STSP, we proposed several scientific applications of very small telescopes to the Deep Impact mission. As evidenced by the success of the American Association of Variable Star Observers (AAVSO, http://www.aavso.org/publications/publications-online.stm) and asteroid observing programs such as Asteroid Photometry at the Parker Observatory (Warner 1999-2001), very small telescopes equipped with CCDs have the potential of contributing data for photometric analysis. In particular, we proposed to perform photometry of the dust coma using broadband-R and broadband-I CCD images. There are little or no gas emissions in these bands. This photometry would allow us to generate a light curve for Tempel 1 and to estimate its dust production rates. We suggested using broadband-V images to aid the calibration process. We considered producing a total visual-magnitude light curve based on unfiltered CCD images.

In addition to photometric analysis, we noted the potential of using the CCD images for rotation rate analysis, jet activity studies, and astrometry. We realized that images of Tempel 1 during 2000 would form a baseline for the impact event and future apparitions.

2.3 Equipment Requirements and Observing Procedures

As a minimum program goal, we wanted the STSP data to have a quality good enough for photometric measurements of the dust coma. Tempel 1 is a relatively faint comet, with a maximum brightness near the 9th magnitude during perihelion (Green 1994a). Therefore, we developed equipment requirements and observing procedures tailored to faint comets and defined data format instructions and observing log guidelines. Our methodology was confirmed at the Pro-Am Session of the 196th American Astronomical Society meeting in June 2000. Several amateur astronomers expressed interest in contributing to astronomical research but stated the need for well-defined observing goals and procedures to ensure the usefulness of their data.

We requested observers to supply raw, unprocessed comet images and calibration frames (dark, bias, and flats) so that data reduction could be uniform. We planned to run all STSP images through the same IDL reduction and photometric routines, in-house, at the University of Maryland.

2.4 Observing Phases

We defined two observing phases based on the observability of Tempel 1. Phase 1 extended from February 2000 through March 2001. The purpose of this phase was to establish the STSP network, gather broadband VRI and unfiltered CCD images of the comet, analyze the images for scientific information, and make recommendations for Phase 2. We also expected this phase to produce a baseline of the post-perihelion activity of the comet.

Phase 2 observing begins in late 2003 and continues through 2005. We will request broadband VRI images of Tempel 1 mainly for photometric purposes. Similar to Phase 1, we expect one result to be a baseline of the pre-perihelion activity of the comet. After the impact in 2005, we expect STSP data to provide a baseline for post-impact behavior and a comparison tool to search for differences from the 2000 post-perihelion baseline.

2.5 Establishing the Network

By late 1999, Gary Emerson, of Ball Aerospace and Technologies Corporation, had gathered 5 amateur astronomers willing to participate in the STSP. To grow the network during early 2000, we issued calls for observations of Tempel 1 to various astronomy-related groups and publications including the Astronomical League, the British Astronomical League, the International Comet Quarterly, and Sky and Telescope's Comet Alert.

Simultaneously, we implemented a website to serve as the communication backbone for the program. We included potential scientific goals, equipment requirements, observing procedures, data-format and transfer instructions, a message board, observers' profiles, and an observers' image gallery. The website is http://deepimpact.umd.edu/stsp. A mirror site is http://deepimpact.jpl.nasa.gov/stsp.

3. PHASE 1 OF THE STSP

3.1 The Network: Observers and Instrumentation

In Table 1, we list the 44 observers who volunteered for the STSP during 2000-2001. They include advanced amateur, student, and professional astronomers. Participants' observing locations and instrumentation are included in the table and illustrate the heterogeneity inherent in such an observing program. All but 4 of the observers were able to obtain CCD images of Tempel 1 during Phase 1. Three observers have not yet transmitted their data.

Table 1.

STSP Observers and Instruments

Observers	Location	Tele[1]	Aper[2] (cm)	Focal Ratio	CCD	X Pix[3] (")	Y Pix[3] (")	Filter[4] Type
Peter Birch (PB)	Perth Obs., Australia	C	61.0	8.0	SBIG ST-7	0.60	0.60	J
Robert Bolster	Hopewell Obs., Virginia, USA	SC	36.0	7.0	SBIG ST-8	1.50	1.50	N
Dennis Borgman, Tracy Knauss, Barbara Wilson (DB, DB2)	George Obs., Texas, USA	C	91.0	7.6	SBIG ST-9 Apogee AP-8	1.19 2.14	1.19 2.14	C C
Maurice Clark (MC)	Chiro Obs., Perth, Australia	N N	20.3 30.5	6.0 6.0	Pictor 416XTE Pictor 416XTE	1.52 1.02	1.52 1.02	N B
Bill Dillon, Cynthia Gustava, Keith Rivich	George Obs., Texas, USA	N	46.0	4.5	SBIG ST-7	2.70	2.70	N
Gary Emerson (GE)	E.E. Barnard Obs., Colorado, USA	C	25.4	4.5	HiSIS-22	3.25 (2x2)	3.25 (2x2)	J
Mike Fuller	Ft. McKavett, Texas, USA	SC	30.5	6.3	Pictor 416XTE	1.02	1.02	N
Gordon Garradd (GG, GG2)	Loomberah, Australia	N	45.0	5.4	Apogee AP-7	2.00	2.00	B
	MDM Obs., Arizona, USA	RC	236.0	7.5	SITe	1.02 (2x2)	1.02 (2x2)	J
Todd Giencke	Wayside Obs., Minnesota, USA	S	20.0	1.93	SBIG ST-237A	3.89	3.89	B
Ian Griffin	Mt. John Obs., New Zealand	C	100.0	7.7	SITe 1024	0.64	0.64	N
Cristovão Jacques, Luiz Duczmal	Wykrota Obs, Brazil	SC	30.5	6.0	MX916	2.44	2.44	J
William Liller	Vina del Mar, Chile	SCam	20.3	1.5	SBIG ST-5	6.80	6.80	J
Hugh Lund	Johannesburg, South Africa	N	31.8	6.0	TC245 chip	2.78	2.16	J

Observers	Location	Tele[1]	Aper[2] (cm)	Focal Ratio	CCD	X Pix[3] (")	Y Pix[3] (")	Filter[4] Type
Rob McNaught (RM)	Siding Spring Obs., Australia	RC	101.6	8.0	SITe	1.21	1.21	J
Gianluca Masi	Belletrix Astro. Obs., Italy	SC	28.0	3.3	SBIG ST-7	2.00	2.00	B
Marko Moilanen, Arto Oskanen (MM)	Nyrölä Obs., Finland	SC	40.0	6.3	SBIG ST-7	1.60	1.60	J
Akimasa Nakamura	Kuma Kogen Obs, Japan	RC	60.0	5.8	Photometrics	1.37	1.37	N
Arvind Paranjpye	TIE, Mount Wilson, California, USA	C	61.0	3.5	SBIG ST-6	2.22	2.60	N
Ralph Pass	Massachusetts, USA	SC	25.4	6.3	Pictor 1616XTE	1.16	1.16	R
Tim Puckett	Puckett Obs., Georgia, USA	RC	61.6	8.0	Apogee AP-7	1.00	1.00	N
Veselka Radeva	National Astro. Obs., Bulgaria	S	50.0	3.5	SBIT ST-8	1.1	1.1	J
Pedro Ré	Cascais, Portugal	SC	35.6	6.0	SBIG ST-7	0.90	0.90	B
S. Sánchez, J. Rodriguez	Mallorca, Spain	S	40.0	2.0	SBIG ST-8E	2.32	2.32	N
Giovanni Sostera	Remanzacco Astro. Obs., Italy	BS	31.0	2.8	HiSIS 24	2.40	2.40	J,C
Paola Tanga, Alberto Cellino	Strada del'Osservatorio, Italy	C	100.0	16.0	SITe 2048	0.31	0.31	B
Andrew Timko	Woomera, Australia	SC	25.4	6.3	SBIG ST-7	1.16	1.16	B
Martin Tschimmel, Christoph Ries, Otto Baernbantner, Heinz Barwig (MT)	University of Munich Obs., Germany	RC	80.0	12.4	"MONICA" with TK1024	0.50	0.05	B

Observers	Location	Tele[1]	Aper[2] (cm)	Focal Ratio	CCD	X Pix[3] (")	Y Pix[3] (")	Filter[4] Type
Craig Tupper, Janet Tupper	Virginia, USA	SC	25.4	5.1	SBIG ST-7	1.3	1.3	N
Paul Weissman (PW)	Table Mountain Obs., California, USA	RC	61.0	16.0	Photometrics	0.52	0.52	J
Maik Wolleben	Hoher List Obs., Germany	C	106.5	14.5	SITe 2048	0.85	0.85	J
Kiriaki Xiluri, Eugene Laura	Fan Mountain Obs., Virginia, USA	C	101.6	13.5	SITe 2048	0.36	0.36	N

[1] Telescope: BS = Baker-Schmidt, C = Cassegrain, N = Newtonian, RC = Ritchey-Chretien, S=Schmidt, SC = Schmidt-Cassegrain, SCam = Schmidt Camera, n/a = not available.
[2] Diameter of the aperture.
[3] Unbinned pixel scale, unless noted.
[4] Filter type: B = Bessell, C = Cousins, J = Johnson, N = None, R = RGB

3.2 Observations

From February 2000 through February 2001, STSP observers took unfiltered and broadband-VRI CCD images of Tempel 1. During late January 2001, several observers (Birch, Emerson, Garradd, private communication) reported imaging was quite challenging because the comet dimmed to an observed magnitude of 19.5_R (from ~18.0_R only one month earlier). By March 2001, observing for Phase 1 ended.

STSP observers produced 445 R-band, 260 V-band, 5 I-band, and 335 unfiltered images of the comet. In addition, we received over 1000 supporting frames for the reduction process. Figure 1 combines the STSP observations with those taken at large telescopes for Deep Impact purposes. From the figure, we see that the STSP observations provided good temporal coverage of Tempel 1.

3.3 Science from STSP Images

After inspecting some of the broadband-R images made during February 2000 through July 2000, we determined that photometric measurements of the dust coma would be the most useful application of STSP data to the Deep Impact mission. Broadband-R filters isolate continuum light scattered by dust in the coma, and there is little or no emission in this band pass. We proposed

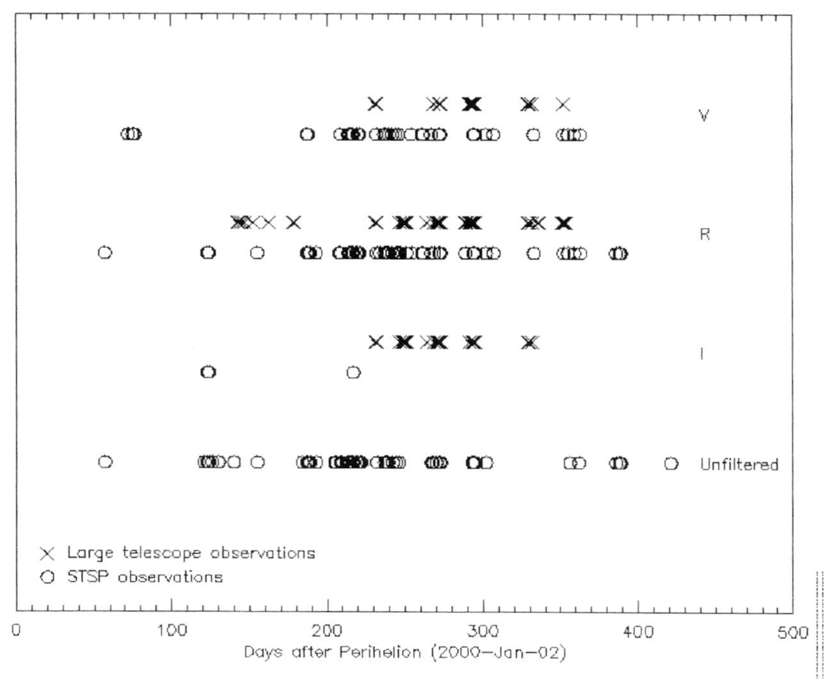

Figure 1. Observations of Tempel 1 taken during 2000 for the Deep Impact mission.

to use the photometric results to construct a light curve for the dust coma in search of brightness variations with respect to heliocentric distance. We planned to estimate the production rate of dust from the photometric results. Broadband-R measurements also take advantage of the facts that CCD chips are most responsive to red light and atmospheric extinction is less at red wavelengths than visual wavelengths. Additionally, we compared the photometry of broadband-R images taken by the George Observatory team on September 30, 2000, to photometry of broadband-R images taken by Karen Meech on the same date with the University of Hawaii's 2.2-m telescope (Lisse et al 2003). The STSP results were comparable to the 2.2-m results.

The uncertainties in the STSP results were small enough that Karen Meech agreed that we should supplement large telescope data with STSP data (Meech 2000). With this encouraging result, we proceeded with photometric analysis of the broadband-R data.

We investigated the possibility of estimating the rotation rate of the nucleus using the broadband-R data. In an active comet, a rotating nucleus is expected to cause periodic changes in the brightness of the coma. This effect can be measured if the dust crossing time in the photometric aperture, t_d, is much less than the rotation rate of the nucleus, T_{best} (Jewitt, 1991). The dust crossing time is estimated using $t_d / T_{best} << 1$, where $t_d \approx 0.35*p$ (arc seconds)

* Δ(AU) * r(AU)$^{0.5}$, p is the aperture used for photometry in arc seconds, Δ is the geocentric distance in AU, and r is the heliocentric distance in AU. For many STSP images, seeing often forced us to use photometric apertures of at least 5 arc seconds. Letting p = 10 arc seconds, Δ = 2.0 AU, r = 2.0 AU, and T_{best} = 21 hours, $t_d / T_{best} \approx 0.5 \approx 1$. We deduced that to detect the effects of rotation in the photometry of Tempel 1's coma, we would need to use a very small photometric aperture, thus requiring excellent seeing. Additionally, images should have temporal coverage that spans several hours and days. With the exception of Birch's images (small pixel scales, good seeing, and good temporal coverage), we determined that STSP images did not meet the requirements for rotational studies.

We established that generating a total visual light curve from the unfiltered CCD images of Tempel 1 was not a viable research goal for Phase 1. Unfiltered CCD images capture visible light scattered by dust as well as gas emissions. Without narrowband observations, it can be very difficult to separate scattered light from gas emissions.

After performing image processing on several frames of Tempel 1, we determined that the relatively large pixel scale of most of the images eliminated processing images in search of dust structure and jet activity.

We decided that photometric measurements of the dust coma in the broadband-R images would be the most scientifically useful information we could extract from STSP data. The following sections present preliminary results of our photometric analysis.

3.4 Analysis of Broadband-R Images

We present preliminary photometric analysis of broadband-R data taken by STSP observers from February 2000 through December 2000. In Table 2, we present the broadband-R data used in our analysis.

3.5 Data Reduction

We developed a set of IDL routines, called "IMPACT" to perform consistent reduction and photometric analysis of Tempel 1 images. We inspected the results of each reduction step for anomalies such as cosmic ray hits, hot pixels, and cold pixels. We applied corrections only if the anomalies existed in areas of the image to be used for photometric measurements.

While taking photometric measurements, we scaled the photometric aperture such that it equated to a diameter of 27,000-km at the comet. We chose this diameter to ensure that most of the coma was integrated when the comet was at small heliocentric distances and that seeing limitations were

Table 2.
Journal of STSP, Broadband-R Observations

Obs #	Obs	UT Date (2000)	UT Start Time	Days After Perihelion	R (AU)	Δ (AU)	α Phase Angle (deg)	Phase Funct	Air Mass	# of Exp	Exp Time (s)	X FOV (')	Y FOV (')
1	RM	02-28	18:36:05	57.145	1.605	2.308	20.89	0.810	3.129	1	40	8.07	8.07
2	PB	05-04	20:45:33	123.235	1.914	2.184	27.50	0.793	1.765	1	1200	5.12	5.12
3			21:28:04	123.264	1.914	2.184	27.50	0.793	1.452	1	1200	5.12	5.12
4	PB	05-05	20:34:57	124.228	1.920	2.181	27.56	0.793	1.859	1	1200	5.12	5.12
5			20:5:057	124.243	1.920	2.181	27.56	0.793	1.646	1	1800	5.12	5.12
6	PB	06-05	20:09:59	155.210	2.096	2.043	28.35	0.793	1.530	1	1800	5.12	5.12
7			21:32:29	155.268	2.097	2.042	28.35	0.793	1.192	1	1800	5.12	5.12
8	DB	07-08	09:48:00	187.778	2.288	1.855	25.80	0.793	1.545	1	240	5.08	5.08
9			10:03:47	187.789	2.288	1.854	25.80	0.793	1.482	1	240	5.08	5.08
10			10:21:20	187.801	2.288	1.825	25.80	0.793	1.421	1	300	5.08	5.08
11	GE	07-13	09:03:40	192.773	2.318	1.825	24.98	0.793	2.112	1	300	20.80	20.80
12			09:47:25	192.778	2.318	1.825	24.98	0.793	2.042	1	300	20.80	20.80
13			09:53:56	192.782	2.318	1.825	24.98	0.793	1.995	1	300	20.80	20.80
14			10:00:20	192.787	2.318	1.825	24.98	0.793	1.994	1	300	20.80	20.80
15	PB	08-03	17:50:15	214.113	2.443	1.716	19.92	0.814	1.265	1	300	5.12	5.12
16			18:11:42	214.128	2.443	1.716	19.91	0.814	1.203	1	300	5.12	5.12
17			18:18:07	214.133	2.443	1.716	19.91	0.814	1.186	1	600	5.12	5.12
18			19:12:38	214.170	2.444	1.716	19.90	0.814	1.095	1	600	5.12	5.12
19	GG	08-04	18:33:46	215.143	2.449	1.712	19.61	0.816	1.082	9	450	17.07	17.07
20	DB	08-05	08:22:39	215.719	2.453	1.710	19.43	0.817	1.509	1	240	5.08	5.08
21			08:59:18	215.745	2.453	1.710	19.42	0.817	1.399	1	240	5.08	5.08
22	PW	08-05	10:41:09	215.815	2.453	1.710	19.40	0.817	1.493	1	300	8.87	8.87
23			10:49:39	215.821	2.453	1.710	19.40	0.817	1.479	1	600	8.87	8.87
24			11:02:41	215.830	2.453	1.710	19.40	0.817	1.461	1	600	8.87	8.87
25	DB	08-06	09:00:30	216.745	2.459	1.706	19.12	0.819	1.391	1	240	5.08	5.08
26			09:37:09	216.771	2.459	1.706	19.11	0.819	1.339	1	240	5.08	5.08
27	GG	08-06	15:30:45	217.016	2.460	1.705	19.03	0.820	1.245	6	360	17.07	17.07
28			15:39:10	217.022	2.460	1.705	19.03	0.820	1.218	1	120	17.07	17.07
29	GE	08-08	08:51:00	218.739	2.470	1.699	18.49	0.823	1.847	1	300	20.80	20.80
30			10:10:00	218.794	2.471	1.699	18.47	0.824	1.632	1	300	20.80	20.80
31	MT	08-26	02:46:09	236.485	2.573	1.664	12.21	0.876	2.311	1	600	8.53	8.53
32	MM	08-27	00:14:35	237.380	2.579	1.664	11.88	0.880	4.187	8	760	10.19	10.19
33	MT	08-27	02:14:09	237.463	2.579	1.664	11.85	0.880	2.211	1	300	8.53	8.53
34			02:26:29	237.472	2.579	1.664	11.85	0.880	2.252	1	600	8.53	8.53
35	MT	08-29	02:45:43	239.485	2.591	1.664	11.10	0.888	2.414	1	600	8.53	8.53
36	GG	08-31	14:32:11	241.976	2.605	1.666	10.20	0.898	1.101	1	120	17.07	17.07
37			15:13:45	242.005	2.605	1.666	10.19	0.898	1.057	1	200	17.07	17.07
38	GG	09-02	15:30:10	244.016	2.617	1.669	9.47	0.904	1.044	1	120	17.07	17.07
39			16:38:14	244.063	2.617	1.669	9.46	0.904	1.070	1	120	17.07	17.07
40	GE	09-03	07:50:00	244.696	2.621	1.670	9.24	0.905	1.775	1	300	20.80	20.80
41			08:13:00	244.712	2.621	1.670	9.23	0.905	1.740	1	300	20.80	20.80
42	PB	09-04	16:33:25	246.060	2.628	1.672	8.78	0.908	1.089	1	300	5.12	5.12
43			19:42:50	246.191	2.629	1.673	8.73	0.908	1.177	1	600	5.12	5.12

Obs #	Obs	UT Date (2000)	UT Start Time	Days After Perihelion	R (AU)	Δ (AU)	α Phase Angle (deg)	Phase Funct	Air Mass	# of Exp	Exp Time (s)	X FOV (')	Y FOV (')
44	MM	09-05	00:35:53	246.395	2.630	1.673	8.67	0.909	4.690	5	300	10.19	10.19
45			00:46:37	246.402	2.630	1.673	8.66	0.909	4.803	5	300	10.19	10.19
46	MM	09-09	23:50:13	251.363	2.659	1.687	7.18	0.919	4.715	9	1080	10.19	10.19
47	MT	09-12	02:27:03	253.472	2.671	1.695	6.68	0.922	3.072	1	120	8.53	8.53
48			02:32:36	253.476	2.671	1.695	6.67	0.922	3.151	1	600	8.53	8.53
49	MT	09-19	22:10:01	261.294	2.715	1.735	5.84	0.928	2.511	1	120	8.53	8.53
50			22:13:53	261.296	2.715	1.735	5.84	0.928	2.494	1	600	8.53	8.53
51	MM	09-20	22:15:01	262.297	2.720	1.741	5.88	0.927	4.970	8	960	10.19	10.19
52	PB	09-25	15:58:24	267.036	2.747	1.775	6.42	0.924	1.037	1	300	5.12	5.12
53			16:13:06	267.046	2.747	1.775	6.42	0.924	1.038	1	1200	5.12	5.12
54	DB	09-30	06:35:40	271.645	2.773	1.814	7.39	0.917	1.484	1	240	5.08	5.08
55			07:51:56	217.698	2.773	1.814	7.40	0.917	1.710	1	240	5.08	5.08
56	DB	10-01	06:03:33	272.622	2.778	1.823	7.63	0.916	1.451	1	240	5.08	5.08
57			06:48:19	272.654	2.778	1.823	7.64	0.916	1.523	1	240	5.08	5.08
58			07:36:16	272.687	2.778	1.823	7.65	0.916	1.699	1	240	5.08	5.08
59	GG2	10-17	04:43:55	288.567	2.866	2.002	11.92	0.879	1.552	1	120	17.32	17.32
60			05:16:24	288.590	2.895	2.003	11.93	0.879	1.516	1	300	17.32	17.32
61	PB	10-22	14:33:16	293.976	2.895	2.076	13.25	0.865	1.050	1	300	5.12	5.12
62			16:00:48	294.037	2.895	2.077	13.26	0.865	1.185	1	600	5.12	5.12
63	PB	10-23	11:42:46	294.858	2.900	2.089	13.45	0.863	1.189	1	300	5.12	5.12
64			15:41:30	295.024	2.900	2.091	13.49	0.863	1.150	1	300	5.12	5.12
65	MC	10-30	17:38:39	302.105	2.938	2.199	14.96	0.847	2.038	1	300	13.02	8.64
66			17:49:44	302.113	2.938	2.199	14.96	0.847	2.200	1	300	13.02	8.64
67			17:55:17	302.117	2.938	2.199	14.96	0.847	2.301	1	300	13.02	8.64
68	MT	11-04	21:17:07	307.257	2.966	2.282	15.85	0.841	2.477	1	600	8.53	8.53
69			21:27:30	307.264	2.966	2.282	15.86	0.841	2.544	1	600	8.53	8.53
70	MT	11-30	19:07:21	333.167	3.100	2.748	18.15	0.826	2.147	1	120	8.53	8.53
71			19:16:52	333.173	3.100	2.748	18.15	0.826	2.176	1	600	8.53	8.53
72			19:28:18	333.181	3.100	2.749	18.16	0.826	2.222	1	600	8.53	8.53
73	DB2	12-19	01:52:26	351.448	3.192	3.101	17.91	0.827	1.431	1	480	12.06	12.06
74			01:28:18	351.487	3.192	3.102	17.91	0.827	1.603	1	480	12.06	12.06
75	MT	12-22	18:24:56	355.137	3.210	3.173	17.72	0.829	2.080	1	600	8.53	8.53
76			18:39:25	355.147	3.210	3.173	17.72	0.829	2.153	1	600	8.53	8.53
77	MT	12-26	17:05:47	359.082	3.229	3.249	17.46	0.830	1.889	1	600	8.53	8.53
78			17:18:51	359.091	3.229	3.249	17.46	0.830	1.903	1	600	8.53	8.53

[1] Dust phase function from Krasnopolsky et al 1986 for comet 1P/Halley. This is the same dust phase function used for Deep Impact targeting. See Appendix A.

[2] Air mass as estimated by JPL's HORIZON'S System (http://ssd.jpl.nasa.gov/cgi-bin/eph) based on the longitude, latitude, and elevation of the observing site.

avoided at large heliocentric distances. We applied this method to all broadband-R measurements.

We planned to calibrate the broadband-VRI images using the University of Hawaii's 2.2-m photometric system (MKO). STSP comet frames contain field stars along the path of Tempel 1. Imaging the field stars on the MKO system would enable us calibrate the STSP images to that system. However, Karen Meech expected the acquisition of calibration images to continue through 2002. In the interim we performed a preliminary photometric analysis of the broadband-R data. For field stars, we elected to use the R magnitudes recorded in the United States Naval Observatory's A2.0 star catalog (USNO-A2.0). We emphasize that the catalog quotes 1-σ, random errors of 0.15 magnitude and 1-σ, systematic errors of 0.50 magnitudes.

To be selected for photometry, we required a field star to have an entry in the USNO-A2.0 catalog, an integrated flux not less than the integrated flux of the coma, and no saturated pixels. The star had to be located outside the coma and tail and far enough from the edge of the frame so that photometry could be performed.

With photometric counts from the coma and field stars in a frame, we calculated the comet's observed magnitude, m_c, outside the Earth's atmosphere (Howell 2000):

$$m_c = m_* + 2.512\log\left(\frac{\text{Flux}_*}{\text{Flux}_c}\right) \tag{1}$$

where m_* = the USNO-A2.0 R magnitude of a field star in the comet frame, $Flux_*$ = photon counts for the field star, $Flux_c$ = the photon counts for the comet. For STSP data, the atmospheric correction term is negligible (less than 0.01 magnitude) because the extinction coefficient is less than 0.3 for the R band and the difference between a field star and the comet is less than 0.03 air masses. We calculated the uncertainty in the magnitude using:

$$\sigma_{mc} = \sqrt{(0.52)^2 + \left(\frac{1.086}{\text{Flux}_*}\sigma_{\text{Flux}_*}\right)^2 + \left(\frac{1.086}{\text{Flux}_c}\sigma_{\text{Flux}_c}\right)^2} \tag{2}$$

where the 0.52 value is the random and systematic uncertainties for the USNO-A2.0 R magnitudes. Most comet frames contained at least 3 field stars. Resulting observed magnitudes were averaged to obtain 1 observed magnitude for each comet frame.

3.6 Photometric Results

3.6.1 Light Curve

In Figure 2, we present the heliocentric magnitudes m_R (r, Δ=1 AU, phase angle α=0 degrees) for Tempel 1 based on photometric analysis of the broadband-R data taken by STSP observers during 2000.

Figure 2 shows an increase of the observed magnitudes when Tempel 1 was at opposition, days 253 to 268 (2.5*log(R) = 1.10). During this period, Tempel 1 had a phase angle less than 7 degrees. For these small phase angles, an increase in brightness is caused by dust particles backscattering the continuum (Swamy 1997). We corrected the data to a phase angle of 0 degrees by applying a dust phase function based on comet 1P/Halley observations (Krasnopolsky et al 1986 and Appendix A). However, the phase correction did not eliminate the brightening. It is possible that the dust phase function needs to be adjusted for Tempel 1.

If the amount of dust within the 27,000-kilometer photometric aperture remains constant over a certain period, the heliocentric magnitudes should follow an inverse square law. Figure 2 plots heliocentric magnitude $m_R(r,1,0)$ versus $2.5\ log(r)$ where $m_R(r,1,0) = m_R(r,\Delta,\alpha) - 5\ Log(\Delta) - f(\alpha)$. At heliocentric distances of 2.096 to 2.966 AU ($2.5\ Log(r)$ = 0.8 to 1.18), the brightness of the coma tends to follow an inverse square law, as indicated by the black line with a slope of –2.0 in Figure 2. This range corresponds to days 155 though 307 after perihelion. Before this period, the comet appears to be more active, which is known from professional observations coordinated by Karen Meech (Lisse et al 2003). After this period, the slope becomes steeper and occurs when the heliocentric distance approaches 3.0 AU. This hints at the cessation of water sublimation. However, we need calibrated results before making a definitive conclusion.

3.6.2 Dust Production Rates

We *calculated* $Af\rho$, an estimate of the dust production rate (A'Hearn et al 1984), using the photometry results from the broadband-R images:

$$\mathrm{Af\rho(cm)} = \frac{2.467\cdot 10^{19}\,\mathrm{r(AU)}^2\,\Delta\mathrm{(AU)}}{\mathrm{a('')}}\,\frac{\mathrm{Flux}_c}{\mathrm{Flux}_{sun}} \tag{3}$$

where A is the Bond albedo, f is the filling factor of the aperture (a), ρ is the radius of the aperture projected at the comet, a is the diameter of the aperture

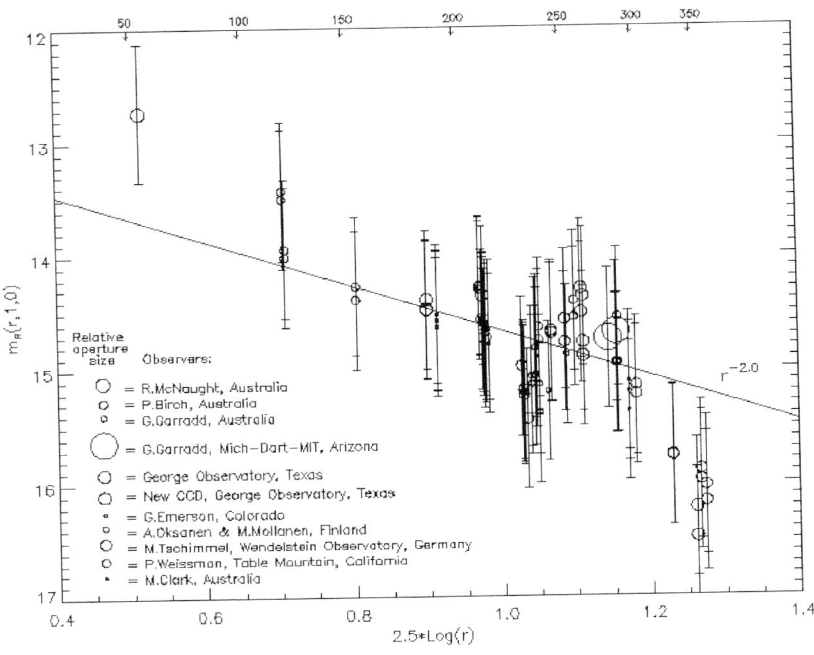

Figure 2. Heliocentric magnitude $m_R(r,1,0)$ versus 2.5 Log(r). Data points are represented by relative telescope apertures (see Table 1). Data points outside of the range 0.8 < 2.5 Log (r) < 1.2 indicate decreasing dust. The large uncertainty in the STSP data is due to the large errors in the R magnitudes reported in the USNO-A2.0 catalog.

used for photometry, $Flux_c$ is the flux from the comet at the Earth and $Flux_{sun}$ is the flux from the sun at the Earth. An $Af\rho$ value of 1000-cm correlates roughly to production of dust in metric tons per second.

We rewrote (3) in terms of phase-corrected observed magnitudes, m_c(r, Δ, phase angle α=0 degrees), using a constant ρ = 27,000 kilometers and m_{Rsun} = -27.10 (Meech et al 1993):

$$\mathrm{Af\rho\ (cm)} = \frac{8.9414 \cdot 10^{26}\, \mathrm{r(AU)}^2\, \Delta\mathrm{(AU)}^2}{\rho\mathrm{(cm)}} 2.512^{(-27.10-\mathrm{m_C})} \tag{4}$$

Then, the uncertainty in $Af\rho$ is:

$$\sigma_{\mathrm{Af\rho}}\mathrm{(cm)} = \frac{\partial \mathrm{Af\rho}}{\partial \mathrm{m_c}} = \ln(2.512) \cdot \mathrm{Af\rho} \cdot \sigma_{\mathrm{mc}} = 0.921 \cdot \mathrm{Af\rho} \cdot \sigma_{\mathrm{mc}} \tag{5}$$

Figure 3 illustrates the dust production curve for Tempel 1 using phase-corrected $Af\rho$ values derived from STSP photometry. The solid line suggests the coma of Tempel 1 contained a constant amount of dust for 155 to 310 days after perihelion, corresponding to r = 2.09 to 2.97 AU. *Afρ* values for days 140 through 260 are consistent with dust production rates derived from visual data taken by Meech on days 142, 231, and 269 after perihelion (Lisse et al 2003). The solid also runs through the Meech data points for days 231 and 269 (Lisse et al 2003). After we calibrate the STSP data, we expect the revised *Afρ* values to supplement and extend the dust production curve for Tempel 1 based on visual and IRAS data from the 1983 apparition and visual data from the 2000 apparition (Lisse et al 2003).

3.7 Discussion of Phase 1 Results

From 155 through 305 days after perihelion, Tempel 1 exhibited a constant amount of dust in its coma, as demonstrated by the similar values for the dust production rate from the large and small telescope data.

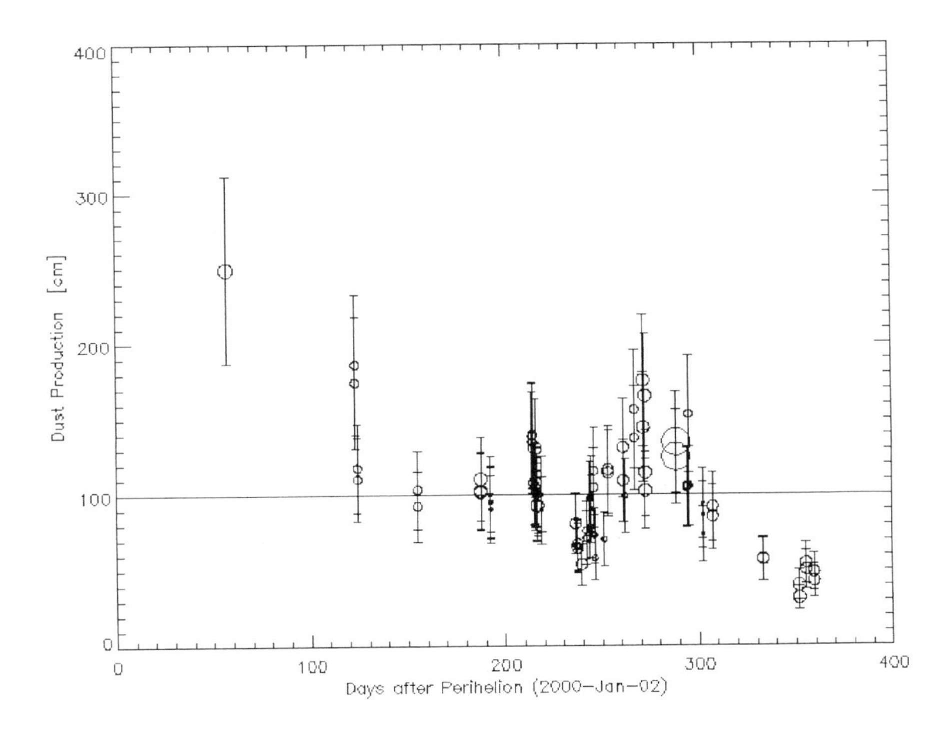

Figure 3. The light curve for dust production, based on data acquired by STSP observers during 2000. Data points are represented by relative telescope apertures (See Figure 2 for the key). The solid line suggests the coma of Tempel 1 contained a constant amount of dust for 155 to 310 days after perihelion, corresponding to r = 2.09 to 2.97 AU. The large uncertainty in the STSP data is due to the large errors in the R magnitudes reported in the USNO-A2.0 catalog.

We plan to investigate the variation in the STSP data for days 235 to 285. Is sky transparency a problem? Will a smaller, projected aperture yield less scatter? Are there short-term variations in the amount or size of dust in the coma? We intend to examine the rapid decrease in brightness after day 310 when Tempel 1 goes through r = 3.0 AU. Did water sublimation turn off?

Must we apply a phase function that includes dust and nuclear contributions? We hope to resolve these issues in 2003, after we have calibrated photometry based on the MKO photometric system.

We could analyze P. Birch's images for the rotation period. The images have a small pixel scale (0.6 arc seconds/pixel), good seeing (2 arc seconds at HWHM), and good temporal coverage (spanning 2-3 hours on 2 consecutive nights).

Based on this preliminary analysis, we expect the data acquired by STSP participants during Phase 1 to be scientifically useful. We plan to insert calibrated STSP results into the light and dust production curves generated from large telescope data (Lisse et al 2003). STSP results will extend these curves to 360 days after perihelion and will provide a baseline for Phase 2 observations.

4. PHASE 2 OF THE STSP

Tempel 1 returns to the inner solar system in late 2003 with an estimated total visual magitude of 21.0. Therefore, in early 2003 we will welcome Phase 1 observers back to the STSP and grow the network by issuing calls for observations to commence in late 2003 and continue through 2005. Observers in the Northern Hemisphere will be able to observe the comet from September 2003 through April 2004 and from November 2004 through September 2005. Observers in the Southern Hemisphere will be able to observe the comet from October 2003 through March 2004 and from December 2004 through December 2005. Phase 1 demonstrated that STSP observations are scientifically useful, especially broadband images, so we will request broadband R images of Tempel 1.

We will generate a pre-perihelion (and pre-impact) light curve to compare with to Karen Meech's pre-perihelion data taken from 1997 through 1999 (Lisse et al 2001). Phase 2 data will also serve as a baseline for future apparitions of Tempel 1. We will merge post-perihelion (and post-impact) data taken during 2005 to the post-perihelion data taken during 2000. Any deviations could give clues about the interior composition of Tempel 1. Phase 2 will complete the characterization of the dust production of comet Temple 1.

5. CONCLUSION AND RECOMMENDATIONS FOR SIMILAR PROGRAMS

Advanced amateur astronomers using very small telescopes are providing scientifically useful data for NASA's Deep Impact Mission. Because of the success of the STSP, several observers from our network supported NASA's Deep Space 1 Mission during August and September 2000 by providing images of comet 19P/Borrelly. Many STSP participants plan to support solar wind research by supplying time-sequenced images of the ion tail of comet C/2000 WM1 (LINEAR) near solar maximum in December 2001 and January 2002 (http://encke.jpl.nasa.gov/request.html).

STSP provided us with valuable insight for future, similar observing program:

- Allow at least 3 months to establish a network of observers.
- Define scientific goals, equipment requirements, and observing procedures. Provide as much detail as possible and make the information available on the web. Require accurate, detailed observing logs. Logs should be in electronic format or entered online by the observer as part of the data transmission process. The quality and accuracy of information in the image headers varies greatly from observer to observer because of the variety of operations software for CCDs.
- Designate a program coordinator to establish the network and maintain communication with the observers.
- Determine if the program benefits most by having the observers reduce their data or by having a research analyst perform uniform.
- Post the status of the data analysis on the web.
- Keep good rapport with the observers and follow their other observing projects when possible. Some observers may ask you to write a paragraph about their contributions to your program to help them win an astronomical research grant.
- Recognize the observers and their efforts. Publish their images on the web. Acknowledge their contributions in all related publications. Refer excellent observers to other observing programs.

6. APPENDIX A

The Deep Impact mission is using the following phase functions for the dust and the nucleus for Tempel 1. The values are based on comet 1P/Halley data (Krasnopolsky et al 1987). The phase function for the nucleus follows a 0.061 m/deg law.

Table 3.

Normalized phase functions for Tempel 1

Phase Angle (deg)	Dust Phase Function (unitless)	Nucleus Phase Function (0.061mag/deg)
0	1.000	1.000
5	0.933	0.755
10	0.900	0.570
15	0.847	0.431
20	0.813	0.325
25	0.793	0.245
30	0.793	0.185
40	0.747	0.106
50	0.720	0.060
60	0.710	0.034
70	0.690	0.020
80	0.667	0.011
90	0.747	0.006
100	0.000	0.004
110	0.000	0.002
120	0.000	0.001
130	0.000	0.001
140	0.000	0.000
150	0.000	0.000
160	0.000	0.000
170	0.000	0.000
180	0.000	0.000

7. REFERENCES

A'Hearn, M.F., Millis, R.L., Schleicher, D.G., Osip, D.J., and Birch, P.V. 1995, Icarus, 118, 223, The ensemble properties of comets: Results of narrowband photometry of 85 comets, 1976-1992.

American Association of Variable Star Observers, Variable Star Research – AAVSO in print, http://www.aavso.org/about/inprint.stm.

Green, D.W.E (Editor) 1994a, International Comet Quarterly (ICQ), 91, 109.

Green, D.W.E (Editor) 1994b, International Comet Quarterly (ICQ), 92, 167.

Green, D.W.E (Editor) 1997, ICQ, Guide to Observing.

Green, D.W.E (Editor) 2000, ICQ, 114, 56.

Hanner, M.S, Tokunaga, A.T., Veeder, G.J., and A'Hearn, M.F. 1984, AJ, 89, 162, Infrared photometry of the dust in comets.

Howell, S. 2000, Handbook of CCD Astronomy.

Jewitt, D. 1991, Comets in the Post-Halley Era, Volume 1, ASSL, Cometary photometry.

Krasnopolsky et al 1986, Nature, 321, 269, Spectroscopic study of comet Halley by the Vega 2 three-channel spectrometer.

Lisse, C.M., A'Hearn, M.F.A., Fernandez, Y.R, McLaughlin, S.A., Meech, K.J., Walker, R.G. 2003 (in preparation), Icarus, The Dust Environment of Comet 9P/Tempel 1.

Meech, K.J. 2000, DPS Workshop: Space Missions to Small Bodies.

Meech, K.J., Belton, M.J.S, and Mueller, B.E.A., Dicksion, M.W., and Li, H.R. 1993, AJ, 106, 1231, Nuclear properties of P/Schwassmann-Wachmann 1.

Monet, D., Bird, A., Canzian, B., Harris, H., Reid, N., Rhodes, A., Sell, S., Ables, H., Dahn, C., Guetter, H., Henden, A., Leggett, S., Levison, H., Luginbuhl, C., Martini, J., Monet, A., Pier, J., Riepe, B., Stone, R., Vrba, F., and Walker, R. 1997, USNO-A2.0, (U.S. Naval Observatory, Flagstaff Station).

Osip, D.J., Schleicher, D.G., Millis, R.L., A'Hearn, M.A., and Birch, P.V. 1991, Asteroids, Comets Meteors, 443.

Osip, D.J., Schleicher, D.G., and Millis, R.L. 1992, Icarus, 98, 115, Comets – Groudbased observations of spacecraft mission candidates.

Swamy, K.S.K. 1997, The Physics of Comets, 2nd Edition, World Scientific Series in Astronomy and Astrophysics, World Science.

Walker, R.G. 1986a, IRAS Asteroids and Comets Survey, IRAS Comets, Compendium of IRAS Comet Observations

Walker, R.G., Matson, D.L., and Veeder, G.J. 1986b, Adv. Space. Res., 6, No. 7, 57

Warner, B., 1999-2001, Minor Planet Bulletin, Asteroid Photometry at the Parker Observatory, Asteroid Lightcurves, http://www.minorplanetobserver.com/htms/pdo_lightcurves.htm.

Chapter 5

Geographical Representation of Large and Small Near-Earth Asteroid (NEA) Discoveries and Observations

John L. Remo
Harvard-Smithsonian Center for Astrophysics
Cambridge, Massachusetts USA

Abstract: Statistics on near-Earth asteroids (NEAs) discovery and follow-up observations as a function of geographical location are presented. The majority of discoveries and follow up observations are made from the northern hemisphere which has the preponderance of facilities. It is suggested that because of the current dearth of observation programs in the southern hemisphere that there are some (high inclination) NEAs less likely to be discovered, tracked, and recovered. To assist in discovering these NEAs and carrying out follow-up (recovery) observations on others, both primary (discovery) and secondary (< 4-m) (follow-up) facilities should be placed in the southern hemisphere and supported by amateur observation follow-up networks. To optimally discover NEAs with orbits inferior to Earth's, an IR observing satellite should be positioned in an inferior orbit, e.g. between Venus and Mercury. To physically characterize NEOs, additional 2- to 4-m telescopes are needed

Key words: Near-Earth Asteroids (NEAs), geographical distribution

1. INTRODUCTION

This chapter's objective is to present available data as of 1 July, 2000 describing the number and geographical locations (latitude, longitude and host country) of the discoveries and follow-up observations of both large (≥ 1-km) in size) and small (< 1-km) near-Earth asteroids (NEAs). From this data some conclusions may be drawn regarding the completeness, or lack thereof, of the current geographical distribution of NEA observational facilities. Recommendations are offered to improve our knowledge of both the

T.D. Oswalt (ed.), The Future of Small Telescopes in the New Millennium, Vol. III, 77-92.

observational and material properties of NEAs using telescopes < 4-m. Data used in this report was extracted from the data-base provided on 1 July, 2000 by G. V. Williams from the Minor Planet Center in Cambridge, Massachusetts (Williams 2000) and is restricted only to those asteroids that have orbits which can possibly intersect Earth's. Comets with potential Earth crossing orbits also represent a significant component of the near-Earth object (NEO) impact hazard, but are not included in this database because the radial distribution of comets is such that very few of these objects, despite their enormous population (whose inclinations encompass the entire sky), are near-Earth objects (NEOs) that can be identified as potentially hazardous objects (PHOs) within the foreseeable future. That is, comet periods can vary from tens to hundreds of years making it all but impossible to predict their (long term) orbits precisely because of the vast number of gravitational interactions that are invariably encountered during these extended orbital periods. To further complicate matters, when comets approach the sun their orbits can be affected by non-gravitational interactions when solar radiation interacts with volatile compounds within the comet. Although one cannot easily locally characterize the population and orbits of comets, to the contrary NEA distributions are primarily confined to the ecliptic in orbits whose radial distributions come very close to intersecting Earth's orbit on a regular basis, providing observational opportunities to assists in precise orbital computations.

2. DISCOVERY RATES

While the discovery rate for NEAs appears to be progressing in accordance with NASA's stated goal of achieving a 90% discovery of > 1 km NEAs by 2008, there are still considerable uncertainties in what this number should be, even though they appear to be confined to within three well-studied dynamical families. That is, while the initial discoveries are generally expected to favor the brightest objects, the time frame for discovering the last 10% might be extended. Even this interpretation is suspect because fundamental uncertainties in the discovery process itself and the true percentage of large NEAs discoveries are exemplified by the recent discovery of some intrinsically bright NEAs corresponding to objects about two to three km in size. These objects which hitherto escaped detection were only recently found, for example the potentially hazardous asteroids (PHAs) 2002 FB3 and 2002 CE. The former discovered on March 18, 2002 turned out to be a larger Aten with (magnitude) M = 16.3; with an aphelion of 1.23 AU and perihelion of only 0.29 AU. The latter was discovered on February 1, 2002 with M = 14.6 and aphelion of 3.13 AU and perihelion of 1.02 AU

(Marsden 2002). A question naturally arises as to 90% discovery of what population number and at what apparent magnitude. Magnitude can, because of uncertainties in knowing the albedo, vary by a factor of five or more. Because of the potentially wide albedo variations, magnitude is only loosely related the NEA mass. In addition, mass estimation is also subject to uncertainties in density which can vary by a factor of three or more.

There is also the problem of discovering NEAs < 1-km. Such objects present a greater collective hazard because of their much greater numbers and smaller sizes which makes them much more difficult to detect. In round numbers one can estimate the number of NEAs 1-km or greater to be roughly 2,000, while the number of NEAs between 0.5 and 1-km are estimated to be about 6,000. These smaller objects can cause considerable regional damage. Of course, with all of these numbers there is considerable guesswork.

3. UNCERTAINTIES IN NEA MASS DETERMINATION

Uncertainties in mass determination underscore another important issue regarding NEA observations which is the lack of observing facilities directed towards characterizing the physical properties of NEAs, such as albedo, masses, sizes, compositions, densities, spin rates, shapes, internal structures, and binary sets. Such observations are critical if a NEA mitigation mission becomes necessary. These major problems in physical characterization are due to both the scarcity of fully-dedicated observing programs with adequate instrumentation and the intrinsic difficulties in measuring many of these quantities, especially mass. For example, because of the very high discovery rate, less than 10% of the cumulative number of discovered NEAs have measured albedos (Cellino 2001). If this trend continues, the rate of follow-up observations to make measurements of physical properties will fall even further behind the discovery rate.

On the other hand, while there are considerable uncertainties regarding the NEA population, even a greater range of uncertainties exists regarding the orbits and number of comets, as well as their varying material composition and luminosity. That is, due to vagaries in past history, solar irradiated surface activity, and potential orbital non-linearities, it would be folly to establish discovery goals based on hypothetical comet population.

4. THE AMOR, APOLLO, AND ATENS GROUPS

Because NEAs can essentially be categorized into three well-studied dynamical groups with relatively short orbital periods, statistics on discovery and recovery can be very helpful for assessing the NEA threat to Earth. Towards this end, the data-base divides the NEAs observed into the three classical (dynamical) groups of Earth-approaching asteroids whose orbits lie outside the main asteroid belt: Amors (semi-major axes greater than the Earth's orbit), Apollos (Earth orbit crossing), and Atens (semi-major axes less than the Earth's orbit). In Tables 1 and 2 respectively, the number of discoveries and observations, at the time of compilation in 7/2000, of the large and small NEAs from each dynamical group are listed below.

Table 1.
Large and small discoveries of NEA Groups; Atens, Apollo, and Amor

	Amor	Apollo	Atens	Totals
Large (> 1-km)	187	153	28	368
Small (≤ 1-km)	222	272	51	545
		Total Discoveries		913

Table 2.
Large and small observations of NEA Groups: Atens, Apollo, and Amor

	Amor	Apollo	Atens	Totals
Large (> 1-km)	3,223	2,991	416	6,630
Small (≤ 1-km)	2,022	2,474	531	5,027
		Total Observations		11,657

Based on numerical integration of NEO orbits, in addition to the well-known Atens dynamical group a new class of objects, IEOs (objects interior to Earth's orbit), with orbits completely within Earth's orbit, are thought to exist. An IEO has yet to be discovered since they are never visible from the ground at large solar elongation.

5. ASTEROID MAGNITUDE, SIZE, AND ABUNDANCE

Since the primary emphasis for most telescope searches has been at those magnitudes associated with the relatively large (≥ 1-km diameter) NEA's, it is appropriate that the delineation between large and small asteroids be set at

1-km. This, of course, sets up a selection effect which presumably substantially undercounts the smaller NEAs, which is suggested by Table 1. Nonetheless, the smaller NEAs (higher magnitude objects), depending on their mass, also pose a serious potential hazard because they are probably far greater in number as well as considerably fainter than the larger objects, and therefore less likely to be discovered. Assuming a density of 3000 kg/m^3, the relationship between absolute magnitude, H, diameter, D, and mass, M, for an asteroid with an average albedo is

$$M \text{ (petagrams)} = 1.6 \text{ D (km)}^3 \quad (1)$$

where

$$D \text{ (km)} = (1.585)^{(18.0 - H)} \quad (2)$$

Equation (2) indicates that an object with a diameter of 1-km a diameter has a magnitude of 18, but a NEA size change from 1 to 0.25-km yields a magnitude change of 3, which can place the smaller objects far beyond the range of many telescopes, i.e. from m = 18 to m = 21. Also, when taking into account NEA distance from Earth completeness of discovery after a given period of time is considerably reduced by smaller size, assuming a constant average albedo. Therefore, some searches should also emphasize these very numerous fainter magnitude objects which can cause considerable damage. For example, a 2,700 kg/m^3, 1 km asteroid with mass = 1.41 x 10^{12} kg impacting Earth at 25 km/s will have an impact energy of 4.4 x 10^{20} Joules. The same impact from a 0.25-km asteroid will have almost two magnitudes (0.0156) less impact energy. The former is thought to be capable of causing regional destruction, while the latter local destruction. Since the region or locale cannot be predicted, both represent substantial threats which must be addressed. These notions of mass-impact destruction have a more realistic meaning in light of the events of 11 September, 2001 which demonstrated how devastatingly catastrophic collisions can be in populated areas, and underscore the need to search out even the smaller objects. So there is little solace in ignoring NEOs with diameters less than 1-km.

To underscore selection effects favoring discovery and observation of the larger NEOs, it is thought by many researchers that there is an approximately linear logarithmic relationship between the size and number of NEAs (Rabinowitz et al 1994). Given an equivalent albedo and observing angle (phase) with respect to Earth, the larger NEAs will appear brighter and therefore more likely to be discovered than smaller NEAs. Because of the large gradient logarithmic relationship there are necessarily far more smaller objects than larger ones, resulting in the similar numbers of large and small objects observed (i.e., there are many more smaller objects with a much lower probability of detection as discussed in the previous paragraph). This selection effect is also the basis for the strong indication, as shown by Tables 1 and 2, that smaller objects, given the putative logarithmic relationship between size and number, are currently underrepresented in the list of observed objects. Recent statistics by Bottke (2000), Rabinowitz (2000), and Harris (2000) indicate that there are probably ~ 25,000 NEOs with diameters greater than or equal to ~ 200-m. This is a very large number for objects that can potentially be so destructive. Presently it is thought by some that there are only ~ 1,300 total (large and small) NEOs known. If this number substantially under-represents the number of smaller NEOs, considerable work is left to be done to discover and characterize fainter objects

6. GEOGRAPHICAL DISTRIBUTION OF NEA OBSERVATIONS AND DISCOVERIES

As is well known to observers and other interested parties, NEO observation facilities are concentrated in three regions of the northern hemisphere (specifically in Europe, Japan, and the USA) and localized in the southern hemisphere in Australia which is now non-operational. Not surprisingly, the discoveries are dominated by observatories in the highly industrialized countries that can afford sophisticated CCD imaging electro-optical systems and the software to automatically operate these facilities and provide the support for highly trained personnel. An excellent example is the LINEAR system that, since it has come on line, has dominated the discovery process.

Since this data has been collected the asymmetry of the discovery and follow up rates between the northern and southern hemispheres has become even more pronounced. Currently, 3/22/02, at least 75% of the discoveries are made by the LINEAR program, with the remainder being made by the NEAT, SPACEWATCH, and LONEOS programs. The follow-up observations are primarily being carried out by amateurs, the overwhelming majority of whom are in the northern hemisphere. Since the Australian facility has been shut down, at the current time there is no survey program in the southern hemisphere and very few amateur observers are available for follow-up work, making it difficult to extend follow-up and recovery observations in the southern hemisphere.

While the following figures are somewhat dated, they nonetheless indicate discovery and follow-up asymmetries which have become even more extreme. Figure 1 shows the total NEA observations as a function of latitude (using five degrees of latitude spacing resolution), with a clustering of observations between twenty and fifty degrees north latitude and at negative thirty degrees (south) latitude. Since some observers may claim that they can effectively

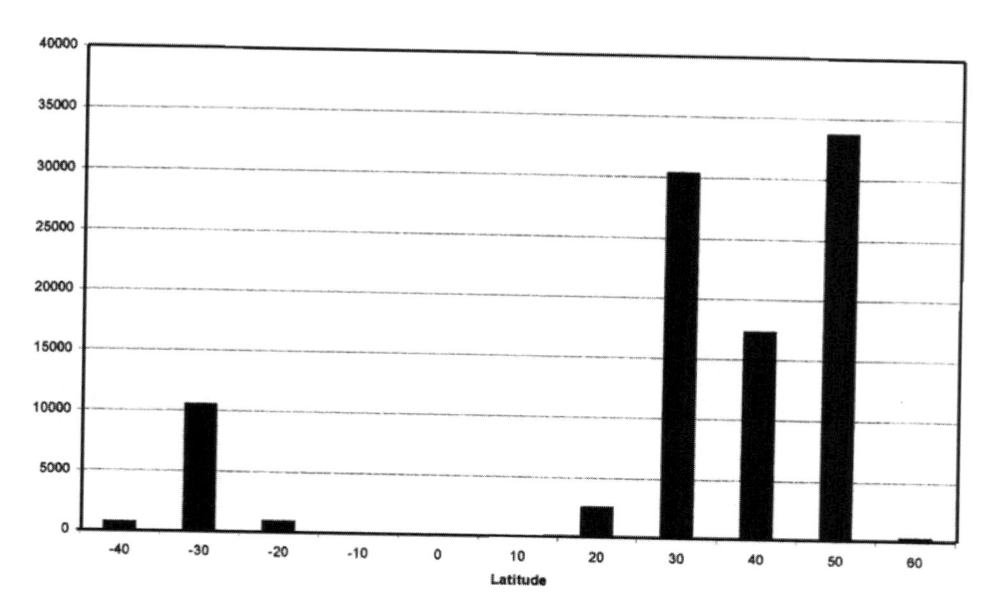

Figure 1. NEA observations as a function of latitude using five degree latitude spacing.

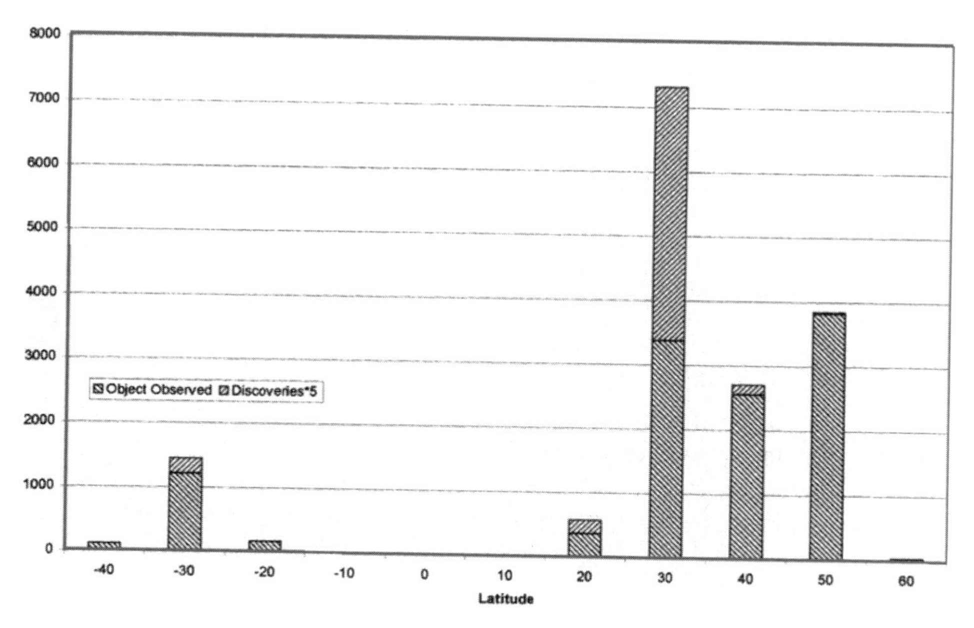

Figure 2. NEA observations and discoveries as a function of latitude.

observe at plus and minus sixty degrees from the zenith, theoretically, at least, the entire sky is covered with the current group of observers in both hemispheres (although the Australian site has been shut down which leaves a large area of space without a survey). However, as Figure 2 shows, the much smaller number of observations and especially discoveries in the southern hemisphere clearly indicates that discoveries and observations are favored in the northern hemisphere. Therefore it is important to make note of the -60 degree off- zenith limit for northern hemisphere observers, since most facilities in the northern hemisphere are above 30 degrees latitude. This limit imposes a constraint of ~ -30 degrees or greater on the declination coordinate of NEAs when observed from the northern hemisphere. This further underscores the need for survey and follow-up facilities in the southern hemisphere.

Figure 3 maps the worldwide distribution of NEA observations and discoveries by country. It shows a disproportionate number of discoveries and observations are made by observers in North America, specifically the United States. Within the US grouping, the vast majority are discoveries and observations from the LINEAR group whose telescope facility is located at a latitude of about 34 degrees in New Mexico. Other major contributors in the US include Spacewatch in Arizona (latitude 32 degrees) and the Palomar Survey (latitude 33 degrees). Contributions from Europe are dominated by a few countries over a latitude range of about 40 to 50 degrees, and Japan (latitude 31-39 degrees). Observations in the Southern Hemisphere, primarily from observers in Australia (when operational), are at a latitude of negative 31 degrees. Because more NEA observers and larger telescopes (i.e., greater

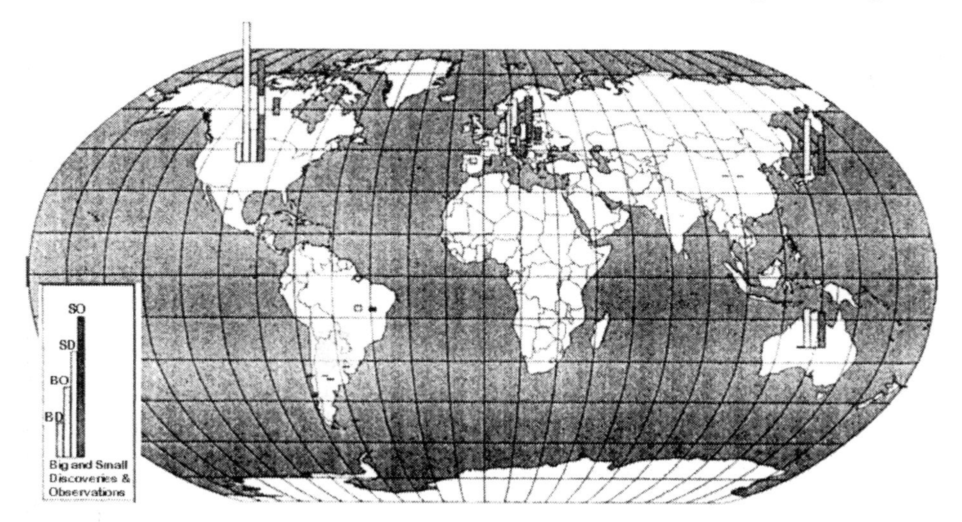

Figure 3. Worldwide distribution of NEA observations. Legend, from left to right: Big Discovery (BD); Big Observations (BO), Small Discovery (SD); Small Observation (SO).

sensitivity) are located there, the northern hemisphere is favored for discovery and follow-up observations. In principal all one needs to do is wait until a NEA is observable from the observatories in the northern hemisphere as long as the declination constraint does not preclude observation from that northern hemisphere position. But there is more to the problem because some NEAs may only be observable from the northern hemisphere in conjunction only over a limited period of time, presenting a potential problem and will be discussed further.

7. DISCOVERY OF BIG AND SMALL OBJECTS

Differences in discovery rates between northern and southern hemispheres become quite stark when considering differences in the big and small discoveries as shown in Figure 4. This suggests that a substantially disproportionate number of many smaller NEAs are not being discovered from the southern hemisphere for an equivalent amount of small observations. Of the 368 big NEA discoveries, only 27 (~7%) were from southern hemispheres (Australian) observers, while of the 545 small NEA discoveries only 13 (~2%) were from southern hemisphere (Australian) observers, at least at the date of this database. Such a paucity of discoveries from southern hemisphere sites especially for the smaller yet still dangerous objects, for reasons alluded to in the previous section, underscores the need for additional (< 4-m) observing facilities in this region. Also, for comets which have a large

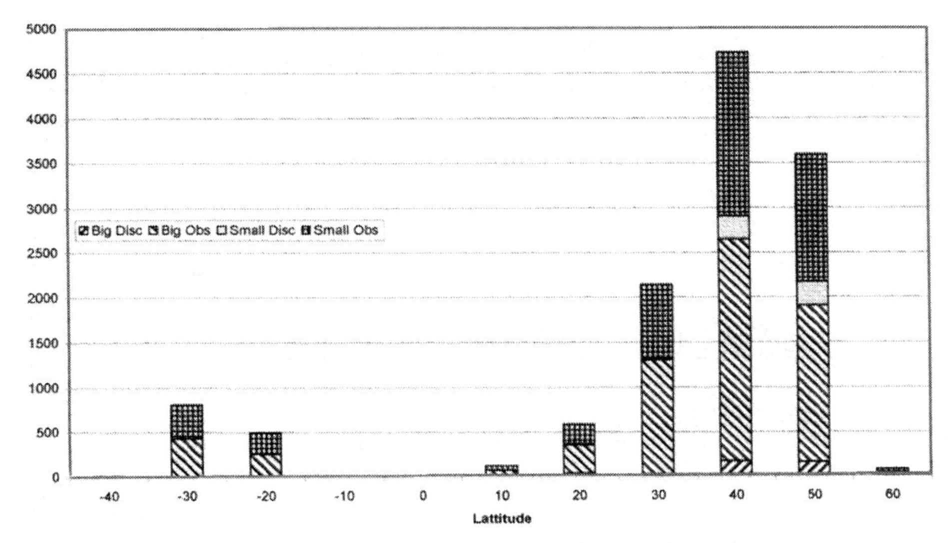

Figure 4. Big and small NEA discoveries and observations.

negative declination, a mid to high negative latitude observing facility would be very helpful in facilitating early discovery and provide invaluable follow-up and recovery data. It is noted that while the impact of a potentially hazardous object, big or small, is not limited by any geographical boundaries or to a single nation, region, or continent, the geographical distribution of NEA discoveries and observations is, incongruously, quite limited.

To further press the above argument, ideally, the position of each NEA discovery/observation (declination) should be displayed as a function of the discoverers/observers latitude. Such a study involves a more extensive analysis using available databases by means of a methodology that establishes correlations of the observers' latitude and longitude with the right ascension and declination of the discovery or follow-up observation. Perhaps a future computer study may carry out this exercise.

8. IMPORTANCE OF FOLLOW-UP OBSERVATIONS

While a NEA discovery provides a great deal of recognition for the discoverer and, is in of itself an important (primary) contribution, one should not underestimate the importance of follow-up observations. A NEA discovery based on a single night's observation cannot provide reliable orbital data. Follow-up observations over additional nights are required to determine accurately and reliably the orbit of the newly discovered NEA. Also, recovery and additional follow-up observations of previously discovered NEAs are necessary to refine the computation of the orbit of some particular NEA. For these reasons it is essential to have regular follow-up observations whenever a new NEA is found in order to optimally predict the discovery coordinates at some time in the (distant) future.

If the discovery rate is low in the southern hemisphere, the recovery rate is also expected to be low. This is especially important when carrying out follow up observations of some NEA orbits that are not necessarily exactly predictable in the long term. A question then naturally arises regarding the likelihood that a recovery would be observed from the southern hemisphere for a PHA in a timely manner. One must understand that for a PHA timely recovery of its orbit is as important as timely discovery, and under some, if not most, circumstances could be vital to a mitigation effort. Consider the case of the very large asteroid Hermes (1937 UB) which is estimated to have a large (800-m) size and was observed for only a few days as it passed by Earth, missing by only 670,000-km, or about 60% further than the Moon. Because of insufficient follow up data, the orbit of this dangerous object was never determined.

Another tool which has proven to be invaluable in establishing NEA orbits is the use of archival photographic plates which, when available, can contribute dynamical data that spans a large period of time thereby providing information on long-term NEA orbital evolution.

9. NEA INCLINATION ANGLE AND DISCOVERY RATE

Since most asteroids have small angles of inclination (less than 15 degrees), observations from the northern hemisphere of NEAs slightly below the celestial equator are both theoretically and practically possible. The question is whether in observational practice can all or even 90% of the asteroids be accounted for by observations from the northern hemisphere? That is, one can make the argument that, all things being equal, if those NEAs which orbit with a small angles of inclinations (<15 degrees) can be (roughly) equally detected from both the northern and southern hemispheres at any time, it is of secondary technical or scientific importance whether the observing facility is in the northern or southern hemisphere mid latitudes even though, ideally, one would like to carry out observations as close to the zenith as possible to minimize atmospheric effects.

However, those NEAs with high angles of inclination (> 15 degrees) favor detection and (follow-up) observation either from the northern or southern hemisphere, depending on the (planetary configuration) phase. This is simply because in general discovery is easier at opposition than at conjunction where the NEA is substantially further from Earth, thereby appearing at a higher magnitude, and where there is less observing time, i.e. observations at conjunction are possible at sunrise and sunset (hardly ideal viewing conditions) and not throughout the night, thereby losing substantial amounts of potentially high quality observation time. On the other hand, while at opposition the NEA can be tracked throughout the night (assuming that the meteorological conditions are favorable) thereby allowing the collection of extended data. This is the reason why, when possible, recovery generally occurs at opposition where long observation times can be carried out.

In addition, if there is a group of high inclination asteroids that have an almost synchronous orbital period with that of Earth's, such that these objects tend to be in opposition when they are below the celestial equator (negative declination) and in conjunction when they are above the celestial equator, then their discovery and recovery will be favored by observations from negative latitudes, i.e., southern hemisphere observatories. This suggests that some NEAs, both big and small, which would have been more likely to be observed had they passed through the observing regions of the northern hemisphere (with more observers), have in actuality not been observed in the southern hemisphere due to the paucity of observers and adequately instrumented observing facilities. Such objects nonetheless exist and can pose an immense threat to Earth and therefore must be discovered in order to develop a reliable statistical database to assist in assessing the possibility (probability) of NEA impacts with Earth (Gehrels 1994). Of course, some of those NEAs with negative declinations (at the time of observation) may still be observable from many sites in the northern hemisphere since most of the inclination angles of the larger Aten, Apollo, and Amors are much less than 20 degrees. However, the inclinations for some NEOs can be as high as, for example, 44.3, 67.9, and 48.5 degrees respectively which effectively puts such NEAs out of viewing from many of the northern hemisphere facilities for a significant portion of their orbit which reduces their overall chance of discovery by the fewer observers in the southern hemisphere. This is partial basis for the selection effect which favors more discoveries and observations through the activities of the greater number of observers in the northern hemisphere.

Clearly, the optimum survey strategy is to extend coverage to as much sky area as possible. Therefore, to a large extent, assuming ideal observing conditions, the NEA discovery probability depends both on the geographical location and the quality (sensitivity) of the telescope electro-optical system of the observer, as well as on the inclination of the NEA orbit and the orbital phase of the NEA during the period of time when the search is being conducted.

10. PHYSICAL PROPERTIES OF NEAs

While the preponderance of the NEA search effort has been directed towards the discovery of large objects as discussed above, a relatively small effort has been directed towards the determination of NEA physical and material characteristics. Indeed, the physical characterization effort is lagging further behind the discovery rate such that less than 10% of the discovered NEAs have measured albedos. This is especially significant because NEA sizes cannot be reliably estimated from the magnitude alone. Visible magnitude depends upon both size and albedo, the latter of which can vary by over an order of magnitude among NEOs. As a consequence, there are currently immense uncertainties about the true NEO size distribution and number. Other important physical characteristics include the NEA density, shape, rotation and topology. The material properties include the (mineralogical) composition of the dominant phases, faults, fractures, craters and related structural characteristics. The physical and material properties will be of primary importance if and when a mitigation operation is required (Remo 1994), and eventually must be well characterized for those NEAs which prove to be potentially hazardous objects (Remo 2002). Spectroscopic and spectrophotometric observations at visible wavelengths can provide some general indications about albedo such as discriminating between S- and C-type NEAs. For these tasks, ground based telescopes with 2-m < aperture < 4-m would be quite helpful. But realistically, because some classes of objects exist that are extremely difficult to observe from the ground, only a limited, but nonetheless critically important, amount of information can be obtained from < 4-m telescopic observations and from studying meteorites. Ultimately, satellite reconnaissance and probes of a range of asteroids are required to reliably characterize the physical and material properties of the NEA population. Projects such as the 3.5-m Herschel Space Observatory at L_2, the GAIA mission, which uses identical twin 1.7 m, telescopes, and the possible Darwin mission which would consist of a constellation of 5 or 6 1-m telescopes operating in the IR would be helpful. But to be most effective, satellite reconnaissance and probe activity will depend on ground-based observations for background and guidance.

11. RECOMMENDATIONS TO IMPROVE NEA DISCOVERY AND OBSERVATION

Recent policy studies on the NEO hazard (Remo 1996, Gehrels 1997, UK Task Force 2000, Govt. Response to UK Task Force 2001, and Remo and Haubold 2001) have not only recognized the striking geographical asymmetry and other limitations in NEO discovery and observation facilities, but have also made recommendations regarding the upgrading of the observational facilities in general. These recommendations include:

1. Upgrading of NEO observational facilities in both the Northern and Southern hemispheres with observation programs that are internationally coordinated with amateur astronomy groups. Specifically recommended is construction of new two to four meter telescopes to search for faint large objects and small objects in both the northern and (especially) southern hemispheres. Using state of the art charge coupled devices (CCD), ideally the sensitivity of these telescopes could achieve sensitivities down to the 23 magnitude and beyond. Observational data from non-dedicated telescopes could be integrated into the dedicated NEO discovery and observation programs when appropriate.

2. Use of space satellite observatories, such as ESA's GAIA (a comprehensive stellar survey telescope) and the BepiColombo mission to Mercury (which has been altered to include a dedicated NEO camera) and NASA's SIRTF, to survey the sky for NEAs, especially for those NEA

orbits which are mostly inferior to Earth's. Success of these missions can pave the way to a dedicated NEO discovery satellite(s).

Related to the above observational recommendations are:

3. Implementation of coordinated space rendezvous missions to different types of NEAs in order to determine a detailed characterization of their physical and material properties. Results of this reconnaissance work combined with laboratory studies can provide a database and be used as a the basis for analytically planning mitigation operations if they are needed.

Since the potential NEO hazard is an international problem, both funding and governance of NEO discovery, observation, reconnaissance, and mitigation efforts should be internationally organized and committed for the long term. Since the entire population of Earth is at risk, there need not be any major political agendas. The threat is from outside Earth and it should not be used as a pretext for political or ideological intrigue. A program to encourage NEA discovery, and follow-up observations in developing countries should be undertaken.

12. CONCLUSION

Based on an analysis of data provided by the Minor Planet Center, it is concluded that the lack of observers in the southern hemisphere seriously affects both NEA discovery rates as well as the recovery capability. Since warning time for a PHA is critical, such a situation is untenable. It is therefore suggested that this situation be minimally rectified with the establishment of additional observing facilities in the southern hemisphere, preferably in the 2 to 4-km aperture range, as recommended by several policy studies.

The immediate issue centers about the question as how to best augment the limited resources to establish the necessary observing facilities with telescopes that are capable of both discovering and observing faint (to the 22, 23 magnitude or beyond) NEAs in the southern hemisphere. Perhaps the most efficient observing strategy would be to have a LINEAR class observation facility to make discoveries and some (critical) follow-up observations, which uses a 1-m class telescope with a 2560 by 1960 pixel format frame transfer CCD (Stokes and Kostishack 1997). Depending on the available electro-optic technology, some carefully selected additional observation facilities using half-meter support telescopes with CCD cameras to carry out recovery observations is suggested. With such a system in place the combined resources of both the northern and southern hemispheres will yield valuable statistics on the NEA population and their orbits. However, to carry out NEO discovery, follow-up, and physical property observations, telescopes with apertures between 2 and 4-m are required at appropriately diversified geographical locations.

Ideally, to provide enhanced levels of planetary security (discovery beyond 90% of the conjectured NEA population and a more extensive recovery capability) for the longer term, a space telescope could be placed in a inferior orbit (semi-major axis smaller than 1 AU) at about the distance of Venus or Mercury in a conjunctive configuration with Earth's orbit to provide sky coverage of NEAs with orbits inferior to Earth's (Atens) as well other potentially hazardous NEOs that would otherwise be observationally inaccessible. A dedicated space based facility would also be the ideal instrument to carry out physical characterizations in conjunction with ground based facilities. It is thought that such a satellite would be placed in an orbit around the L2 Lagrangian point of the Sun-Earth system which is a relatively low radiation environment. In this way, space based discovery and observation systems can provide substantial advantages for the discovery of potentially hazardous NEOs. A modest, dedicated space based IR telescope with an aperture ≤ 1-m operating above Earth's atmosphere can measure many of the important physical characteristics of NEAs. One concept (Cellino 2001) is for a 70-cm aperture Ritchey-Chrétien telescope with both a visible CCD and IR array (6 - 11 μm), each of which with a six channel filter spectrophotometer wheel. However, there are other options for a space telescope design. Whatever particular design is chosen, to reliably discover and thoroughly characterize potentially hazardous comets, several space-based telescopes will be required.

13. ACKNOWLEDGMENTS

I wish to thank B. G. Marsden and T. Spahr for helpful comments, and G. Williams for making orbital data from the Minor Planet Center in Cambridge available for analysis.

14. REFERENCES

Bottke, W. F., Jedicke, R., Morbidelli, A., Petit, J. M., Gladman, B., Understanding the Distribution of Near-Earth Asteroids, Science 288, 2190-2194, 2000.

Cellino, A., "Physical Characterization of Near-Earth Objects," International School of Space Chemistry, Erice, Italy June 17-25, in press 2001.

Gehrels, T. (ed.), "Hazards Due to Asteroids and Comets," Univ. Ariz. Press, Tucson 1994.

Gehrels, T., A proposal to the United Nations regarding international discovery programs of near-Earth asteroids, 603-605, in "Near-Earth Objects; The United Nations International Conference, " Annals of the New York Academy of Sciences, 822, J. Remo (ed.) 1997.

Harris, A. W., Searching for NEAs from Earth or Space in "Highlights of Astronomy, II-A, 257-261, J. Andersen ed. 2000.

Harris, A. W., "Evaluation of Ground-Based Optical Surveys for Near-Earth Asteroids," Planet. Space Sci. 46 283-290, 1998.

Marsden, B. G., Personal communication, 2002.

Rabinowitz, D. L., Bowell, E., Shoemaker, E. M., and Muinonen, K., Population of Earth Crossing Asteroids, in "Hazards Due to Asteroids and Comets," Univ. Ariz. Press, Tucson , Gehrels, T. (ed) 1994.

Rabinowitz, D. Bowell, E., Lawrence, K., and Pravdo, S., A Reduced Estimate of the Number of Kilometer Sized Near-Earth Asteroids, Nature 403, 165-166, 2000.

Remo, J. L., Classifying and Modeling NEO material properties and interactions, in "Hazards Due to Asteroids and Comets," 537-550, Univ. Ariz. Press, Tucson, Gehrels, T. (ed) 1994.

Remo, J. L. Policy Perspectives from the UN International Conference on Near-Earth Objects, Space Policy 12, 13-17 1996.

Remo, J. L. (ed.) Near-Earth Objects; The United Nations International Conference, "Annals of the New York Academy of Sciences 822, 1997.

Remo, J. L. and Haubold, H. J., NEO Scientific and Policy Developments: 1995 – 2000, Space Policy 17, 213-218, 2001.

Remo, J. L. 2002, Characterizing the Near-Earth Object Hazard and Its Mitigation, Acta Astr. 50, 737-746.

Stokes, G. H. and Kostishack, D. F., Technology for the Detection of Near - Earth Objects, in Near-earth Objects, The United Nations International Conference, "Annals of the New York Academy of Sciences 822, J. L. Remo (ed.) 1997.

UK Task Force on Potentially Hazardous Near-Earth Objects, 2000.

US Govt. Response to the UK Task Force on Potentially Hazardous Near-Earth Objects, 2001.

Williams, G. V., Personal communication, July, 2000.

Chapter 6

Finding High Priority Projects for Stellar Physics Through Spectroscopy

C.J. Corbally
Vatican Observatory
University of Arizona
Tucson, Arizona USA

R.F. Garrison
David Dunlap Observatory
University of Toronto, CANADA

R.O. Gray
Appalachian State University
Boone, North Carolina USA

Abstract: The question, "What are the scientifically interesting objects?" finds a ready answer through surveying and monitoring of stars. In such reconnaissance spectral classification remains an effective tool for the smaller telescopes, while current computing power allows the same spectra that are suitable for classification to provide good stellar parameters. The methods and possibilities for automated spectral classification are outlined, as well as suggestions of how to interface these to spectra obtained with large telescopes. Some priority projects are suggested, both for cluster and for field stars. In all this the complementarity between data obtained with small and large telescopes is stressed, since both kinds are needed to understand the most interesting stars.

Key words: stars, peculiar stars, spectral classification, spectral analysis, automation

T.D. Oswalt (ed.), The Future of Small Telescopes in the New Millennium, Vol. III, 93-109.

1. INTRODUCTION

We might think that we know much about stars, but there are tantalizing gaps in our understanding. Just browsing through recent conference proceedings will confirm this statement. These gaps concern our understanding of the evolution of stars and of the relation between stars, gas, and dust in the Galaxy. In turn, if we understand the components of our Milky Way Galaxy, we stand a better chance of understanding other galaxies. There is nothing new in this, but it does remind us why, in the quest via the largest of today's telescopes for the most distant and faintest of galaxies, the study of relatively nearby and seemingly prosaic stars remains essential.

"Stars are very complex," as Jeff Linsky (2000) quotes Sylvie Vauclair in the 1999 conference on stellar clusters and associations at Palermo, and consequently "More data are needed" (Sofia Randich's understatement of the same meeting). What kinds of data are these and how can the smaller telescopes gather some of them with greater efficiency than large telescopes? Let us first review how the MK System of spectral classification is so very suited to providing initial data on stars, then we shall look at how modern computing methods, both by analysis and by automation, are extending the possibilities of these spectra, and finally we shall pick out some areas where all these methods can best contribute in the coming decade.

2. THE MK PROCESS AND THE MK SYSTEM IN THE NEXT DECADE

The MK Process is an inductive methodology involving the use of specimens. An important part of this process is the separation of the classification from the calibration. In the case of the MK System of Stellar Classification, this independence gives the MK type permanency as a description of the appearance of the blue-violet spectrum and therefore as a fundamental property of the star. The type does not depend on an ever-changing calibration or on the theory of stellar atmospheres. A carefully determined MK type from 1943, based on high quality photographic spectra and a stable set of standards, is as valid today as a modern type, in most cases, even though the detector and optics may be quite different. Refinements, corrections, revisions and extensions have been made to the MK System since its publication in 1943 (Morgan et al. 1943), but the basic structure and the grid of standard stars have been relatively stable, much more so than theory or calibration. For a given bright star, the spreads among various determinations of T_{eff}, log g, and $[M/H]$ are much greater than the spread of spectral-type determinations by different classifiers.

The MK System is used for reconnaissance (surveys) and spectroscopic parallaxes, as well as the isolation and study of peculiar stars. Small telescopes are preferred for these tasks, in most cases. A very large telescope would only be needed for follow-up, e.g., to study very faint peculiar stars at significantly higher resolution and S/N.

In this chapter, we will be discussing mainly high-quality spectral types (S/N $\geq$ 100), recognizing that it is possible for an experienced classifier to get a very rough spectral type with poor-quality data and few or no standards.

Because the resolution required by the MK Process of Stellar Classification is moderate (1-3 Å, or R~5000), very large telescopes (4-m and larger) are not needed except for the very faintest stars, which may be intrinsically faint (e.g., L stars and brown dwarfs) or very distant. Even so, most can be observed with a smaller telescope, though it may take longer. In any case, several millions of stars are within reach of a small telescope (e.g., 60-cm) and a 2-m telescope is capable of reaching several billions of stars.

It even can be argued that very large telescopes are first-class instruments which are being used to do second-class science. The reason is that it is very difficult to observe enough standards to make first-class determinations of the MK type.

1. There is no time to spare for taking an adequate grid of standards when the TAC only gives a few nights. Clearly, it is a pure waste of time to use a very large telescope on bright stars and TACs are justified in not giving enough time for standards. One solution is to have a small telescope feed into the spectrograph when it is not being used on the large telescope, but this would have to be designed into the spectrograph from the beginning.

2. The bright standards are too bright to observe with a very large telescope. A few decades ago, we formed a committee to recommend a set of faint secondary standards for use with very large telescopes and with the Hubble Space Telescope, which did not have any neutral-density filters. There was to be a set at about 10th magnitude and another at 15th. We soon discovered that the first problem was to find candidates. It was easy to find candidates for the lower Main Sequence, using clusters, but that didn't help with the upper main sequence OB stars or with the supergiant regime. The second problem was that the high-luminosity, faint stars were either heavily reddened or way out beyond the Milky Way in other galaxies and therefore were not suitable standards since their overall abundance ratios differed.

3. Multi-object spectrographs are of little or no use because there are generally no standards within the field of view. Even if there is a standard, it will be far brighter than the program objects.

One of us (Garrison) has compiled a database of all the stars that have ever been proposed, at one time or another, as reliable standards by Morgan or Keenan or one of their former students (Garrison, Lesh, Walborn, Gray). The table contains data for 630 standard stars, for all of which there are both photographic and CCD spectra. Some of the standards were later discarded or slightly revised. At the 1993 MK Workshop in Tucson, Garrison (1994) outlined a hierarchical structure for the standards in the sample. Included in that paper is a list of Anchor Points, for which the classifications have remained the same since 1943. The purpose of this list is to prevent drifting in the system. However, the Anchor Points may not be the best or the most conveniently located standards. Garrison is now working on a set of Primary Standards, which are the best and most conveniently located standards. This list of Primary Standards is meant to be the one most often used at the telescope. Secondary, local, and special standards can be set up by anyone, using the Primary and Anchor-Point sets.

In this maintenance and extension of spectral standards small telescopes excel due to the large time available on them, their low cost of operation, and their ease of use. Exactly the same can be said with respect to the other kinds of stellar standards, whether spectrophotometric, radial velocity, polarization, etc.; and the same inability to get adequate observations of these kinds of standards on large telescopes pertains. For all work in standards, without which data are scientifically meaningless, small telescopes remain essential.

3. NEW TECHNIQUES FOR LOW-RESOLUTION SPECTROSCOPY

Before the advent of digital detectors, stellar spectroscopy on small telescopes was largely limited to survey and classification. During the last decade, the dramatic increase in computing power, plus the near ubiquity of relatively inexpensive digital detectors on the spectrographs of small telescopes has opened the door to the application of new techniques and methods to the low to medium resolution (classification–resolution) spectra obtainable on small telescopes. These new applications can be summarized under two headings: 1) the use of the powerful technique of spectral synthesis to derive basic physical parameters from classification–resolution spectra and 2) the use of AI techniques to classify and characterize stellar spectra. Both of these techniques are currently in the stage of development, and small telescopes are needed to bring these tools to perfection. Once perfected, these tools can be applied to spectra from the largest telescopes, thus extending our knowledge of the most distant stars observable with these instruments.

3.1 The Analysis of Low-Resolution Stellar Spectra

The derivation of the basic physical parameters of stars (T_{eff}, log g, ξ_t and [*M/H*]) has traditionally been carried out, except for the bright stars, using high-dispersion spectra obtained on medium to large telescopes. Indeed, for fine analyses in which the abundances of individual elements are derived, this is still the case. However, it is now possible to use classification--resolution spectra to obtain good estimates of the basic physical parameters, including an estimate of the overall metal abundance, using techniques which take advantage of the power of modern desktop computers.

At the core of these techniques lies the tool of stellar spectral synthesis. A stellar spectral synthesis program (many are now available that can run on desktop computers, including SYNTHE, Kurucz 1993; SYNSPEC, Zboril 1996; SYNTH, Piskunov 1992; SPECTRUM, Gray & Corbally 1994; plus others) begins with a model stellar atmosphere (those easily available include Kurucz ATLAS9 models and NextGen models of Hauschildt & Allard 2001) and numerically integrates the radiative transfer equation to yield the emergent synthetic spectrum. These synthetic spectra, once convolved with the line–spread function of the spectrograph, may be compared directly with observed spectra. By searching in the 4-dimensional space of the basic physical parameters T_{eff}, log g, [*M/H*] and ξ_t (the microturbulent velocity), it is possible to come up with a good match between the synthetic spectrum and the observed spectrum, and thus obtain reliable estimates of the basic physical parameters.

Two to three decades ago, when spectral synthesis was just being introduced into stellar astronomy, the computing resources of the time limited the synthesis of stellar spectra to very small spectral ranges of 10 - 20Å. Now, with powerful modern desktop computers, it is possible to synthesize the full spectral range of a classification–resolution spectrum ($\geq$ 2000Å) with a step size of 0.01Å in a few minutes. Thus it is now possible to take full advantage of the large amount of information in the form of hydrogen-line profiles, metallic-line blends, molecular bands, etc. present in classification–resolution spectra.

To make this technique practical, however, some automatic or semi–automatic means of finding the best fit between synthetic and observed spectra must be employed, else finding the best match can be very time consuming indeed. A number of such tools have been described in the literature including Cayrel et al. (1991), Cuisinier et al. (1994) and Jones et al. (1996).

We briefly describe here a technique by Gray et al. (2001) designed to extract the basic physical parameters by fitting synthetic spectra and fluxes to observed spectra and fluxes (from Strömgren *uvby* and/or Johnson–Cousins BVRI photometry) which is suited to the analysis of classification–resolution spectra. This technique utilizes the multidimensional downhill SIMPLEX algorithm (Press et al. 1992) to search an interpolated four dimensional grid of synthetic spectra and fluxes to find the best fit. Once the four dimensional grid (which contains a few thousand synthetic spectra and flux distributions) has been computed (a process which can take a week or more of desktop computer time), the actual fitting procedure to an observed spectrum can be carried out in minutes. The errors in the fits are typically comparable to or better than ±80K in T_{eff}, ±0.1 in log g, ±0.1 in $[M/H]$ and ±0.5km/s in ξ_t. An example of such a fit can be found in Figure 1. The technique is best suited to stars with a solar abundance mix, but the method has been used with success on mild λ Boötis and Ap stars (Gray & Kaye 1999, Koen et al. 2001, Breger et al. 2002) as well as Ae stars (Miroshnichenko et al. 2001), and, as is shown in Koen et al. and Breger et al., gives reliable results even in cases where purely photometric methods fail.

Once one uses a technique such as the SIMPLEX method to find the basic parameters of a star or a set of stars, then for what can these parameters be used? In the case of variable stars, such as δ Scuti stars (cf. Breger et al 2002), γ Doradus stars or Cepheids, knowledge of the basic parameters is useful in the identification of modes of oscillation and in the location of the relevant instability strips in the theoretical HR diagram. If one knows the distance to the star, then the basic physical parameters permit the derivation of other parameters of interest, such as the mass and the radius (see, for instance, the recent determination of the masses of blue horizontal branch stars with Hipparcos parallaxes, de Boer 1997). Good estimates of the basic parameters enable, with the aid of stellar evolutionary tracks, a determination of the evolutionary stage of the star in question. Knowledge of the basic parameters for a large set of stars can sometimes yield interesting astrophysical insights. For instance, Gray et al. (2001) were able to demonstrate that the MK luminosity class in late A, F and early G--type stars is more strongly correlated with the microturbulent velocity than with the surface gravity. All of these investigations have been carried out on and are accessible to spectrographs on small telescopes.

A similar effort, but using a quite different approach with Artificial Neural Nets (ANN), to determine the basic physical parameters of stars from the comparison with synthetic spectra has been carried out by Bailer–Jones et al. (1997). These authors used this technique to determine T_{eff} and log g for 5000 spectra extracted from objective prism plates.

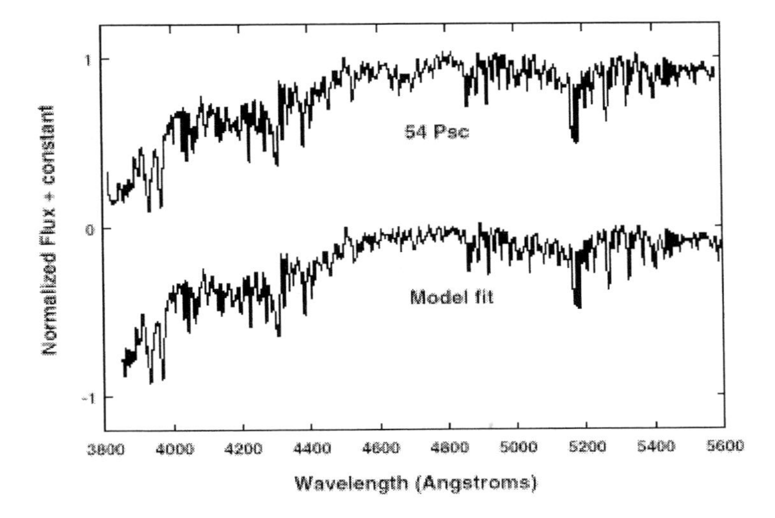

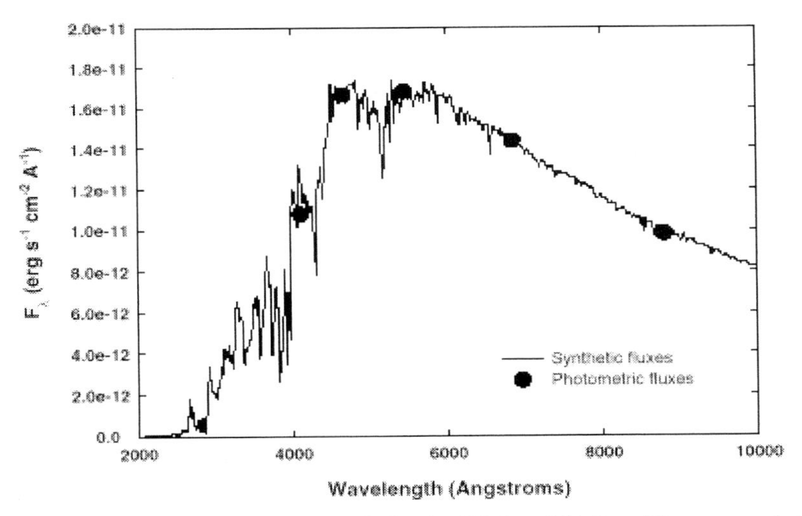

Figure 1. An illustration of a SIMPLEX fit for the K0 dwarf 54 Psc. The top panel shows the comparison between the observed classification spectrum and the synthetic spectrum. The bottom panel shows the synthetic fluxes compared with observed fluxes (large dots) from Strömgren *vby* photometry and Johnson *RI* photometry.

We will discuss below the automatic classification of stellar spectra using ANNs, but it should be kept in mind that this technique can be used to give estimates of the physical parameters as well.

Such techniques will assume increasingly greater importance in the future, especially with ground–based and space–based missions which are returning and will return large numbers of spectra. Small telescopes have an important role in developing these techniques.

3.2 Automatic Classification of Stellar Spectra

MK spectral classification is a powerful tool which is able to place a star in the context of a broad population of stars, as well as single out astrophysically interesting stars for further study. Traditionally, MK spectral classification has been carried out visually by comparing the program star with spectra of standard stars. The technique has been successfully transferred to modern digital spectra, but the process can still be very time consuming. For instance, the authors of this chapter are collaborating on an NStars (nearby stars) project in which we are obtaining classification–resolution spectra (on a number of small telescopes with aperture size from 0.6 - 2.3-m) of the nearly 3600 stars earlier than M0 within 40pc of the sun (see http://stellar.phys.appstate.edu/). We are classifying these spectra visually on the computer using XMK18, an X-windows graphics program designed specifically for the purpose of MK classification. The first group of 250 stars has taken nearly 2 months to completely classify (each of us classifies the spectra independently, then these classifications are collated and we iterate until we reach agreement). Our mode of operation guarantees very precise and accurate spectral types, but for even this relatively small project a reliable and accurate automatic technique to deal with these spectra would be very desirable. With the advent of new surveys, both ground–based and space–based, and the introduction of multiple-object spectrographs, the flood of stellar spectra is likely to accelerate, and some practical way will need to be found to categorize all of these stellar spectra in an automatic way. Ideally, a technique should be devised that will classify the star in essentially real time, i.e., at the telescope as the spectra are being acquired. Small telescopes can play an important role in the development of such a tool both by providing spectra and by acting as a test bed for techniques.

Some progress has been made in the last decade and a half in devising automatic classification techniques. Essentially two approaches have been investigated in detail — the "metric–distance technique" (cf. Kurtz 1984 and LaSala 1994) and techniques based on Artificial Neural Networks (ANN) (cf. von Hippel et al. 1994, Bailer–Jones 1997, Weaver 1994, Gulati et al. 1994 and Singh et al. 1998). As was outlined in Section 2, the advantage of using such techniques to determine spectral types as opposed to physical parameters (see paragraphs above) is that the spectral type is independent of theoretical stellar atmosphere models, and thus will not change as the models evolve. Indeed, there are a large number of peculiar stars which cannot yet be adequately analyzed using theoretical models and thus are most usefully characterized using classification techniques.

Both techniques show promise even though they are based on quite different philosophies. The metric–distance technique regards the digitized spectrum as an n-element vector (where n refers to the number of resolution elements in the spectrum) and defines the metric distance between a program spectrum X and a standard S as

$$d_{xs} = \frac{1}{n}\sqrt{\sum_{i=1}^{n} \alpha^2 (X_i - S_i)^2} \tag{1.1}$$

where the weighting factor α^2 is defined in such a way that the highest weights are given to those features which most strongly discriminate the final spectral type from other surrounding spectral types (cf. LaSala 1994). One advantage of this method is that it is quite close to the original philosophy of the MK system (e.g., that spectral types are determined by comparison with MK standards). A disadvantage is that the method requires that the spectra should be accurately rectified (i.e., the shape of the continuum should be removed). This is straightforward in the early type stars, but is problematical in the late–type stars (say later than K0), as in these stars there are no true continuum points. This problem becomes extreme in the M- and the C-type stars. This should not be an insurmountable problem, but so far this method has only been applied to stars earlier than K0.

Artificial Neural Networks (ANN) have been used quite extensively to distinguish, on digital images, between stars and galaxies and to classify galaxies, and so it is quite natural to apply the technique to stellar classification. We refer the reader to the references listed above for details on the method, but in brief, the ANN method uses an artificial neural network consisting of a number of layers of nodes, including a layer for input of the data (in this case, the spectral data), a layer for the output, and one or more "hidden layers" of summation nodes together with weighted connections between the layers (see Figure 2). The most popular configuration used for automatic spectral classification uses a gradient-descent, non-linear minimization ANN technique, known as back propagation, with a single hidden layer. The ANN must first be trained on a set of spectra with well–determined spectral types; this training changes the weighting used in the connections between the layers. Once trained, the ANN can be used to classify new data. Errors involved in temperature classification are typically on the order of 1.5 - 2 spectral subtypes when classifying across the entire HR diagram. One of the advantages of this method is that rectification of the spectrum is not as critical as in the metric-distance method. Indeed, the ANN can, in principle, be trained with non-rectified spectra with known reddenings, and can be designed to output not only the spectral type but also the reddening. A disadvantage is that the ANN method does not, contrary to the principles of MK classification, classify a star against the standards, but is instead an "expert" system which effectively reproduces the results of an expert classifier (in the case of the work by von Hippel's group, Nancy Houk, 2001), and thus classifies

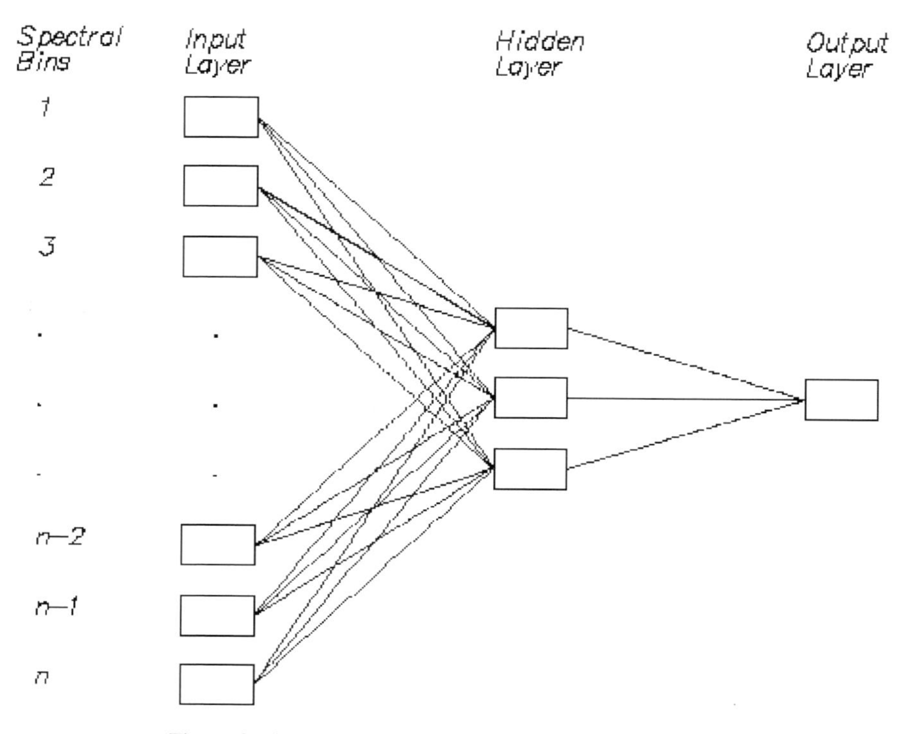

Figure 2. A sample ANN architecture with one hidden layer.

stars against the "cloud" of stars classified by that expert as a single spectral type.

Both of these techniques show adequate maturity with respect to their theory (Zhu 1995), but refinement of their details is much needed. In our opinion, both methods need more work to improve the accuracy of temperature classification. After all, skilled visual classifiers, such as Garrison and Houk, can get to within 0.6 spectral subtypes of each other. Of greater concern is that little work has been done on the much more subtle process of luminosity classification (see, however, the work of Weaver, 1994, on the ANN luminosity classification of A–type stars). Both techniques need considerably more work before they can usefully identify peculiar spectra (bear in mind the vast range of stellar peculiarities, from Ap and Am stars to emission-line stars to carbon stars) and deal with composite spectra (see recent work by Weaver, 2000). And, considerable work will be required to make these methods "robust" so that they can be used on large quantities of real data. Small telescopes can be used to advantage in developing these techniques by supplying libraries of spectra. For instance, Garrison has a spectral library of more than 30,000 spectra of 10,000 stars, obtained on identical Garrison spectrographs on small telescopes in both hemispheres.

Gray has a library of over 12,000 CCD spectra of about 4000 individual stars obtained on the Gray/Miller spectrograph on the Dark Sky Observatory 0.8-m telescope. These and other spectral libraries (such as the Coudé Feed Spectral Library) could be excellent resources for the development of automatic classification techniques.

Until these very large datasets can be utilized, an interim solution would be to make up an observed set of spectra of the Primary Standards (Section 1) at relatively high resolution. This grid would be available to all, and especially to the large telescopes. Software would be written to adapt the grid to the spectral resolution and configuration being used on a particular night, perhaps through the use of a standard specimen. At the end of each exposure, a rough preliminary spectral type would be generated automatically. In the case of multi-object spectrographs, it may take a little longer and be accessible at the end of the night. Our guess is that using this methodology, more peculiar objects would be discovered than is the case now and the information efficiency of very large telescopes would increase by a significant fraction; and this would come thanks to the smaller telescopes.

4. SURVEYING AND MONITORING THE MOST ASTROPHYSICALLY INTERESTING STARS

4.1 Surveys of Cluster and Association Stars

Most stars in clusters lack spectral classification. Mermilliod's (2001) WEBDA database currently gives 5700 stars in 291 open clusters having MK types. This is up from the 4000 stars in 225 clusters recorded in 1984 (Levato 1986), but it is still a woefully small number of the stars reachable spectroscopically with small telescopes. A significantly increased database of classified cluster stars will provide the subjects for research into the following astrophysical problems.

4.1.1 The Distances to Clusters and Associations

While Hipparcos distances are accepted as the best available, there is still need for checking them, even that to the Pleiades. Double-lined spectroscopic binaries are very effective for making such an independent check (Torres et al. 1997, and similar papers by the same authors), but more are needed. Indeed, only some 486 spectroscopic binaries, single and double-lined, are currently identified in clusters (WEBDA database, Mermilliod 2001), despite their value in testing models of stellar evolution (Andersen 1991; Lastennet et al. 1999).

Even when one falls back on main-sequence fitting for distances, rogue stars that do not quite fit will be found. Classification spectroscopy is essential for determining why these in fact appear rogue.

4.1.2 Stellar Evolution

As just indicated, this major concern of astrophysics finds a perfect test ground in clusters and associations by matching isochrones to real data once the distances to the clusters are established, preferably independently of main-sequence fitting. One problem in such matching is to understand how rapid rotation in high-mass stars can affect their scatter on the upper main sequence of clusters. This broadening in their HR diagram positions can be due to changes in their interior structure and their surface-layer abundances, and it is also affected by whether these stars are viewed pole-on or equator-on.

An ongoing problem for stellar evolution theory has been the precise modeling of convection. It is true that the relatively recent inclusion of overshooting has successfully lowered the evolutionary ages of clusters, but it is still a matter of choice over just how much overshooting is used. We need better constraints from a larger database, even for a cluster as near as the Pleiades. Essential first steps in that database are good classification and parameter definition for candidate stars. This is most efficiently done by small telescopes.

4.1.3 Abundances and Ages

Element mixing in the atmospheres of stars certainly occurs through rotation, convection, and diffusion, but its importance in the overall scheme of a star's evolution is disputed. What this mixing certainly does do is to change the rate of depletion of lithium with the age of a star (Jeffries 2000), so our calibration of the lithium chronometer is in question until mixing is better understood. Again, while lithium analysis needs the larger telescopes, more candidates from clusters for research on mixing need to be found in the coming decade.

We also need more data on the overall metallicities of stars within clusters. Section 3.1 on the analysis of low-resolution stellar spectra shows how this can be done very efficiently with the smaller telescopes. Since metallicity affects the star's HR diagram position, which in turn affects its assigned age, such information is essential for sorting out both the overall age of a cluster and the spread of age within a cluster. The latter, if it exists, is a fascinating record of the star formation sequence within a particular cluster.

4.1.4 Peculiar Stars

Most stars are normal. Where then do the stars with peculiarities fit into the general scheme of stellar evolution and environment? Clusters are clearly places ripe for studying this relationship between normal and peculiar stars, but the latter have first to be identified. This is where low-resolution spectroscopy excels, in complementary partnership with photometry. (The WEBDA database again shows that the number of peculiar stars known in clusters is quite limited.)

Among the peculiar stars thus selected we might highlight the pre-main-sequence stars, always needed for studying the development of circumstellar disks and their effect on stars. The presence of disks is obviously best determined through infrared observations, but the youth of stars can been seen in their spectra, particularly from emission-line features. So, Be and Ae stars come in here, while the still enigmatic λ Boötis stars may indeed be related to the latter and affected by the fate of their circumstellar disk material (Gray & Corbally 2001; for a review, see Gray 1997). Studying the development with stellar age of peculiarities such as Am and the broader Bp/Ap phenomena is also much promoted by cluster stars, provided they are first identified using small telescopes and the new techniques described in Section 3.

4.2 Surveys of Field Stars

It might seem that there is not much left to be done among nearby field stars that is astrophysically interesting, at least at low to medium dispersion. After all, isn't this covered by the Michigan Spectral Catalogue project (Houk 2001)? The answer is definitely “no”. NSF funding for that project has, at least temporarily, dried up when there is still almost all of the northern sky to cover. But even if this survey were complete, there would yet be much to do among field stars, both those of the HD catalog and fainter.

4.2.1 Variable Stars

The monitoring of spectroscopically variable stars, whether in clusters or in the field, is still going to produce interesting astrophysics. For instance, there is variability among both Be and Ae stars that we do not understand (Marlborough 2000). How is the variability connected to the stars' pulsation, to their magnetic fields, or modulated by their rotational velocity? What is the interaction of the star with its surrounding shell (Abt 1997)? How much difference does environment make to the emission phenomenon of, for instance, the forbidden lines of the B[e] type stars?

We have been fortunate to have a multitude of variable stars discovered by extensive surveys such as Hipparcos, MACHO, and OGLE. These are all candidates for obtaining parameters by the method described in Section 3.1 or for monitoring the spectrum variables, such as the Mira and VV Cephei stars for which each cycle is unpredictably different. This is another niche for small telescopes.

4.2.2 Peculiar Stars

The stars that do not fit neatly into our classification schemes are not confined to clusters. Among such stars in the field are the Am and Ap stars. Morgan (1984; Morgan et al. 1978) highlights the Mn-Hg stars, which are localized in the MK plane, and the Sr II stars, which have a considerable range in spectral type. Dworetsky (1993) was convinced that Hg-Mn stars were "telling us something quite fundamental about stellar interiors and evolution, and about star formation," and we still have not unraveled their story completely. This problem is within easy range of small telescopes.

4.2.3 Population II Stars

These are sensitive tracers of the history of galactic nucleosynthesis. Even at low to medium resolution, spectra of these stars will yield quite accurate abundance information on Ca (related to the most massive stars; see Maeder 1992), on C and N (related to the intermediate mass stars), and on Fe (related to the SN Ia contributions). There is significant work to be done in refining the precision of these abundances, both from the aspect of defining grids of spectral standards for weak-lined stars (Corbally 1994, Gray 1989) and from the application of spectral synthesis to these spectra (see Section 3.1). Then, at a comparatively small outlay in observing resources, the history of our Galaxy can be better understood.

4.2.4 Limited Samples

Getting precise, homogeneous data on a set of stars is always an adventure in astrophysics. You never know what treasures you are going to discover. The authors' NStars project (Section 3.2) was conceived to help refine a list of nearby stars suitable for the Space Interferometry Mission and the Terrestrial Planet Finder mission, but serendipity will turn up interesting objects. Such are the young binary system, HD 37572 plus Hip 26369, and the two other highly chromospherically active stars that are reported by Gray & McFadden (2001). This is a volume-limited survey, thanks to Hipparcos data, but other surveys could well be limited to specific regions of the HR diagram, using the Michigan Spectral Catalogue as their starting database, all within reach of small telescopes.

5. CONCLUSIONS

It is clear that for understanding the complexity of stars, we need a balance of available telescope sizes to be used in a complementary way. Low and medium resolution spectroscopy has brought serendipitous discoveries to stellar astrophysics, and these come when small telescopes are doing what they do best, surveying and monitoring. So, for the next decades spectroscopic surveys, such as currently for the nearby stars, will still be key in picking out the interesting objects. If these are spectroscopic variables, then follow-up monitoring by small telescopes becomes essential before the right questions can be posed for the high-resolution, large telescope work. What is the relationship between the normal and the peculiar stars we find? Does each peculiar star have a perfectly normal counterpart on the HR diagram? How does galactic environment affect the peculiarities we observe?

Traditionally, stellar classification has been the prime way both to pick out and monitor peculiarities in stars. Since this will still be foremost in storing in a convenient form the description of the spectra of billions of stars, the next decade must see the standards for this technique maintained and refined. Refinements are particularly needed in describing the varieties of peculiar stellar spectra.

Other tools are coming to low resolution spectroscopy through an increase in computing power. There are new and very efficient ways to get good stellar parameters from such spectra, and so to improve both the discovery and monitoring aspects of this kind of astronomy. The methods of automatic spectral classification will give as yet unrealized potential for surveys. In this development small telescopes are needed to bring the tools to perfection. When perfected, our reach towards the most distant stars with the largest telescopes will be extended as well. We must conclude that small telescopes are indispensable accessories to very large telescopes.

6. REFERENCES

Abt, H.A. 1997, ApJ 487, 365.

Andersen J. 1991, A\&AR 3, 91.

Bailer-Jones, C.A.L. 1997, PASP 109, 932.

Bailer-Jones, C.A.L., Irwin, M., Gilmore, G., von Hippel, T. 1997, MNRAS 292, 157.

Breger, M., et al. 2002, MNRAS 329, 531.

Cayrel, R., Perrin, M.N., Buser, R., Barbuy, B., Coupry, M.F. 1991, A&A 247, 122.

Corbally, C.J. 1994, in ``The MK Process at 50 Years," eds. C.J. Corbally, R.O. Gray & R.F. Garrison, ASP Conference Series Vol. 60, p. 237.

Cuisinier, F., Buser, R., Acker, A., Cayrel, R., Jasniewicz, G., Fresneau, A. 1994, A&A 285, 943.

de Boer, K.S., Tucholke, H.-J., Schmidt, J.H.K. 1997, A&A 317, L23.

Dworetsky, M.M. 1993, in "Peculiar versus Normal Phenomena in A-Type and Related Stars," eds. M.M. Dworetsky, F. Castelli, and R. Faraggiana, ASP Conference Series Vol. 44, 1.

Garrison, R.F. 1994, in "The MK Process at 50 Years," eds. C.J. Corbally, R.O. Gray & R.F. Garrison, ASP Conference Series Vol. 60, p. 3.

Gray, R.O. 1989, AJ 98, 1049.

Gray, R.O. 1997, in "The Third Conference on Faint Blue Stars," eds. A.G.D. Philip, D.S. Hayes and J.W. Liebert (Schenectady: L. Davis Press), p. 237.

Gray, R.O., and Corbally, C.J. 1994, AJ, 107, 742.

Gray, R.O., and Corbally, C.J. 2002, AJ 124, 989.

Gray, R.O., Graham, P.W., Hoyt, S.R. 2001, AJ 121, 2159.

Gray, R.O., and Kaye, A.B. 1999, AJ, 118, 2993.

Gray, R.O., and McFadden, M.T. 2001, IBVS #5055.

Gulati, R.K., Gupta, R., Gothoskar, P., Khobragade, S. 1994, ApJ 426, 340.

Houk, N. 2001, Michigan Spectral Catalogue, http://www.astro.lsa.umich.edu/users/hdproj/

Hauschildt, P., and Allard, F. 2001, http://dilbert.physast.uga.edu/~yeti/mdwarfs.html

Jeffries R.D. 2000, in "Stellar Clusters and Associations: Convection, Rotation, and Dynamos," eds. R. Pallavicini, G. Micela, and S. Sciortino, ASP Conference Series Vol. 198, 245.

Jones, J.B., Gilmore, G., Wyse, R.F.G. 1996, MNRAS 278, 146.

Koen, C., Kurtz, D.W., Gray, R.O., Kilkenny, D., Handler, G., Van Wyk, F., Marang, F., Winkler, H. 2001, MNRAS 326, 387.

Kurtz, M.J. 1984, in "The MK Process and Stellar Classification," ed. R.F. Garrison (Toronto: David Dunlap Observatory), p. 131.

Kurucz, R.L. 1993, CD-ROM 13, ATLAS9 Stellar Atmosphere Programs and 2 km/s Grid (Cambridge: Smithsonian Astrophys. Obs.)

LaSala, J. 1994, in "The MK Process at 50 Years," eds. C.J. Corbally, R.O. Gray & R.F. Garrison, ASP Conference Series Vol. 60, p. 312.

Lastennet, E., Valls-Gabaud, D., Lejeune, Th., and Oblak, E. 1999, A&A, 349, 485.

Levato, H. 1986, in IAU Symposium 118, eds. J.B. Hearnshaw & P.L. Cottrell (Dordrecht: Reidel), p. 359.

Linsky, J.L. 2000, in "Stellar Clusters and Associations: Convection, Rotation, and Dynamos," eds. R. Pallavicini, G. Micela, and S. Sciortino, ASP Conference Series Vol. 198, 565.

Maeder, A. 1992, in "The Stellar Populations of Galaxies," eds. B. Barbuy and A. Renzini, IAU Symposium 149 (Dordrecht: Kluwer), p. 109.

Marlborough, J.M. 2000, in "The Be Phenomenon in Early-Type Stars," eds. M.A. Smith, H.F. Henrichs, J. Fabregat, ASP Conference Series Vol. 214, 743.

Mermilliod, J.-C. 2001, WEBDA database, http://obswww.unige.ch/webda/

Miroshnichenko, A.S., Bjorkman, K.S., Chentsov, E.L., Klochkova, V.G., Gray, R.O., Garcia-Lario, P., Perea Calderon, J.V. 2001, A&A 377, 854.

Morgan, W.W. 1984, in "The MK Process and Stellar Classification," ed. R.F. Garrison (Toronto: David Dunlap Observatory, Univ. of Toronto), p. 18.

Morgan, W.W., Abt, H.A., and Tapscott, J.W. 1978, "Revised MK Spectral Atlas for Stars Earlier than the Sun," (Chicago and Tucson: Yerkes and Kitt Peak National Observatories).

Morgan, W.W., Keenan, P.C., and Kellman, E. 1943, "An Atlas of Stellar Spectra" (Chicago: Yerkes Obs).

Piskunov, N.E. 1992, in Proceedings of the International Meeting on "Stellar Magnetism", eds. Yu.V. Glagolevskij and I.I. Romanuk, (St. Petersburg, Special Astrophys. Obs., Nauka), 92.
Press, W.H., Teukolsky, S.A., Vetterling, W.T., and Flannery, B.P. 1992, Numerical Recipes in C (2d ed.; Cambridge: Cambridge Univ. Press).
Singh, H.P., Gulati, R.K., Gupta, R. 1998, MNRAS 295, 312.
Torres, G., Stefanik, R.P., Latham, D.W. 1997, ApJ, 485, 167.
von Hippel, T., Storrie-Lombardi, L.J., and Storrie-Lombardi, M.C. 1994, in "The MK Process at 50 Years," ed. C.J. Corbally, R.O. Gray and R.F. Garrison ASP Conference Series Vol. 60, p. 289.
Weaver, Wm.B. 1994, in "The MK Process at 50 Years," ed. C.J. Corbally, R.O. Gray and R.F. Garrison ASP Conference Series Vol. 60, p. 303.
Weaver, Wm.B. 2000, ApJ 541, 298.
Zboril, M. 1996, in "Model Atmospheres and Spectrum Synthesis," eds. Saul J. Adelman, Friedrich Kupka, and Warner W. Weiss, ASP Conference Series Vol. 108, p. 193.
Zhu, M. 1995, M.Sc. Thesis, University of Toronto.

Chapter 7

The NStars Project and Small Telescopes

Todd J. Henry
Georgia State University
Atlanta, Georgia USA

Dana E. Backman
Franklin & Marshall College
Lancaster, Pennsylvania USA

Jerry Blackwell
Nearby Star Observers
Federal Way, Washington USA

Takeshi Okimura
Symtech, NASA-Ames Research Center
Moffett Field, California USA

Sharon Jue
Johns Hopkins University
Baltimore, Maryland USA

Abstract: The NStars Project is investigating all stellar, substellar, and planetary objects within 25 parsecs of the Sun. This effort involves collecting data from the astronomical community into a comprehensive and quality-controlled master compendium (the NStars Database) available over the worldwide web (http://nstars.arc.nasa.gov, at the time of this writing). Unlike earlier general and "passive" stellar databases that amass published data, NStars is also actively carrying out research to characterize known members of the sample as well as to

T.D. Oswalt (ed.), The Future of Small Telescopes in the New Millennium, Vol. III, 111-121.

discover new members, with special emphasis on properties of astrobiological interest. Much of the fundamental astrometric, photometric, and spectroscopic research needed to develop a complete picture of the Sun's neighbors is being carried out on telescopes with apertures less than 2 meters. Small telescopes continue to provide the backbone of stellar astronomy, just as they have for decades.

Ultimately, NStars will provide target lists of stars to be closely examined by NASA Origins missions and programs such as SIRTF (Space InfraRed Telescope Facility), SOFIA (Stratospheric Observatory For Infrared Astronomy), SIM (Space Interferometry Mission), and TPF (Terrestrial Planet Finder). Thus, much of the fundamental work required before any of these missions can be successful takes place at small telescopes on the ground. We will discuss the crucial role that these small telescopes play in the NStars Project, and how they provide the foundations for these billion-dollar missions.

Key words: nearby stars, parallaxes, astrometry, photometry, spectroscopy

1. THE NSTARS PROJECT

The Exozodiacal Dust Workshop held at NASA-Ames in October 1997 (Backman et al. 1998) was driven by a need to understand the effects of exozodiacal dust clouds on the proposed Terrestrial Planet Finder (TPF). Both TPF's ability to detect Earth-sized life-bearing planets around nearby stars and the optimization of TPF's design were addressed. A main recommendation of the workshop was development of a master database regarding stars in the solar neighborhood.

The Nearby Stars (NStars) Project was established in April 1998 at NASA-Ames to create the Database, promote new research on nearby stars, and ultimately produce a carefully selected list of target stars for TPF. The first NStars Workshop focusing exclusively on nearby stars, sponsored by the NStars Project and its Science Working Group, was held at NASA-Ames in July 1999 (Backman et al. 2001).

The NStars Database is presently located at http://nstars.arc.nasa.gov and will be moving to a site at Northern Arizona University in early 2003. As of January 1, 2003, there are 2633 stars in 2029 systems in the Database. The total number of stars expected within 25 pc (scaled from the better-known sample within 5 pc) is roughly 10000 in 5500 systems, indicating that nearly three-quarters of the sample is "missing" (see also section 2 below). In addition, substellar objects are being found rapidly enough that their total space density may be comparable to stars; brown dwarfs and confirmed extrasolar planets will be included in the NStars Database.

In contrast to the European general stellar database SIMBAD and the nearby stars database CNS which are compendia of published data on their catalog objects, NStars includes only quality-filtered data. Also unlike these general and "passive" stellar databases, the NStars Project is focused on:

- critically examining completeness of the sample, with active research carried out by the NStars team aimed at adding to the list of nearby stars,
- determining which stellar characteristics need further definition and measurement for a comprehensive understanding of stellar astrophysics, and fostering research by the astronomical community in those directions,
- identifying crucial gaps in the NStars Database that must be filled in order to provide a robust set of data suitable for astrobiology research and support of NASA Origins missions, with active research by the NStars team aimed at filling those gaps.

In parallel to NASA's NStars Project centered at NASA-Ames (moving to NAU) which has the primary goal of database development, the National Science Foundation in cooperation with NASA announced a separate funding opportunity called the NStars Program, which should not be confused with the NStars Project. This cooperative effort between NSF and NASA is specifically for research programs on nearby stars (NSF) and to support the upcoming space missions that will target nearby stars (NASA). Ultimately, results from the NStars Program will be included in the NStars Database. The NStars Project has agreed to specify formats and quality criteria for incoming data and to help coordinate the target lists and observations of the NSF and NASA funded teams.

2. SMALL TELESCOPES AND THE NSTARS SAMPLE TODAY

The defining measurable characteristic of a nearby star is its trigonometric parallax. Here we provide a few details about the parallax work that has allowed us to measure the third dimension of our night sky and thereby constrain the NStars sample.

Productive ground-based astrometric parallax programs require consistent scheduling of observing time through several years to allow untangling of a star's parallax from its proper motion. Because observing time, not telescope aperture, is crucial, small telescopes are the champions in the business of stellar parallaxes. Examination of the telescope apertures for some of the more famous parallax programs shows this clearly: Allegheny Observatory (0.76-m refractor, University of Pittsburgh), Cape Observatory (0.66-m reflector), McCormick Observatory (0.66-m refractor, University of Virginia), Sproul Observatory (0.61-m refractor, Swarthmore College), U.S. Naval Observatory (1.55-m reflector, Flagstaff), Van Vleck Observatory (0.51-m refractor, Wesleyan University), Yale University (0.66-m refractor).

To date, the NStars Project team has carefully combined ground-based (van Altena et al. 1995) and spaced-based (HIPPARCOS, ESA 1997) trigonometric parallaxes to develop the definitive NStars membership list. It is important to point out that nearly all of the trigonometric parallaxes determined from the ground, including targets as faint as V = 20, were done using telescopes with primary mirrors less than 2 meters in size. Even HIPPARCOS, which determined parallaxes for roughly 120,000 stars to V ~ 13, had only a 0.29-m primary mirror.

The first stellar parallax (of 61 Cygni) was determined in 1838 yet the construction of a comprehensive three-dimensional map of the solar neighborhood is far from complete. Although the NStars sample is arguably the most complete list of nearby stars ever constructed, straightforward analyses of sub-sample densities reveal obvious shortfalls in the present database. The sample is incomplete in two senses: (1) the density of known stars decreases with distance, and (2) a disproportionate number of stars remain unidentified in the southern sky. Figure 1 illustrates the first point clearly. Using the NStars list we estimate that 62% of the stellar systems within 25 pc are missing from the NStars sample. Table 1, also derived using the current NStars list, illustrates the second type of incompleteness -- more stars are missing in the southern sky than in the north. We estimate that more than 70% of the southern systems remain undiscovered.

Table 1.
Number of Stellar Systems within 25 Parsecs

Region of Sky	# Systems Known	Fraction of Sample	Total Predicted	Fraction Missing
+90 to +30	603	29%	1375	56%
+30 to +00	591	29%	1375	57%
-00 to –30	479	23%	1375	65%
-30 to –90	403	19%	1375	71%
TOTAL	2076	100%	5500	62%

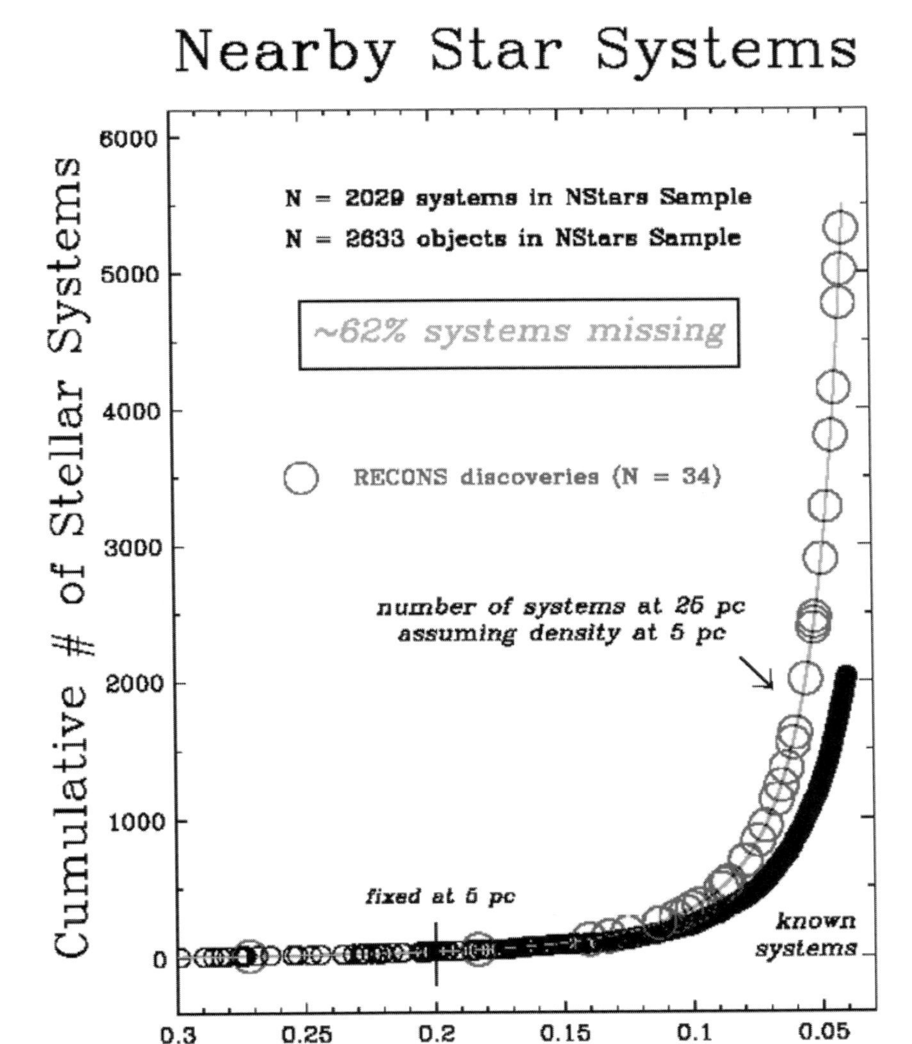

Figure 1. The cumulative number of stellar systems known within 25 pc is shown as a function of parallax. Assuming the density of stars within 5 pc is constant to 25 pc, roughly 3500 systems are estimated to be missing from the present NStars sample. Large circles are new nearby systems discovered by the RECONS team, which is supported in part by the NStars effort.

3. DISCOVERING NEW NEARBY STARS

Three aggressive parallax programs are currently being carried out to reveal the Sun's hidden neighbors. All use small telescopes, and some of the recent results are striking. In the northern hemisphere, the U.S. Naval Observatory program is continuing using optical and/or infrared imaging cameras. Phil Ianna leads a program using the Siding Spring 1.0-m telescope in Australia. Here we discuss the third and youngest program of the three, which is supported in part by NStars.

The Research Consortium on Nearby Stars (RECONS, Henry et al. 1997) has the goals of discovering and characterizing all stars, brown dwarfs, and planets within 10 parsecs. Established in 1994, RECONS can be considered an early NStars effort. In 1999 RECONS initiated a parallax program in Chile. Under the auspices of the NOAO Surveys Program, Henry and co-investigators received more than 200 nights from August 1999 through July 2002 on the CTIO 0.9-m and 1.5-m telescopes to obtain parallaxes, proper motions, and photometry of southern white, red, and brown dwarf candidates in the NStars volume. The program is called CTIOPI, for Cerro Tololo Interamerican Observatory Parallax Investigation. CTIOPI's primary goal is to measure definitive parallaxes accurate to 3 milliarcseconds for 150 new members of the NStars sample in three years. As of January 1, 2003, preliminary results from CTIOPI's 0.9-m program include 10 new systems containing 13 stars within the RECONS horizon, and an additional 24 systems containing 29 stars within the NStars horizon.

Closing of the CTIO 0.9-m in 2003 and "privatization" of much of the observing time on the 1.5-m is motivating development of a second-generation CTIOPI effort. The plan is to continue the southern parallax program for several more years using the CTIO 0.9-m telescope, as part of a new consortium that will operate the CTIO 1.5-m, 1.3-m, 1.0-m, and 0.9-m telescopes. Plans are also underway to place an infrared camera on one of those telescopes, from which nearby brown dwarfs can be efficiently targeted in an infrared astrometric program.

Measuring the first accurate parallaxes for hundreds of white, red, and brown dwarfs is not simply a matter of knowing who your neighbors are. These comprehensive parallax programs will also provide

- direct measurements of luminosities, colors, and temperatures for these faint but numerically dominant members of the solar neighborhood,
- detection of new astrometric binaries that will become targets for crucial mass determinations,
- data allowing determination of the initial mass function down to a few x 10 Jupiter masses,

- a robust database for stellar kinematic and multiplicity studies, and
- the best targets for NASA's planned Origins space missions, in particular SIM and TPF.

4. CHARACTERIZING THE NEARBY STARS

Although most of the nearby stars are faint red dwarfs, many of the more famous members of the immediate solar neighborhood, such as Sirius, Procyon, alpha Centauri A and B, Fomalhaut, Altair, Vega and many others, are difficult or impossible to observe with today's largest telescopes — they are simply too bright. Nonetheless, these prominent members of the nearby sample are worthy of detailed scrutiny, if we are to develop a comprehensive assessment of our Sun's place in the Galaxy.

NStars is collecting published photometry and making new observations to build a photometric matrix that includes 22 optical/infrared bands for each star: Johnson UBV, Cousins RI, Sloan ugriz, 2MASS JHK, CIT LL'MNQ, IRAS 12, 25, 60, and 100 microns. These data allow astronomers to calibrate absolute magnitudes and colors for all types of stars in the solar neighborhood, discover hidden multiple systems that can be used for crucial mass determinations, and estimate photometric parallaxes of candidate nearby stars that can be targeted for trigonometric parallax measurements.

Small telescopes can be used to fill in the many photometric "holes" in the NStars Database. To ensure high quality data in NStars, only efforts that have targeted hundreds of nearby stars and which have been fully evaluated for photometric consistency have been used. The most reliable optical photometric work includes that of Bessell (1990), Cousins (several references in the Bessell paper), Leggett (1992), which is compendium of many smaller efforts each of which has been painstakingly and very helpfully calibrated to a common photometric system, and Weis (1996), who examined nearly 1200 high proper motion stars. Infrared work includes Leggett (1992) and 2MASS. All of these studies were done using telescopes with apertures smaller than 2 meters.

Similarly, even the most straightforward spectroscopic data — spectral types — are missing for many NStars members. Currently, only about half of the stars within 25 pc have spectral types on a standard system. The standard system presently adopted for the brighter AFGK type stars is that of Houk (Houk and collaborators 1975-1999), which are based on objective prism spectra of HD stars taken with the 0.6/0.9-m Michigan Curtis Schmidt telescope at CTIO. Many of the missing spectra for these bright stars are currently being acquired by Richard Gray and colleagues using the 0.8-m Appalachian State and CTIO 1.5-m telescopes. For the fainter M stars, the standard system is that of RECONS (Henry et al. 2002), supplemented with data from the Palomar/Michigan State University effort (Reid et al. 1995 and Hawley et al. 1996) which obtained spectra for ~2000 stars and is directly linked to the RECONS system. In the latter case, the spectra were all taken using the Palomar 1.5-m or CTIO 1.5-m telescopes.

In contrast to the short-term photometric and spectroscopic work so far described, long term monitoring of nearby stars for variability, flaring, stellar activity cycles, and eclipses/transits is in its infancy. Only small percentages of the nearest stars have been monitored for any significant time period, although radial velocity programs searching for planets are making headway for that particular application. In most cases, however, it is again large blocks of time on small telescopes that are required to complete comprehensive surveys for long-term changes in stellar surfaces.

Finally, determining the multiplicity of comprehensive samples of stars is best accomplished using small telescopes on which sufficient observing time can be acquired to survey hundreds or thousands of stars. Wide companions are typically found by taking astrometric frames with large field sizes, while close companions are revealed at finer scales. Unseen companions can also be found in astrometric frames as periodic perturbations that are detected after modeling the long-term image centroids for proper motion and parallax.

5. NSTARS CONTRIBUTIONS TO FUNDAMENTAL ASTROPHYSICS

The NStars Project performs "in-house" analyses of information in the Database to determine the most likely values for basic quantities such as luminosity, mass, radius, and age for all stars within 25 pc. Here we discuss only a few of the results that will be forthcoming from the NStars Database.

Luminosities and Masses: Luminosities (typically using M_V and M_K) can be determined via photometry and parallaxes from small telescopes. Through a long-term study of stars in various mass regimes, T. Henry is leading an effort to calibrate the mass-luminosity relation from 30 $M_\odot$ to the end of the main sequence at 0.08 $M_\odot$ (Henry et al. 1999). Accurate luminosities combined with this comprehensive empirical mass-luminosity relation allow estimation of mass — the single most important stellar parameter — for every star within 25 pc.

Radii: Available radii determinations from eclipsing binaries can be supplemented using direct measurements from the Georgia State University CHARA Array, which has a maximum baseline of 330-m and a resolution limit of 0.3 milliarcsecond in the V band. Granted, the effective baseline is not representative of a small telescope, but the six individual telescopes are only 1-m in size.

Ages: The age of individual field (non-cluster) stars are notoriously difficult to determine, requiring use of a number of correlated indices such as chromospheric emission intensity and rotation speed. An example of this type of investigation was a spectroscopic survey of a large sample of nearby solar-type stars (Henry et al. 1996) that measured Ca II emission and used the results to estimate stellar ages (Figure 2). The observations were made at the CTIO 1.5-m telescope. The effort is being expanded by David Soderblom using the KPNO Coudé Feed Telescope with a 1.5-m mirror and the CTIO 1.5-m to include ~5000 solar-type stars within 50 pc.

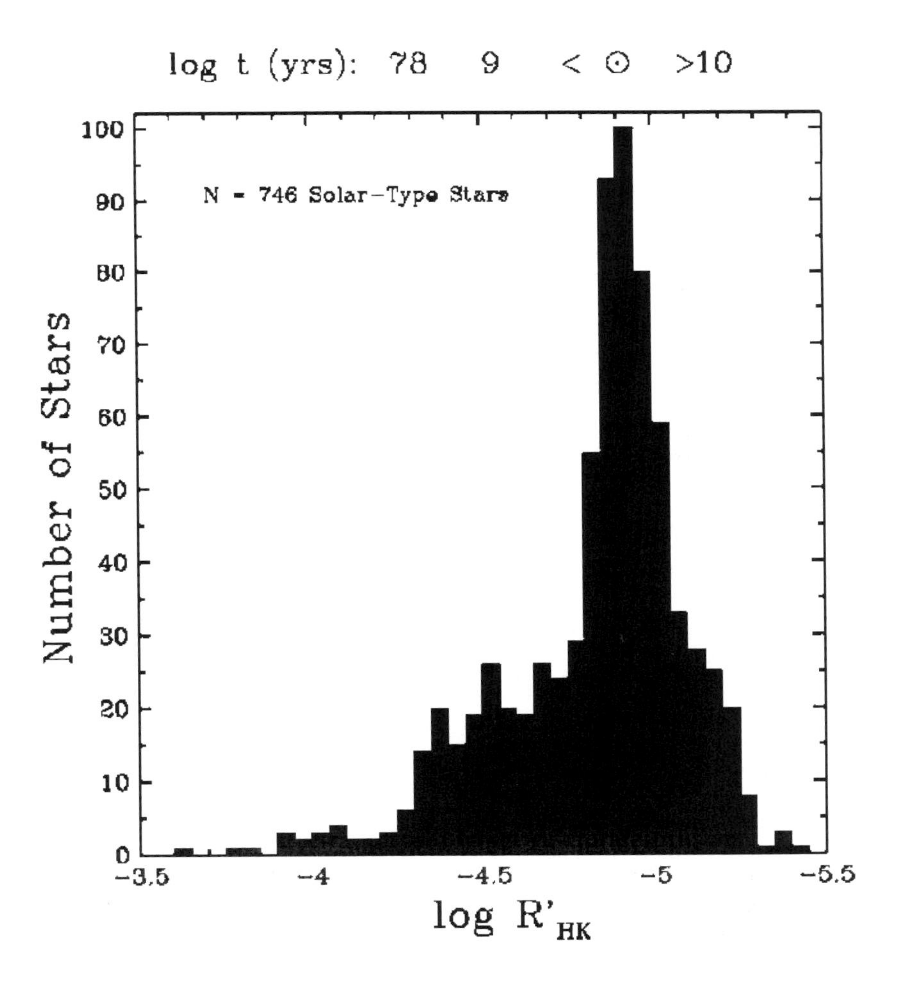

Figure 2. Histogram of log R'HK for a large sample of southern, solar-type stars. R'HK is an indicator of chromospheric emission which is linked to age. Representative ages derived using the chromospheric-age relation of Donahue (1993), based on a large body of cluster star spectroscopy, are shown along the top of the figure. The range in log R'HK values of the Sun throughout its activity cycle is indicated by the brackets around the Sun symbol.

6. SMALL TELESCOPES IN THE BIG PICTURE OF NSTARS

We have merely touched on some of the attributes of the nearest stars that can be addressed by small telescopes, including everything from their initial discovery to detailed studies of their behavior over time. Some of the most fundamental of all astronomical questions can be addressed by small telescopes examining the nearest stars: What is the true nature of the Galaxy's population? How many stars travel through the Galaxy alone, compared to those that travel in pairs or higher-order multiple systems? What are the masses of stars? How might environmental conditions on extrasolar planets change because of stellar variability?

For example, most astrobiological thought inclines toward the notion that only stars older than a certain age (in the few-Gyr range) can have planets where life has progressed enough to produce non-equilibrium atmospheric compositions detectable by TPF. Thus, TPF should be aimed first at the older nearby stars; their identities are mostly unknown, waiting to be determined and collated via the NStars Project.

Our ongoing comparisons of optical and infrared photometry have already revealed close, hidden multiple systems that will complicate SIM's and TPF's interferometric fringe patterns and point spread functions. Long-term astrometric series will reveal additional low mass companions that will affect target suitability for both SIM and TPF. Moreover, the NStars sample will be critical for terrestrial planet searches because the closest targets will exhibit the largest astrometric perturbations for a given planetary companion mass (for SIM) and present the largest separations on the sky for a planet at a given distance in AU from a star (for TPF).

Completing the sample of nearby stars down to low-luminosity M dwarfs has a special application to defining the TPF target list. Another example of astrobiological "common wisdom" has it that M stars don't host Earth-like planets because such planets would need to be so near their primaries for water to be liquid that tidal friction would halt or slow planetary rotation, rendering them sterile. However, recent research indicates that a substantial atmosphere can redistribute heat around an otherwise too-slowly-rotating planet (Joshi et al. 1997). Furthermore, an M0 star has fully 60% of the Sun's mass and would heat a planet to terrestrial temperatures at the position of Mercury's orbit, where tidal braking around a 0.6 $M_\odot$ star may not be severe. In addition, in our solar system, Europa is an example of an environment containing liquid water outside the conventional habitable zone. Thus, blanket exclusion from TPF's target list of the 70% of nearby stars that fall into the arbitrarily-defined spectral class M might be a mistake. NStars Project studies of M dwarf metallicities, ages, variability, etc. could lead to substantial broadening of the TPF target list and a determination of the local prevalence of planets less biased toward solar-type stars.

It should come as no surprise that modest, but consistent, investments of observing time on small telescopes are of tremendous benefit to scientific inquiries that ultimately require further work with large ground-based telescopes or expensive space-based missions. The NStars efforts described here, nearly all of which are carried out with small telescopes, are a prime example. We have come full circle and returned to the question with which we started this essay — the goal of finding other Earths orbiting our galactic neighbors, and perhaps even life on those Earths. We will only be fully prepared to carry out what is perhaps the most fundamental of all astronomical searches — the search for life on other planets — once the homework is done on the nearest stars. Most of that homework can and should be done using small telescopes.

7. REFERENCES

Backman, D. E., Caroff, L. J., Sandford, S. A., & Wooden, D. H. 1998, editors, "Exozodiacal Dust Workshop", NASA Conference Proceedings (Moffett Field: NASA-Ames)
Backman, D. E., Burg, S. J., & Henry, T. J. 2001, editors, "Nearby Stars (NStars) Workshop", NASA Conference Proceedings (Moffett Field: NASA-Ames)
Bessell, M.S. 1990, A&ASup, 83, 357
Donahue, R. A. 1993, Ph.D. thesis, New Mexico State University
ESA 1997, The HIPPARCOS and Tycho Catalogues, ESA SP-1200
Hawley, S.L., Gizis, J.E., & Reid, I.N. 1996, AJ, 112, 2799
Henry, T. J., Franz, O. G., Wasserman, L. H., Benedict, G. F., Shelus, P. J., Ianna, P. A., Kirkpatrick, J. D., & McCarthy, Jr., D. W. 1999, ApJ, 512, 864
Henry, T.J., Ianna, P.A., Kirkpatrick, J.D., & Jahreiss, H. 1997, AJ, 114, 388
Henry, T. J., Soderblom, D. R., Donahue, R. A., & Baliunas, S. L. 1996, AJ, 111, 439
Henry, T. J., Walkowicz, L.M., Barto, T.C., & Golimowski, D. A. 2002, AJ, 123, 2002
Joshi, M. M., Haberle, R. M, & Reynolds, R. T. 1997, Icarus, 129, 450
Leggett, S.K 1992, ApJS, 82, 351
Reid, I.N., Hawley, S.L., & Gizis, J.E. 1995, AJ, 110, 1838
van Altena, W. F., Lee, J. T., & Hoffleit, E. D. 1995, The General Catalogue of Trigonometric Stellar Parallaxes, Fourth Edition (L. Davis Press: Schenectady)
Weis, E.W. 1996, AJ, 112, 2300

Chapter 8

Speckle Interferometry with Small Telescopes

Brian D. Mason and William I. Hartkopf
U.S. Naval Observatory
Washington, DC USA

Abstract: Speckle interferometry is a mature technique first utilized by Antoine Labeyrie over a quarter century ago. Unlike conventional interferometry, speckle is filled-aperture interferometry utilizing the turbulence cells of the atmosphere over the telescope to constructively interfere and produce diffraction limited information. Despite an initial flurry of activity, however, work at the present time is limited to a few groups, typically utilizing telescopes of 0.7 to 4.0 meters. Although actual images have been difficult to obtain and are rather expensive (both in computer and telescope time) to produce, speckle interferometry is well suited to simple morphologies, such as relative astrometry of binary and multiple stars. The vast bulk of work done in speckle interferometry is observations of binary stars. These observations can be used to: determine stellar masses (although this is also dependent on the availability of other data, e.g., spectroscopic or parallactic), provide an independent check on proper motions determined for close doubles by other techniques, provide verification or confirmation of close binaries found by other techniques (e.g., Hipparcos), or, due to the rapid operation of most speckle observing, provide information as to the multiplicity characteristics of a large sample of stars over a large regime of ρ-Δm space. Speckle investigation can also aid in the study of stars in planetary searches by culling from targeted searches stars unlikely to have life harboring planets, or to ensure that "planets" detected spectroscopically are not (due to the *sin i* dependence) pole-on binary stars. Many of these known binaries in need of investigation can be observed with moderate or smaller sized telescopes.

Key words: techniques, Interferometry, visual binaries

T.D. Oswalt (ed.), The Future of Small Telescopes in the New Millennium, Vol. III, 123-138.

1. THE GENESIS OF SPECKLE INTERFEROMETRY

As a technique, interferometric measurement of binary stars is over one hundred years old. Karl Schwarzschild made the first measure of a binary star in late May, 1895 (Schwarzschild 1895) only four years after the technique was first used to measure the angular diameters of the Galilean satellites by Michelson (1891). Speckle interferometry is a much younger child of this far older method, having only been in use for the last thirty years. First proposed by Antoine Labeyrie (1970), speckle was applied to measuring angular properties, both binary star separations and stellar diameters, about one year later (Gezari et al. 1972). Programs using speckle interferometry were soon initiated at CERGA in France (Blazit et al. 1977), at Imperial College in England (Morgan et al. 1978), at the 6m telescope in Russia (Balega & Tikhonov 1977), and at Kitt Peak and Georgia State University in the United States (McAlister 1977a) with binary stars as the primary area of study. Other work in binary star speckle interferometry has been done at the University of Erlangen-Nurnberg in Germany (Weigelt 1978), at the University of Sydney in Australia (Tango et al. 1979), and at Steward Observatory in the United States both in the visual (Hege et al. 1982) and in the infrared (McCarthy et al. 1987). It was recognized quite early that speckle provided a means to determine orbital parameters for systems that can also be studied spectroscopically (McAlister 1976a). Some of the first results from speckle were of spectroscopic binaries (Labeyrie et al. 1974, McAlister 1976b), where this new technique quickly proved its worth in determining fundamental stellar properties (Popper 1980).

2. THE THEORY BEHIND SPECKLE

As light passes through the Earth's atmosphere it encounters turbulence cells. The density fluctuations causing the turbulence cells distort what would be an Airy pattern by providing some tilt within each cell, as well as some group delay to the entire series of parallel rays within each of these patches. The diameter of each of these patches, or coherence cells, is defined in terms of r_0, their characteristic size. A telescope of this size could produce an instantaneous image of the Airy disk. Over time these r_0 patches move due to wind and may even grow in size or dissolve due to atmospheric turbulence. With a telescope of diameter (D) larger than r_0, the image becomes more complex as the image now contains $(D/r_0)^2$ coherence cells (McAlister 1992). However, some of these patches will have the same tilt and group delay. As long as light from the binary system is passing through the same set of coherence cells (a condition known as isoplanicity), there will be a correlation in the speckle pattern produced by the components of the system. Indications are that this condition is met for stars closer than about 2" to 3" (Labeyrie 1970). The stellar light in these coherent patches will then interfere to produce dark and light patches that are known as "speckles." To capture or "freeze-out" these speckle patterns requires very short exposure images, but the result is that each isolated set of speckle pairs represents a diffraction-limited image of the object. Labeyrie demonstrated that the convolution of the object intensity distribution $[O(\alpha,\beta)]$ and the speckle image of a point object $|p(\alpha,\beta)|^2$ (i.e., the point-spread function) is what is observed in the image intensity distribution $[I(\alpha,\beta)]$:

$$[I(\alpha,\beta)] = [O(\alpha,\beta)] \odot |p(\alpha,\beta)|^2.$$

The Fourier transform of this intensity (in itself a convolution),

$$i(x,y) = o(x,y) \cdot A[P(x,y)],$$

and

$$|i(x,y)|^2 = |o(x,y)|^2 \cdot |A[P(x,y)]|^2,$$

is equal to the product of the Fourier transform of the two convolved quantities. Here, the autocorrelation function of $P(x,y)$ is A and the squared modulus of A is the transfer function. Dividing the Fourier transform of the speckle images by the Fourier transform of a single star then is the essence of speckle interferometry. A simple and efficient alternative to full power spectrum analysis is the vector-autocorrelation method. A vector autocorrelogram is produced by plotting all the pairings of speckles that occur within the speckle image in a diagram where each point for each member of the pair is translated to the origin. The actual binary pairing of speckles occurs with much greater frequency than random pairings; hence, the binary star geometry will dominate the autocorrelogram.

2.1 Resolution limits

One of the most frequently discussed parameters in dealing with interferometry, whether it be speckle, long baseline optical or even radio is the diffraction limit. We typically quote a 4-m resolution limit of 30 milliarcseconds (mas), which is arrived at through the Rayleigh criterion ($= 1.22\lambda/D$). More classical interferometers typically discuss the Michelson limit ($= \lambda/2D$). The crucial point to make here is that diffraction limited optics implies different things depending on wavelength. While adaptive optics and other imaging are possible with ~8-m telescopes and larger, it is usually done in K or other long wavelength bands where the atmosphere is more forgiving in terms of both coherence length r_0 and time (τ_0) The Rayleigh limit of a 10-m telescope in K is much larger than that of a 4m telescope in V doing optical speckle.

3. FROM TEST-BED TO INDUSTRY: THE GROWTH OF SPECKLE

Table 1 lists the more long-lived (5 years or more) programs in speckle interferometry or similar techniques, together with the most recent entries in the field. Names and dates are based on references in Hartkopf et al. (2001) and are approximate. (Due to changes in equipment, telescopes, and personnel, it is often difficult to precisely define a specific observing program; our apologies to any efforts neglected or given short shrift).

Table 2 gives the total number of high-resolution astrometric measurements in the current *Fourth Catalog of Interferometric Measurements of Binary Stars* (Hartkopf et al. 2001) obtained since the birth of speckle in 1970 (excluding unresolved observations and a handful of measures for which no telescope aperture was provided). Figures 1 – 3 illustrate the distribution of these observations with time, separation, and declination. Data are broken down by telescope aperture, with observations obtained from space (Hipparcos/Tycho and HST) listed separately. Data obtained using speckle interferometry (or similar techniques) are also listed, highlighting the dominance of speckle with small telescopes—91% of all ground-based observations—in this field. Finally, the contributions of the CHARA and USNO speckle efforts illustrate both the success of these projects and the potential danger of dominance by one or two groups. These two programs are responsible for over 83% of the published speckle data. Equipment and reduction techniques used in both programs are similar, and the principal scientists of the USNO program held similar roles in the now-closed CHARA effort. Despite extensive calibration efforts the worry persists that any systematic error would be hard to detect due to the dominance of this (essentially) single source of data. The lack of any apparent systematic offset between CHARA/USNO and Hipparcos measures of close slow-moving systems in encouraging; however, more observations by a variety of observers, cameras, and reduction methods are still needed. The recently begun efforts by the Docobo and Marchetti groups, as well as that of the talented amateur Chris de Villiers, are all the more welcome developments, in this regard.

Table 1.
Major speckle programs, past and present

Observer(s)	Active Dates
Labeyrie, Blazit, Bonneau, et al.	1970 –
CHARA: McAlister, Hartkopf, Mason, et al.	1975 – 1998
Balega, Balega, et al.	1977 –
Morgan, Vine, Argue, et al.	1978 – 1991
Weigelt et al.	1978 – 1994
Tokovinin	1978 – 2000
McCarthy, Henry, et al. (IR)	1983 – 2000
Karovska et al. (IR)	1985 – 1997
Leinert, Zinnecker, et al. (IR)	1986 –
Ghez et al. (IR)	1989 – 1998
USNO: Worley, Douglass, Mason, et al.	1990 –
Isobe, Miura, et al.	1990 – 1995
Scardia et al.	1991 –
Horch, Ninkov, van Altena, et al.	1992 –
Ageorges et al. (IR)	1997 –
Docobo et al.	2001 –
Marchetti et al.	2001 –
de Villiers	2001 –

Table 2.
High-Resolution Observations 1970 – 2001

	Telescope Aperture			
	<2-m	2-4-m	>4-m	Space-Based
All techniques	18,690	19,544	1,638	52,007
Speckle techniques	18,567	19,401	1,048	—

Program	$N_{observations}$	Percentage
CHARA	18,555	47.6
USNO	14,044	36.0
Other	6,417	16.4

4. WHAT SPECKLE DOES WELL: BINARY STARS

4.1 Resolved Spectroscopic Binary Stars

Some of the most fertile ground for speckle interferometry has been in the resolution of spectroscopic binaries, especially double-lined systems, where no further information is needed in order to determine individual masses and a geometric parallax. Catalogs of spectroscopic binaries (e.g., Batten et al. 1989) were quickly exhausted in searches for resolved systems. Cross-correlation techniques and dedicated campaigns led to successful analyses of more difficult and/or longer-period spectroscopic systems (Griffin 1992), many of which were also resolvable by speckle. However, it can be said that the period of rapid success by either speckle interferometry or spectroscopy

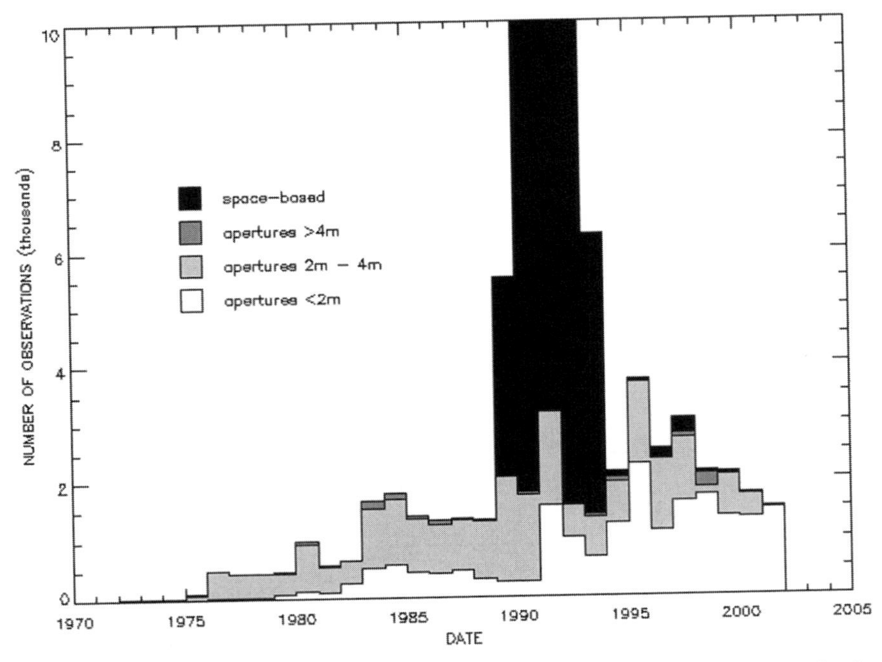

Figure 1. Distribution of interferometric observations with time. In this and the following two figures, all data in the current *Fourth Catalog of Interferometric Measurements of Binary Stars* (Hartkopf et al. 2001) obtained since 1970 are tallied. Observations are categorized by telescope aperture—"small" (aperture < 2-m), "medium" (2 to 4-m), and "large" (> 4-m)—and shown as white to medium grey shaded regions of the histogram. Observations made from space (shown in dark grey) are dominated by the Hipparcos/Tycho mission from 1989 – 1993; later observations are from HST. Over 97% of the ground-based interferometric data were obtained by speckle or similar techniques, and over 91% of these data were obtained with telescopes of aperture less than 4-m.

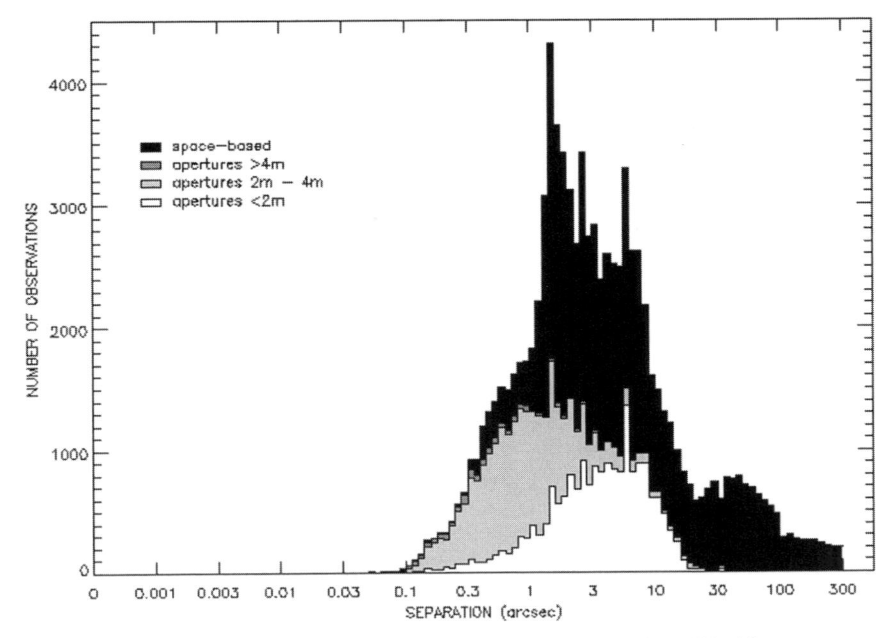

Figure 2. Distribution of interferometric observations with separation. The histograms are skewed as expected, with most of the wider binaries being observed using small apertures.

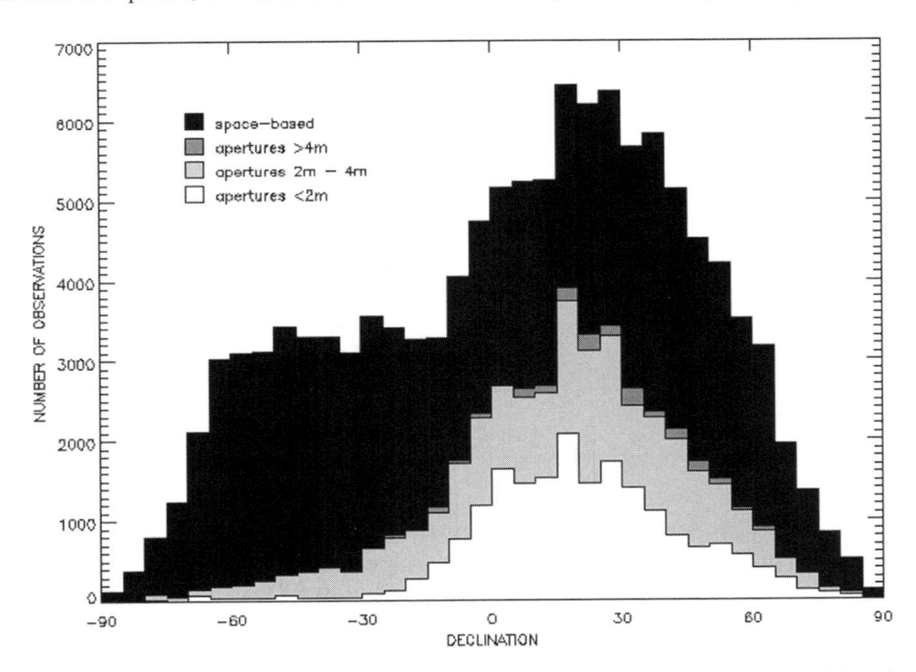

Figure 3. Distribution of interferometric observations with declination. Most ground-based interferometry data are still from the Northern Hemisphere, although some increased effort has been made to survey the southern skies, especially by Horch and colleagues.

alone has mostly passed and the improvement in the regime in common between them will be made by continued efforts with both techniques. Most of the rapid advance in this area will be made by optical interferometers. While those that are currently operational use small telescopes or siderostats, an instrument with a resolving power measured in microarcseconds is not small by most conventional uses of the word.

However, as mentioned above, results in this area continue to be made. Shown in Figure 4 is a calculation of the relative orbit of μ Ori (= A 2715 = HD 40932) from a recently submitted combined spectroscopic/astrometric solution by Fekel et al. (2002).

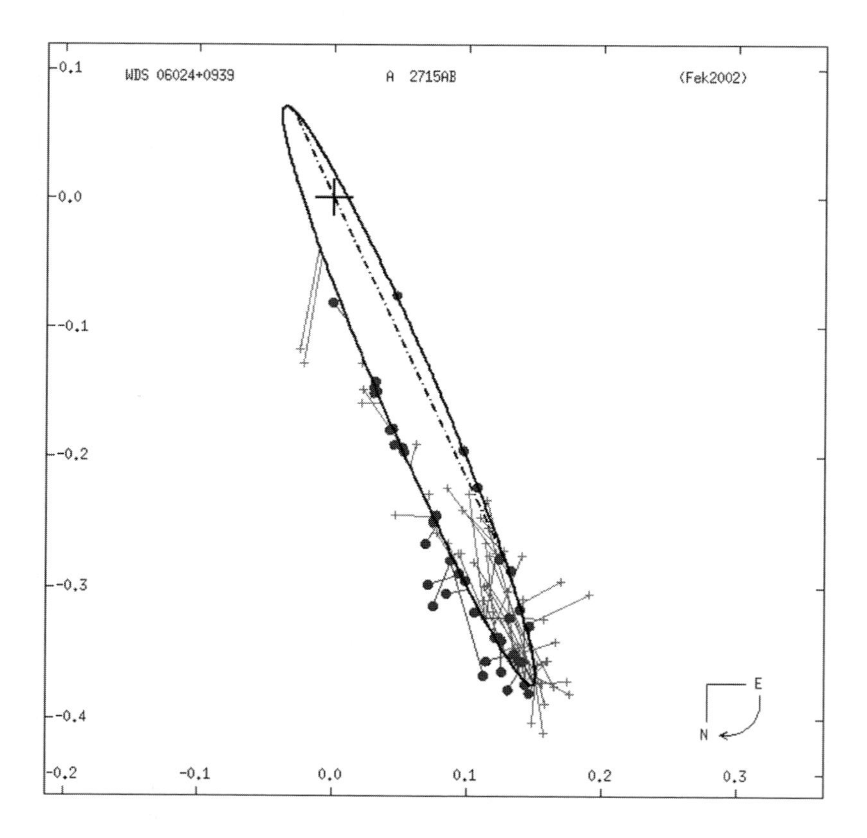

Figure 4. Astrometric orbit for the close visual binary A 2715 = μ Ori. The location of the primary star is indicated by a large '+' sign, the relative orbit of the secondary by the heavy ellipse. Individual observations are coded by observing method (+ = visual micrometry, ● or ★ = speckle interferometry); "O – C" lines connect each observation to its predicted position along the orbit. The line of nodes is indicated by a dot-dash line. Scales along the x and y axes are in arcseconds. Discovered by Aitken in 1914, this system has also been the subject of extensive spectroscopic study, and a combined orbital solution has recently been completed by Fekel et al. (2002).

4.2 Other Binary Stars

Each technique used in the study of binary stars is amenable to a different spectral class, magnitude and Δm range, separation regime, etc., and in turn yields its own bit of information about the object being studied. Observations of the same star using more than one technique can produce a powerful synergy, yielding far more information than either technique alone. Some examples of techniques which complement visual (and speckle) astrometry include:

- spectroscopy $\rightarrow$ component masses and distance. This is described in greater detail in the section above;
- parallax or other distance determination $\rightarrow$ mass sum, via the formula $M_{1+2} = a''^3/\pi^3P^2$;
- occultation $\rightarrow$ duplicity confirmation and magnitude difference (a quantity which is very difficult to obtain through autocorrelation methods, see Mason 1996, 1997). Lunar occultation can detect extremely small angular separations, although this is only a vector projection of the true separation; and
- astrometry from space $\rightarrow$ duplicity confirmations. Many binaries first resolved by Hipparcos or Tycho have been confirmed using speckle interferometry from the ground, see Mason et al. 1999, 2001a, Horch et al. 1999, 2000).

Indeed, in the final case above, many new discoveries of both Hipparcos and Tycho have been confirmed using speckle interferometry with the USNO 26-inch telescope in Washington. While conventional wisdom prior to Hipparcos would have been to save all survey work for larger telescopes (with a smaller resolution limit), a not insignificant number of new Hipparcos pairs could have been discovered with a small telescope at an urban site.

Observation of "garden variety" visual binaries can result in confirmation and verification where observations are lacking. These can also aid other techniques where only a photocenter is seen (e.g., spectroscopy) or where the point-spread function may be modified by a double. This can have implications for future space missions (e.g., FAME, see Johnston et al. 2000) or pending ground-based projects which are capable of observing faint systems but whose resolution capabilities are rather limited (e.g., UCAC, described elsewhere in this publication; see also Zacharias et al. 2000). Indeed it is these astrometric projects or missions where unresolved doubles, physical or optical, cause significant problems. It is clearly not possible for speckle programs to observe all objects in projects with $>10^6$ objects, but it is possible to utilize speckle and other inexpensive ground-based techniques with small telescopes to form a library of physical (rather than software) templates for use in deconvolution algorithms.

Another example of the utility of speckle interferometry is the solving of high-resolution problems such as is posed by the close binary designated "Tweedledum and Tweedledee" by the great visual interferometrist William S. Finsen (1954). STF2375 is a wide binary where each component is itself a close binary. Each of the closer pairs was of approximately the same separation and position angle at the time of discovery, and all four stars were of approximately the same brightness. Indeed, the pairs were often mistaken for each other. That, coupled with the quadrant ambiguity inherent in most interferometric results, has made this a longstanding problem in interferometric circles. A complete analysis of this system, shown in Figure 5, is in progress.

Historically, most binary star research has been done from the ground, so it is hardly surprising that the majority of binary stars can be observed with small telescopes. It is perhaps more instructive to look at the new discoveries of Hipparcos and Tycho and see how many of their detections can be confirmed and eventually have their motion characterized from the ground. For speckle interferometry a Δm limit of 3 is often quoted (though clearly, greater values can be reached depending on seeing and reduction algorithms; see Horch et al. 1999), and V limits of ~11 are typical. While separation is often a factor, it doesn't enter into these arguments. Only a few of the new Hipparcos binaries (ESA 1997) are closer than 100 mas, and the Tycho-2 Catalogue (Høg et al. 2000a,b) limited double star processing to binaries of 800 mas or wider; none approach the resolution limit of speckle with a 4m telescope of 30 mas. Consequently, the parameters of significance are the magnitudes. Considering the quoted H_p and V_T magnitudes (taking V_T when both are given or when V_T is available elsewhere, see Fabricius & Makarov 2000) and given a Δm limit of 3 and a primary magnitude limit of 11, 69% of the 4,636 binaries first resolved by Hipparcos and Tycho are observable with speckle from the ground (15% have a primary which is too faint, while 16% have a measured Δm which is too large).

Of the 84,486 systems in the Washington Double Star Catalog (Mason et al. 2001b), perhaps a dozen are too close for speckle interferometry with a 4-m class telescope (though this is a selection effect). A much larger number of binaries are too wide, so their components don't fall within the same isoplanatic patch.

5. WHAT SPECKLE DOES WELL: SURVEYS

Speckle interferometric observations can typically be made with a minimum of data (typically 1-3 minutes' worth), and observations can usually begin as quickly as the telescope can point; i.e., there is very little setup and a low duty cycle. This can result in a significant number of observations being

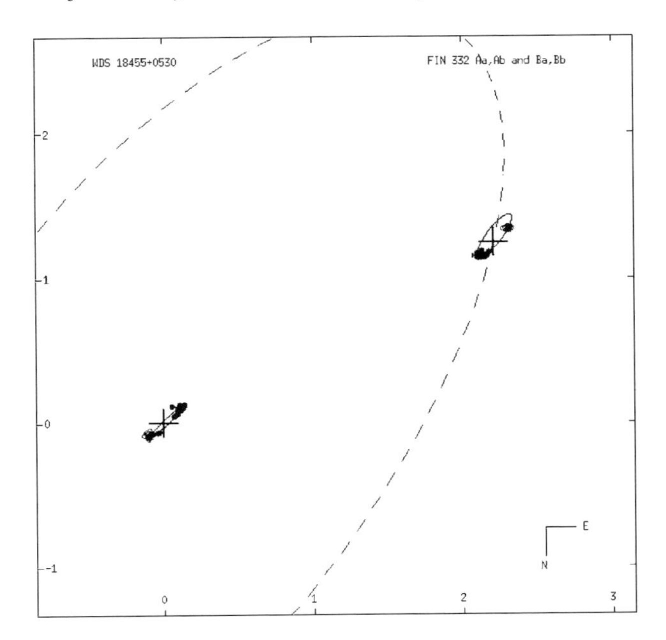

Figure 5. Astrometric orbits for the quadruple system WDS18455+0530. The large dashed ellipse indicates the (very poorly determined) orbit of the wide pair STF 2375 AB, which has moved barely 10° since its discovery by F.G.W. Struve in 1825. Finsen resolved both the A (lower left) and B (upper right) components in 1953, using his eyepiece interferometer. After nearly a half-century of observation, coverage is now sufficient to allow determinations of the orbital elements for both these close pairs (Mason et al., in prep.).

obtained in a typical night (usually around 300 depending on the pointing speed of the telescope). While writing a concise scientific justification for so many objects is often not possible giving the space requirements of telescope allocation committees, the speed of operation lends itself quite well to doing surveys of large samples of stars. Examples of survey topics undertaken using speckle interferometry include:

- groups sorted by proximity or their coeval nature (e.g., Hyades, Pleiades, Praesepe and IC 4665 open clusters; thus see Mason et al. 1993a,b),
- stars of a given spectral classes [e.g., white dwarf stars (McAlister et al. 1996), Be stars (Mason et al. 1997), O stars (Mason et al. 1998a), WR Stars (Hartkopf et al. 1999), or nearby G dwarf stars (Mason et al. 1998b)],
- stars with certain kinematic characteristics (e.g., high-velocity stars, see Lu et al. 1987), and
- stars sorted by other characteristics (e.g., bright stars, see McAlister et al. 1987, 1989, and 1993).

Determining the multiplicity characteristics for different subsets of stars, by type, proximity, age, etc., has important implications for stellar evolution and galactic dynamics. For example, preliminary results of the G dwarf survey noted above suggest a decrease in duplicity fraction with age. While smaller telescopes (i.e., 2 meters or less) can certainly be used, it is only when you get to the resolutions possible with 4-m class instruments that you approach separations where astrophysical phenomena are observed by other techniques. The following specific system analyses came about as a result of survey work described above: HR 1071 (McAlister et al. 1992), 15 Mon (Gies et al. 1993, 1997), and HR 7767 (McKibben et al. 1998). It is certainly correct to say that none of these studies would have been done without the survey work which prompted them.

As important as searches for duplicity may be, perhaps of equal importance is the search for "non-duplicity". Photometric or spectroscopic variability may be mimicked by a nearby star occasionally falling within the photometer's aperture or the spectrograph's slit. (For example, there is some speculation that the entire class of λ Boo variable stars may be normal binaries; the variability and observed abundance anomalies may simply be due to blended spectral features; see Farraggiana & Bonnifacio 1999). The point-spread function of a "standard" star near an adaptive optics target may be distorted by a stellar companion. The meters/second or tens of m/sec radial velocity variations thought to be due to an extrasolar planet may instead be the residual radial velocity curve of a binary star viewed nearly pole-on. The accuracy of a celestial reference frame may be compromised by the presence of too many "slightly resolved" doubles. With binaries comprising at least 2/3 of the stars visible to us, these occurrences are hardly rare; indeed for many astronomers binaries have earned the pigeon-worthy sobriquet "vermin of the sky".

While no single technique can survey all of separation (or equivalently, period) space, speckle neatly covers the gap between the spectroscopic and wide-field astrometric regimes. Coupled with its efficient use of telescope time, this makes speckle interferometry an attractive first-cut technique for winnowing out these "vermin".

6. WHAT SPECKLE DOES WELL: OTHER PROJECTS

Most of this chapter has stressed the contributions of speckle interferometry to binary star astrometry and duplicity surveys, as these have been the major areas of research to date. Speckle techniques have been utilized in other projects as well, however — both stellar and non-stellar. A few examples are listed below.

6.1 Globular Cluster Proper Motions

Speckle data were taken for several fields in the globular cluster M5 in April, 1987 and for M15 in both October, 1987 and November, 1991 (mostly non-overlapping fields). The high precision of speckle measurements will allow, with one more set of observations, differential proper motion determinations to be made with an expected accuracy of less than 1 mas/yr. This has great potential to look for ejected members or possibly investigate the virialization of these clusters.

6.2 Submotions Due to Unresolved Companions

The high precision of speckle interferometry makes it suitable for examining submotions due to unresolved companions of wide pairs. Cole et al. (1992) discovered a 4-mas companion to the wide binary ADS 784, also detected as a 5-year perturbation in the spectra of a 4-day spectroscopic component. Mason et al. (1995) derived an orbit for the 669-day astrometric subsystem of ADS 8119 = ξ UMa, based on residuals to speckle observations of the 60-year visual system. Finally, a program was conducted at the Lowell Observatory 1.8-m telescope from 1982 - 1989, attempting systematic monthly speckle observations of about 60 wide binaries and a similar number of calibration pairs. The goal of the project was to search for possible submotions due to any planets which may be orbiting one component of the wide pair. Results of the project were presented by Al-Shukri et al. (1996) and Fu et al. (1997).

6.3 Mutual Events of the Galilean Satellites

Close pairings of the Galilean satellites occur in series that are determined by mutual resonances, within a geometric envelope defined by the apparent inclination of the orbital planes (i.e., Jupiter's equator) and distance. Observations have been made of these mutual events in an effort to refine the ephemerides of the orbits of these satellites. While experimentation is still underway involving image reconstruction, progress thus far has been quite favorable, and it is anticipated that there will be significant work in this area in 2003 (the next passage of the Earth through the Jovian orbital plane).

6.4 Asteroid Duplicity

During the early 1980's, a total of 57 asteroids were observed (most on multiple occasions) for possible companions (See Roberts et al. 1995). No companions were detected within the separation/Δm regime accessible by speckle, but the project did provide limits to the separations and sizes of possible secondary components.

6.5 Diameters and Images

Although data reduction is considerably more complicated, speckle interferometric techniques can also be used to determine diameters and even diffraction-limited images of some objects. The available targets are mostly limited to supergiants and their dust shells and mass-loss envelopes (Wittkowski et al. 1998, Gauger et al. 1999) and a few active galaxy cores (Wittkowshi et al. 1999), although some image reconstruction of binary stars has been used to determine accurate magnitude differences (Schoeller et al. 1998), and Schertl et al. (1995) have also made images of a few asteroids.

7. CONCLUDING REMARKS

Speckle interferometry is a well established and surprisingly versatile technique, whose equipment can be designed and constructed at fairly low cost (and is indeed now within the reach of serious amateurs), and whose data-reduction techniques can mostly be implemented in a quite straightforward manner. Its ability to make robust detections of close companions (or set useful detection limits), determine object diameters, and produce diffraction-limited images allows for synergy with many other techniques in the investigation of source structure in astronomical objects. As a result, this "poor man's adaptive optics" is particularly well suited to small- and medium-sized telescopes as a complement to an observatory's usual retinue of photometric and spectroscopic equipment.

The authors wish to acknowledge the continued support given by the U.S. Naval Observatory to its double star speckle interferometry project. Thanks also to Georgia State University and the National Science Foundation for over twenty years of support to the CHARA speckle interferometry program. Finally, our thanks to Dr. Harold McAlister, who started the most productive program of speckle interferometry to date, is an active and accommodating collaborator, and provided the authors with stimulating careers.

8. REFERENCES

Al-Shukri, A., Hartkopf, W.I., & McAlister, H.A., 1996, AJ 111, 393
Balega, Y.Y. & Tikhonov, N.A. 1977, SvAL 3, 272
Batten, A.H., Fletcher, J.M., & MacCarthy, D.G. 1989, Eighth Catalogue of the Orbital Elements of Spectroscopic Binary Systems, Victoria, Dominion Astrophysical Observatory
Blazit, A., Bonneau, D., Koechlin, A. & Labeyrie, A. 1977, ApJL 214, 79
Cole, W.A., Fekel, F.C., Hartkopf, W.I., McAlister, H.A., & Tomkin, J. 1992, AJ 103, 1357
ESA 1997, The Hipparcos and Tycho Catalogues (ESA SP-1200) (Noordwijk: ESA)
Fabricius, C. & Makarov, V.V. 2000, A &AS 144, 45
Farraggiana, R. & Bonnifacio, P. 1999, A &A 349, 521
Fekel, F.C., Scarfe, C.D., Barlow, D.J., Hartkopf, W.I., Mason, B.D., & McAlister, H.A. 2002, AJ 123, 1723
Finsen, W.S. 1954, Observatory, 74, 41
Fu, H.-H., Hartkopf, W.I., Mason, B.D., McAlister, H.A., Dombrowski, E.G.,Westin, T., & Franz, O.G., 1997, AJ 114, 1623
Gauger, A., Baleha, Y.Y., Irrgang, P., Osterbart, R., & Weigelt, G. 1999, A &A 346, 505
Gezari, D.Y., Labeyrie, A. & Stachnik, R.V. 1972, ApJL 173, 1
Gies, D.R., Mason, B.D., Bagnuolo, Jr., W.G., Hahula, M.E., Hartkopf, W.I., McAlister, H.A., Thaller, M.L., McKibben, W.P., & Penny, L.R. 1997, ApJL 475, 49
Gies, D.R., Mason, B.D., Hartkopf, W.I., McAlister, H.A., Frazin, R.A., Hahula, M.E., Penny, L.R., Thaller, M.L., Fullerton, A.W., & Shara, M.M. 1993, AJ 106, 2072
Griffin, R.F. 1992, Spectroscopic Studies of Binary and Multiple Star Systems, in Complimentary Approaches to Double and Multiple Star Research, ASP Conference Series, Vol. 32, IAU Colloq. 135, H.A. McAlister and W.I. Hartkopf, Eds., p. 98
Hartkopf, W.I., Mason, B.D., Gies, D.R., ten Brummelaar, T., McAlister, H.A., Moffat, A.F.J., Shara, M.M., & Wallace, D.J. 1999, AJ 118, 509
Hartkopf, W.I., McAlister, H.A., & Mason, B.D. 2001, AJ, 122, 3480
Hege, E.K., Hubbard, E.N., Strittmatter, P.A. & Cocke, W.J., 1982, Opt. Acta 29, 701
Høg, E., Fabricius, C., Makarov, V.V., Bastian, U., Schwekendiek, P., Wicenec, A., Urban, S., Corbin, T., & Wycoff, G. 2000b, A &A 357, 367
Høg, E., Fabricius, C., Makarov, V.V., Urban, S., Corbin, T., Wycoff, G., Bastian, U., Schwekendiek, P., & Wicenec, A. 2000a, A &A 355, L19
Horch, E., Franz, O.G., & Ninkov, Z. 2000, AJ 120, 2638
Horch, E., Ninkov, Z., van Altena, W.F., Meyer, R.D., Girard, T.M., Timothy, J.G. 1999, AJ 117, 548
Johnston, K., Gaume, R., Harris, F., Monet, D., Murison, M., Seidelmann, P.K., Urban, S., Johnson, M., Horner, S., & Vasser, R. 2000, A &AS 197, 1403
Labeyrie, A., 1970, A &A 6, 85
Labeyrie, A., Bonneau, D., Stachnik, R.V. & Gezari, D.Y. 1974, ApJL 194, 147
Lu, P.K., Demarque, P., van Altena, W., McAlister, H., & Hartkopf, W. 1987, AJ 94, 1318
Mason, B.D. 1996, AJ 112, 2260
Mason, B.D. 1997a, AJ 114, 808
Mason, B.D., ten Brummelaar, T., Gies, D.R., Hartkopf, W.I., & Thaller, M.L. 1997b, AJ 114, 2112
Mason, B.D., Gies, D.R., Hartkopf, W.I., Bagnuolo, Jr., W.G., ten Brummelaar, T., & McAlister, H.A. 1998a, AJ 115, 821
Mason, B.D., Hartkopf, W.I., Holdenried, E.R., & Rafferty, T.J. 2001a, AJ 121, 3224
Mason, B.D., Hartkopf, W.I., McAlister, H.A., & Sowell, J.R. 1993b, AJ 106, 637

Mason, B.D., Henry, T.J., Hartkopf, W.I., ten Brummelaar, T. & Soderblom, D.R. 1998b, AJ 116, 2975
Mason, B.D., Martin, C. Hartkopf, W.I., Barry, D.J., Germain, M.E., Douglass, G.G., Worley, C.E., Wycoff, G.L., ten Brummelaar, T., & Franz, O.G. 1999, AJ 117, 1890
Mason, B.D., McAlister, H.A., Hartkopf, W.I., & Bagnuolo, Jr., W.G. 1993a, AJ 105, 220
Mason, B.D., McAlister, H.A., Hartkopf, W.I., & Shara, M.M. 1995, AJ 109, 332
Mason, B.D., Wycoff, G.L., Hartkopf, W.I., Douglass, G.G., & Worley, C.E. 2001b, AJ 122, 3466
McAlister, H.A., 1976a, PASP 88, 317
McAlister, H.A., 1976b, PASP 88, 957
McAlister, H.A., 1977a, ApJ 215, 159
McAlister, H.A., 1977b, Sky & Telescope 53, 346
McAlister, H.A., 1992, Encyclopedia of Physical Science and Technology, Vol. 15, Academic Press, San Diego, 668
McAlister, H.A., Hartkopf, W.I., & Mason, B.D. 1992, AJ 104, 1961
McAlister, H.A., Hartkopf, W.I., Hutter, D.J., Shara, M.M., & Franz, O.G. 1987, AJ 92, 183
McAlister, H.A., Hartkopf, W.I., Sowell, J.R., Dombrowski, E.G., & Franz, O.G. 1989, AJ 97, 510
McAlister, H.A., Mason, B.D., Hartkopf, W.I., Roberts, Jr., L.C., & Shara, M.M. 1996, AJ 112, 1169
McAlister, H.A., Mason, B.D., Hartkopf, W.I., & Shara, M.M. 1993, AJ 106, 1639
McCarthy, D.W., Cobb, M.L. & Probst, R.G., 1987, AJ 93, 1535
McKibben, W.P., Bagnuolo, Jr., W.G., Gies, D.R., Hahula, M.E., Hartkopf, W.I., Roberts, Jr., L.C., Bolton, C.T., Fullerton, A.W., Mason, B.D., Penny, L.R., & Thaller, M.L. 1998, PASP 110, 900
Michelson, A.A. 1891, Nature 45, 160
Morgan, B.L., Beddoes, D.R., Scaddan, R.J. & Dainty, J.C. 1978, MNRAS 183, 701
Popper, D.M., 1980, ARA &A 18, 115
Roberts, L.C., Jr., McAlister, H.A., Hartkopf, W.I., & Franz, O.G. 1995, AJ 110, 2463
Schertl, D., Grieger, F., Hofmann, K.-H., Mauder, W., Reinheimer, T., Weghorn, H., & Weigelt, G. 1995, P &SS 43, 313
Schoeller, M., Balega, I.I., Balega, Y.Y., Hofmann, K.-H., Reinheimer, T., & Weigelt, G. 1998, SvAL 24, 283
Schwarzschild, K., 1895, Astr. Nach. 139, 23
Tango, W.J., Davis, J., Thompson, R.J. & Hanbury Brown, R., 1979, Proc. Astron. Soc. Australia 3, 323
Weigelt, G.P., 1978, A &AL 67, 11
Wittkowski, M., Balega, Y., Hofmann, K.-H., & Weigelt, G. 1999, annual scientific meeting of the Astronomische Gesellschaft, Goettingten Germany, 20-25 Sept 1999
Wittkowski, M., Langer, N., & Weigelt, G. 1998, A &A 340, L39
Zacharias, N., Urban, S.E., Zacharias, M.I., Hall, D.M., Wycoff, G.L., Rafferty, T.J., Germain, M.E., Holdenried, E.R., Pohlman, J.W., Gauss, F.S., Monet, D.G., & Winter, L. 2000, AJ 120, 2131

Chapter 9

"Visual" Binaries in the Twenty-First Century

Elliott Horch
University of Massachusetts at Dartmouth
North Dartmouth, Massachusetts USA

Abstract: For thirty years, speckle interferometry has continued and extended the work of the visual binary star observers of the past century. What does it offer as we look to the future? Because of the rise of space astrometry and the much more precise parallaxes that these satellites are likely to achieve, one of the most difficult aspects of determining reliable stellar masses from visual orbits (*i.e.* the system distance) will most likely fall in the next decade. In addition, Hipparcos has discovered more than 3400 new double star systems, some fraction of which is certainly gravitationally bound. These and other recently discovered systems give us a chance to significantly improve our knowledge of stellar masses, the mass-luminosity relation, and binary statistics, if their orbits can be determined in sufficient detail. Regular speckle and spectroscopic observations at smaller telescopes, in combination with ultra-precise parallaxes, therefore hold the promise of providing the fundamental empirical calibration of stellar structure and evolution theory that the first visual binary star observers had envisioned.

Key words: close binaries, visual binaries, interferometric techniques, photometric techniques

1. INTRODUCTION

Visual (and by extension interferometrically resolvable) binary stars yield stellar mass information through Kepler's harmonic law, usually written:

$$m_1 + m_2 = \frac{a^3}{\pi^3 P^2} \qquad (1.1)$$

T.D. Oswalt (ed.), The Future of Small Telescopes in the New Millennium, Vol. III, 139-153.

where m_1 and m_2 are the masses of the components, a is the semi-major axis, π is the absolute trigonometric parallax, and P is the period of the system. Defining the total mass of the system as $m_{tot} \equiv m_1 + m_2$ and computing the fractional uncertainty in this quantity, one obtains

$$\frac{\delta m_{tot}}{m_{tot}} = [3(\frac{\delta a}{a})^2 + 3(\frac{\delta \pi}{\pi})^2 + 2(\frac{\delta P}{P})^2]^{1/2} \tag{1.2}$$

Of the three observables on the right-hand side of the above equations, parallax has traditionally been the limiting factor in the uncertainty of the total mass for nearly all relevant systems. In turn, this has been a fundamental constraint in the usefulness of these systems for increasing our knowledge of stellar astrophysics.

The Hipparcos mission and subsequent release of the Hipparcos Catalogue (ESA, 1997), have substantially improved parallaxes of stars within a couple of hundred parsecs of the sun. Hipparcos parallaxes have typical uncertainties in the range of 2 to 5 times smaller than ground based parallax results (*i.e.* from 1 to 2 milliarcseconds [mas]). As impressive as this achievement is, the dominant source of mass uncertainty remains the parallax for most visual and speckle binaries. For example, the well-known binary Bu 151AB (= WDS 20375+1436 = HIP 101769) has Hipparcos parallax of 33.49 mas, and the latest orbital parameters are those of Hartkopf (private communication), where a = 439.4 ± 0.6 mas and P = 26.6327 ± 0.0110 years. The fractional uncertainties in the observed quantities are therefore $\delta a/a$ = 0.00137, $\delta\pi/\pi$ = 0.02628 $\delta\pi$ = 0.88 mas in this case), and $\delta P/P$ = 0.00041. The parallax uncertainty is nearly a factor of 20 (!) larger than the other two contributors, even for this system which is only 30 parsecs from the sun, and even post-Hipparcos.

However, three major satellite missions to be launched in the coming years will totally transform our idea of what systems are "close" enough to be astrophysically interesting. These are: the Full-Sky Astrometric Mapping Explorer[1] (FAME; as of this writing the future of the mission is unfortunately in doubt), the Space Interferometry Mission[2] (SIM; planned launch 2009), and the Global Astrometric Interferometer for Astrophysics[3] (GAIA; planned launch 2010-2012). All three of these missions plan to determine parallaxes to *at least* the tens-of-microarcseconds μas level, some 50 to 100 times smaller than Hipparcos. For Bu 151AB, the fractional error in parallax using a conservative estimate of $\delta\pi$ = 50 μas becomes 0.00149, and the overall error budget from the three observables would be less than 0.4%, over 10 times better than the current value (using the Hipparcos parallax) of 4.6%. Table 1 gives similar results for a representative set of well-known visual/speckle binaries. Column headings are: (1) the Discoverer Designation from the Washington Double Star (WDS) Catalog (Worley and Douglass, 1997[4] except for γ Virginis = StF 1670; (2) the Hipparcos Catalogue number; (3) the Hipparcos parallax; (4) the B - V color listed in the Hipparcos Catalogue; (5) the current uncertainty in the total mass using Equation 2, the latest orbital elements, and the Hipparcos parallax results; (6) the total mass uncertainty assuming no change in orbital parameters, but a reduction in parallax uncertainty to 50 μas; and (7) the (current) uncertainty in the mass fraction. "Unknown" indicates that no mass fraction exists in the literature (mass fraction information courtesy of William van Altena and Reed Meyer, Yale University). In all cases, the parallax error post-FAME is reduced at least to the current level of uncertainty in the semi-major axis and in several cases, well below.

Table 1.
Current & Expected Mass Uncertainties for Several Well-Known Visual & Speckle Binaries

Star Name	HIP	HIP π (mas)	HIP B – V (mag)	Current $\delta m_{tot} / m_{tot}$ (%)	Post-FAME $\delta m_{tot} / m_{tot}$ (%)	$\delta f/f$ (%)
Bu 101	38382	59.98	0.600	2.8	0.42	4.1
Fin 347 Aa	45170	48.83	0.371	4.4	3.0	2.7
Kui 48AB	49658	15.80	0.469	9.0	0.83	unknown
γ Vir	61941	84.53	0.368	2.5	0.69	unknown
StF 1937AB	75312	53.70	0.577	4.0	0.29	5.0
Kui 79AB	84140	158.17	1.485	3.6	0.37	2.0
Kui 94	96907	12.00	0.182	9.0	2.9	unknown
Bu 151AB	101769	33.49	0.425	4.6	0.36	4.4

The above exercise is illustrative of the importance of space-based astrometry in unlocking the power of the speckle binaries for astrophysics. There is still much work to do, but we owe a debt of gratitude to the visual observers, such as Wulff Heintz, Charles Worley, Geoff Douglass, Paul Couteau, their contemporaries and predecessors, as well as the sustained as well as the sustained speckle efforts of the Center for High Angular Resolution Astronomy (CHARA) group at Georgia State University and the U.S. Naval Observatory (USNO), who have already provided us with a wealth of orbital information: they knew long ago that one day the demon of parallax would be subdued.

But total mass determinations are the beginning of what can be done with these stars, not the end. What else is needed? Certainly, as Table 1 demonstrates, precise mass fractions will be necessary in many cases to convert the total masses into individual masses. This can be approached with spectroscopic observations. However, even individual masses are significant only if they can be related to other *observed* stellar parameters, such as luminosity, effective temperature, metallicity, age, helium abundance, and so on. For the classic sample of visual and speckle binaries, regular speckle and spectroscopic observations on small telescopes can provide new and extremely important information in these areas. In this chapter, we focus primarily on speckle observations, but also appeal to spectroscopists, in particular to make the radial velocity observations of these objects necessary for determining high-quality mass fractions.

2. SPECKLE OBSERVATIONS

By far, the most successful use of speckle has been resolving close binary stars, where despite the advent of adaptive optics, it continues to be used regularly. The CHARA and USNO speckle programs have been the most productive; their measures account for nearly half of the 70,000 measures in the *Third Catalogue of Interferometric Measurements of Binary Stars* (Hartkopf, McAlister & Mason, 2000). Speckle has been shown to give excellent relative astrometry, allowing over 100 orbits of close binaries to be determined and/or significantly refined (many from the series of orbit papers by Hartkopf, Mason, and their collaborators, *e.g.* Mason, Douglass & Hartkopf 1999 and references therein). Speckle is an extremely efficient observing technique; 200 objects per night or more is routine for the best speckle observers. The standard measurement techniques are stable down to about 10th magnitude, where decreasing signal-to-noise ratio renders precise astrometry too difficult with current detectors (a new detector now being used at the USNO may be capable of going up to two magnitudes fainter). A nagging problem even for bright stars has always been the inability to determine the relative photometry of the two stars in a binary system, sometimes referred to as the "Δm" problem. This is the simplest non-trivial photometry problem possible, and yet seasoned speckle observers, such as Hartkopf et al. (1996), generally assign uncertainties of 0.5 mag for any magnitude difference estimates.

The root causes of the Δm problem for speckle observing are non-trivial; for example, detector effects can play a significant role. Although these effects generally do not strongly degrade the astrometry, they make photometry very problematic in the case of microchannel-plate detectors such as intensified-CCDs (ICCDs) and other photon counting cameras. These devices are susceptible to a detector non-linearity known as microchannel saturation, which has been discussed by e.g. Sams (1991). Microchannel saturation leads to a loss of counts in locations on the detector where several photons are detected in rapid succession. Speckles are by definition localized areas where several photons are detected in a short amount of time, so these devices are particularly limited in determining the brightness of speckles. This results in a disastrous effect for recovering photometric information from speckle data. Other effects, such as pixel sampling, detector field of view, and atmospheric effects, also play significant roles in the photometric error budget of the speckle technique.

However, microchannel saturation can be eliminated using normal, bare CCDs to record speckle patterns, since they are not susceptible to these effects. Tyler and Matson (1993) were among the first to suggest the use of CCDs over intensified imaging systems for speckle; their work was revived and extended by Horch, Ninkov & van Altena (1998). As CCDs continue to improve in terms of read noise and readout speed, their use as speckle imaging devices becomes feasible and indeed advantageous, especially if they can retrieve the photometric information for close binaries that has been elusive with ICCDs. In the limit of zero read noise and large bandwidth, CCDs effectively become *linear* photon-counting cameras with very high quantum efficiency. CCDs have been employed in observations of α Orionis (Klückers et al. 1997) and binary stars e.g. Horch et al. 1999, 2001, both types of observations benefiting from the linearity of CCDs compared with ICCDs and other photon counting cameras.

Some recent speckle results using a fast readout CCD have been compared with space-based magnitude differences derived from Hipparcos (Horch, Ninkov and Franz, 2001), where the speckle data were obtained at the Lowell-Tololo 24-inch Telescope at the Cerro Tololo Inter-American Observatory (CTIO). The results indicated good agreement between the space-based and speckle results. The speckle data were also compared with the adaptive optics (AO) differential photometry of ten Brummelaar et al. (2000), which has also yielded encouraging results, with typical uncertainties in the magnitude differences of 0.05 to 0.10 mag. Good agreement was again found, although the number of systems observed in common was small. So far, the CCD approach appears to be a step towards solving the long standing "Δm" problem of speckle imaging. This fact may allow for the component magnitudes and colors of all speckle binaries to be determined in the coming years, either with AO or as a part of routine CCD-based speckle observations. Effective temperature and luminosities follow directly, which in combination with masses, begin to give us new astrophysical data to compare with theory. By using bare CCDs, we can do more with speckle now than in the past.

3. THE MASS-LUMINOSITY RELATION AND ASTROPHYSICS

For a resolvable binary, the fundamental observational data are of course the position angle and separation measurements, which when continued over a substantial part of one revolution can then be analyzed to yield the true orbit. As previously mentioned, the total mass is obtained by combining the orbital information with the trigonometric parallax of the system and simple two body mechanics. The individual masses can be derived once the mass ratio of the system is found, either from the same observations that yielded the trigonometric parallax or from spectroscopic observations of the radial velocity of the system.

Radial velocity observations of double-lined spectroscopic binaries can be used to provide minimum masses. If the system is also an eclipsing binary (or is visually or interferometrically resolvable) the inclination can be obtained, yielding individual masses and the distance of the system. The observation of spectroscopic and eclipsing binaries have traditionally been most effective at providing information for high mass stars, while visual binaries provide more of the information for low mass stars. However, this separation of technique for the different mass ranges, initially bridged by speckle in the 70's and 80's (See McAlister 1985 for a review), has effectively disappeared as a result of the large improvement in measurement accuracy available from both high-precision radial velocity measurements and long baseline optical interferometry (LBOI) e.g. Armstrong et al. 1998; Hummel et al. 1998; Hummel et al. 2001. However, LBOI is currently limited to very bright objects at present (brighter than ~5th magnitude, with improvements to perhaps 7th or 8th magnitude in the next few years).

In terms of the current understanding of the empirical mass-luminosity relation (MLR), the study of eclipsing and spectroscopic binaries has yielded excellent results for stars more massive than the sun. However, the relation is on much less firm ground for lower mass stars (See e.g. Henry & McCarthy, 1993). Henry and McCarthy have produced some of the best infrared speckle data of low mass stars on the systems that are closest to the solar system. However these data are comprised of observations of surprisingly few systems; when limited by ground-based observations as they were, there were few systems that were both bright enough and close enough to observe with their speckle camera. Recent improvements in the MLR have been made with the incorporation of Hipparcos parallaxes (Martin et al. 1998, Söderhjelm 1999), and recent Space Telescope results (Henry et al. 1999), and again these are primarily new mass determinations of well-known systems. Henry et al. gave us a significantly clearer picture of the lowest mass systems, but the range 1.0 $M_\odot$ to 0.2 $M_\odot$ still remains extremely uncertain. More data are still needed to improve the low end of the diagram, especially to investigate so-called “second parameter” effects in the MLR such as metallicity and age. Better distances and magnitude differences of known binaries are will significantly tighten the existing MLR, but more objects are also needed. Nonetheless, CCD-based speckle and AO photometry can now give us the luminosities and effective temperatures of the existing sample.

Andersen (1991) discussed what is needed observationally before meaningful comparisons with stellar theory can be made. Measures of mass should have uncertainties of less than 2%; effective temperature uncertainties, less than 2%; radius uncertainties, less than 1% (note that the radius can obtained from the luminosity and effective temperature via $L = 4\pi R^2 \sigma T_{eff}^4$); and metallicity uncertainty, less than 25%. In a few cases, *e.g.* Lydon, Fox & Sofia (1993), extremely good observational parameters have provided a basis for evaluating stellar models, but in general the sample remains meager. Even α Centauri, which was the focus of the Lydon et al. work, has had a recent revision in the masses (Pourbaix et al. 1999), highlighting the difficulty of the observational problem. More and better observations are needed. For very bright spectroscopic binaries, observations with the LBOI arrays will provide excellent information. However, speckle has the unique role to play for objects in the magnitude 6 to 12, which includes many more objects, and many important sub-solar mass systems.

4. HIPPARCOS BINARIES

In addition to deriving the trigonometric parallax for 118,000 of the nearest stars, Hipparcos also discovered some 3400 new double star systems and flagged thousands of other stars as “suspected double.” Magnitude differences in the so-called H_p filter, with typical uncertainties in the range 0.1 to 0.15 magnitudes (Mignard et al. 1995) are also listed in the Hipparcos Catalogue, as well as color information for binaries of wider separation in the Tycho-2 Catalogue. In view of the dramatic improvement in the distance measures to these stars that can be expected over the next 10 to 15 years, the Hipparcos discoveries are a treasure trove of information concerning stellar masses and binary statistics. Most importantly, these systems are bright and relatively easy to observe. Figure 1 shows both a magnitude difference versus total magnitude plot for the subset of the Hipparcos stars that are within 200 pc of the solar system, a total of approximately 1500 objects. (A parallax uncertainty of at worst ~1% would be obtained at 200 pc with FAME, for example.) The total magnitude range places most of these objects too faint for long baseline optical interferometry (LBOI) at present, and the magnitude differences are generally less than 4, perfect for speckle observing. In Figure

2, a color-magnitude diagram of the same subsample is plotted. Many of these systems are blue and more luminous than the sun, indicating that, if gravitationally bound, they are probably higher mass systems and therefore they may in general have less to contribute to the understanding of the MLR. Of the red stars, some are luminous and probably giants, but some are also faint. If bound, these are probably low mass dwarfs that would be very important objects for refining the lower end of the MLR. Figure 3 illustrates that speckle observations at even a 1-m class telescope would be capable of resolving the vast majority of Hipparcos discoveries.

In the case of the low mass stars used by Henry and McCarthy to construct their MLRs and many other well-known binary stars, a wealth of position angle and separation measures is available from the WDS, including a large amount of high-precision speckle data. Observations of those systems are ongoing through the USNO and other speckle efforts. However, the new double stars discovered by Hipparcos currently have little or no orbital data. If just a small percentage of these systems turn out to be true binaries with reasonable periods, the number of points in the most recently published MLRs would be significantly improved, probably doubled or tripled.

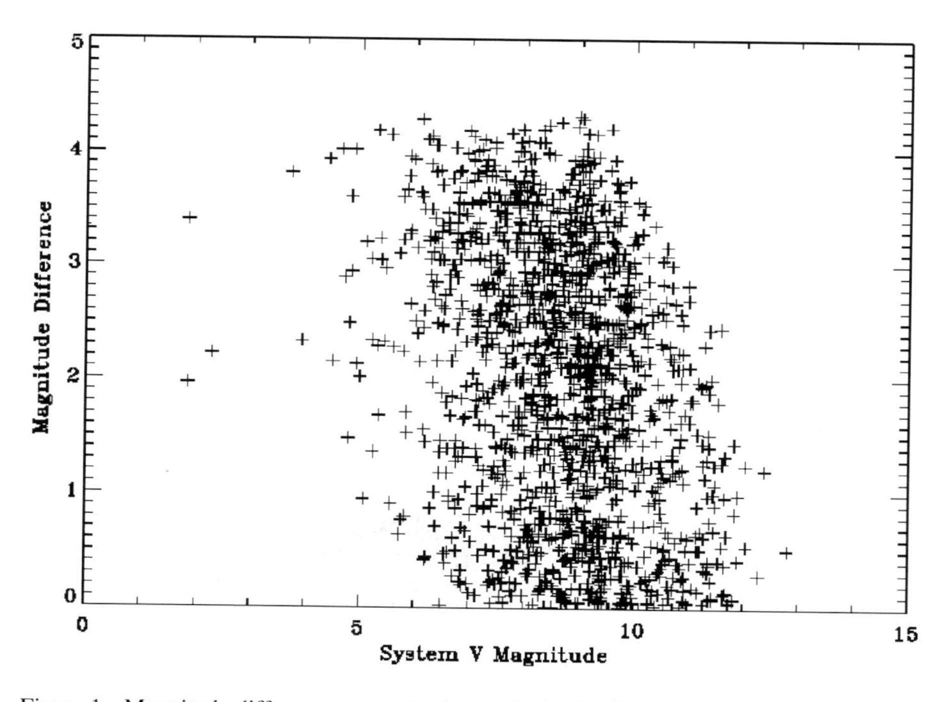

Figure 1. Magnitude difference versus total magnitude plot for the sample of "firm" Hipparcos discoveries within 200 pc of the sun.

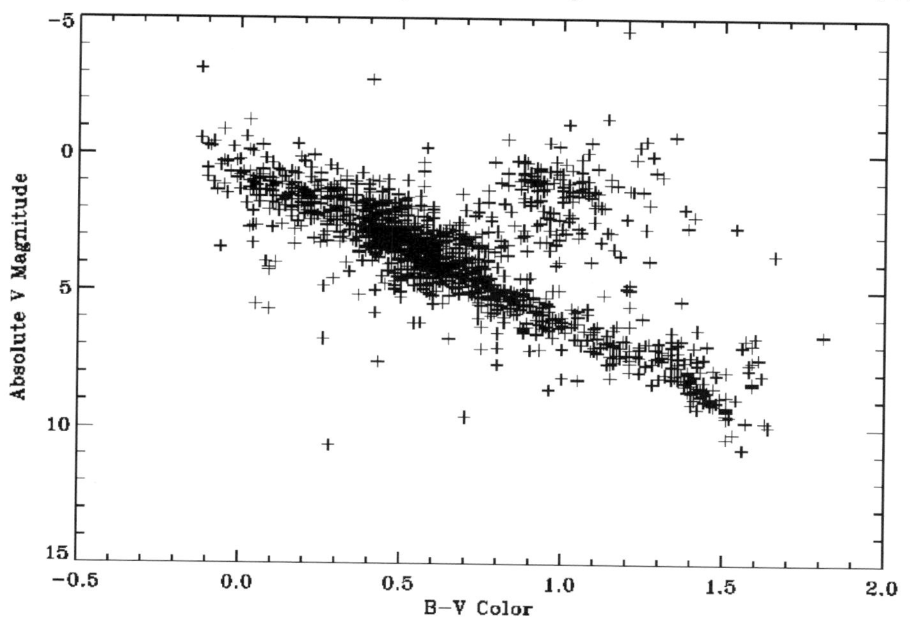

Figure 2. A B-V color versus absolute magnitude plot for the same sample as in Figure 1. A number of these are redder than the sun (B-V ≥ 0.6) and intrinsically fainter, indicating that, if bound, they are probably low-mass dwarf systems.

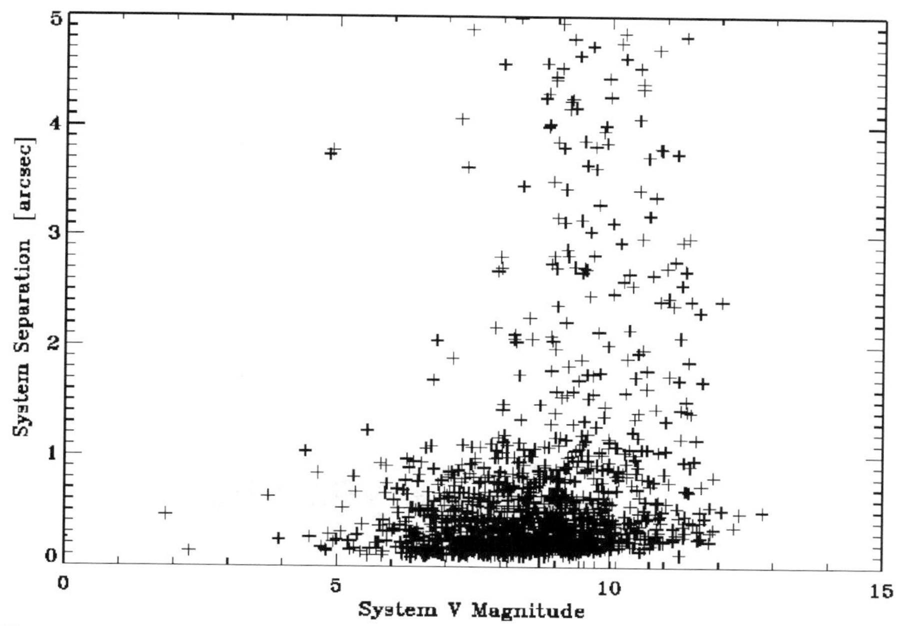

Figure 3. A separation versus system magnitude plot for the same sample. The great majority of these systems are well-suited to speckle observations, and indeed some systems (not shown here) have separations larger than 5″, and can easily be observed by amateurs. The smallest separations (~ 0.1″) are easily resolvable by even a 1.5-m telescope.

While decades of observations may be needed in order to determine high-quality orbits and total masses of some of the slow-moving Hipparcos discoveries, there are some systems for which mass determinations can be made over the next few years. Objects such as CAR 1 (= HIP 17891 = WDS 03496+6318) and HDS 318 (= HIP 11352 = WDS 02262+3428) have already shown substantial orbital motion since the Hipparcos observations (> 270° change in position angle, indicating periods in the 10-15 year range, based on speckle observations by the RIT-Yale group). With a 10-year baseline already since the Hipparcos discovery epoch, fast moving candidates for orbital motion can be identified with one observation. In many cases, a follow-up observation one or two years later would be sufficient to confirm orbital motion. It would then be relatively easy to produce a list of true binaries that could be studied in greater detail and with greater precision to determine the orbital parameters, as well as component magnitudes and colors. For the objects that prove especially interesting, the forthcoming space-based astrometry will complete the path to a precise total mass for the system. Individual masses could be obtained by initiating spectroscopic observations once orbital motion is identified.

5. OTHER SURVEYS

In addition to the Hipparcos discoveries, Mason et al. are currently surveying approximately 3600 stars with B-V colors in the range 0.5 to 1.0 for duplicity via ICCD-based speckle at the Kitt Peak and CTIO 4-m telescopes. The colors place these objects in the range of interest for the lower MLR. Mason expects roughly 3% of the targets (approximately 100 objects) to be discovered to be double (i.e. no previous mention of duplicity in the literature) (Mason, private communication). These objects, with separations presumably too small to be detected by Hipparcos, may also have shorter periods. Because of their composite spectral types and presumed mass range, this sample will have obvious value to the MLR, once their orbits, magnitude differences, and component colors are known.

The RIT-Yale speckle effort, based at the WIYN 3.5-m telescope at Kitt Peak, has started to observe the Hipparcos discoveries and the objects listed as *suspected double* in the Hipparcos Catalogue. Hipparcos itself did not have a large aperture, and so some of the suspected doubles are resolvable from 2 to 4-m class apertures on the ground using diffraction-limited techniques. As of this writing, over 100 suspected doubles have been observed, and 17 have been resolved. The group is also beginning spectroscopic observations of the Hipparcos discoveries that show clear orbital motion, such as CAR 1 and HDS 318, previously mentioned.

6. BINARY STATISTICS AND STAR FORMATION IN THE GALACTIC DISK

In the last decade, some theories of binary formation have emerged that can be tested observationally. For example, some fragmentation models (e.g. Clarke, 1999) predict that the properties of the secondary in a binary system should be independent of the primary mass, while in capture models (McDonald & Clarke, 1995), the higher mass stars have companions with greater frequency. These different predictions can be evaluated by observing the companion star fraction (CSF) for a sample of appropriate stars. In general, imaging methods (including speckle) can be used to determine the CSF of a sample, if orbital motion is confirmed or may reasonably be assumed. If orbit determinations are possible, then further statistics such as the mass ratio distribution, the period distribution, and correlations between the orbital parameters can be observationally determined. Spectroscopic binaries have provided much of this information to date due to their generally shorter periods. If orbits are not determined, the mass ratio distribution can still be obtained indirectly, if the magnitude differences of the systems are known and the evolutionary status of the companions is known or assumed.

Speckle interferometry has already been a powerful method for examining the CSF in clusters and star forming regions. In this case, all stars in the sample may be thought of as coeval, having the same metallicity, and equidistant from the solar system. In recent years there has been a substantial amount of activity in this area including the infrared work of Ghez and her collaborators (Ghez et al. 1993, 1997) the European effort of Leinert et al. (1993), as well as optical speckle interferometry (e.g. Mason et al. 1993). The Space Telescope Fine Guidance Sensors have also been used in the study of cluster binaries (e.g. Franz et al. 1992). These studies have generally shown that young stars have a very high CSF, statistically consistent with the proposition that most if not all stars form in multiple systems.

More recent work by Patience et al. (1998) has given a significantly clearer picture of the CSF and mass ratio distribution in the Hyades cluster, allowing for a meaningful comparison of theory and observation. In particular, they ruled out capture and disk assisted capture as primary mechanisms of binary formation on the basis of mass ratio statistics of the cluster binaries. This and other studies point to fragmentation as the primary mechanism of binary formation. However, Abt & Willmarth (1999) have recently published spectroscopic results indicating evidence for 3 body capture processes as the primary binary formation mechanism in some open clusters.

Our knowledge of the field population in the solar neighborhood has been primarily determined through spectroscopic means. The paper of Abt & Levy

(1976) pointed out that binary and multiple stars systems are more common than single stars, a conclusion that has been confirmed by Duquennoy & Mayor (1991) and Udry et al. (1998). However, the field CSF seems to be considerably below that measured in the star forming regions, leading to the notion that many stars in the field population have been stripped of their companions. Mason et al. (1998) compared chromospherically active and non-active stars in the field via speckle interferometry to gain further insight on the CSF as a function of age. They found that the active (and presumably younger) stars have a higher CSF than the non-active stars. The study of Patience et al. (1998) of the Hyades further corroborates the decrease in the CSF as a function of age, as the Hyades are much older than the star forming regions studied by Ghez et al. (1993) but younger than the general field population, and had a CSF value that was intermediate between the other two samples. However, Duchêne (1999) recently re-analyzed some of the previous studies of star forming regions and concluded that some regions do not have binary excesses. Thus, while there has been an evolving discussion in the field in recent years, there is no firm consensus yet on what the basic properties of the binary population tell us about formation scenarios.

Once it can be determined which of the Hipparcos and other recent double star discoveries are gravitationally bound, it will be possible to obtain a better picture of the field CSF and the mass ratio distribution for these new binaries. Incorporation of these objects into the field star database along with their revised space-based parallaxes will provide a significant new resource to answer questions such as: Is the field CSF dependent on age and spectral type? Does the (presumed) decrease in CSF with age match theoretical expectations? Is the formation mechanism the same for field stars and clusters? What, if any, implications does this have on brown dwarf and planetary formation?

As noted by Lebreton (2000), large samples of binaries can also provide more information on the variation of helium abundance and age with metallicity, a key area of interest for galactic evolution studies. At present, the lack of binaries with well known T_{eff}s and abundances is a limiting factor in making progress. Both issues can be addressed by a combination of speckle and spectroscopic measures of the Hipparcos discoveries at small telescopes. In particular, space astrometry, speckle and/or AO will provide the necessary information for component luminosities and colors, so that these components can be placed on an HR diagram. Using stellar evolution theory and metallicities from spectroscopic work, age estimates can be derived for all systems. These age estimates in combination with their space velocities could provide a basis for a clearer chronology of star formation and helium enrichment in this region of the disk.

7. CONCLUSION

Significant improvements in the distances to objects in this region of the galaxy are expected from upcoming space astrometry missions. In terms of known visual and speckle binary systems, we stand to gain considerably in terms of total mass determinations, and individual mass determinations where the mass fraction is known to sufficient precision. In addition, new "nearby" binaries discovered by the Hipparcos satellite and other survey work need astrometric measures in order to determine orbital parameters, as well as component magnitudes and colors. Once periods and semi-major axes are known in these cases, these objects will provide a wealth of new information about the mass-luminosity relation, binary statistics, and the star formation history in this region of the galaxy.

While differential photometry of close binaries has been difficult with speckle in the past, both adaptive optics and CCD-based speckle observations are now obtaining reliable results in this area. This opens the door to obtaining component magnitudes and colors, effective temperatures and luminosities, key parameters in astrophysical studies. Together with more precise masses, the future appears bright for finally mining these important objects of the astrophysical information they possess.

8. REFERENCES

Abt, H. A. and S. G. Levy. (1976). "Multiplicity Among Solar-Type Stars," Astrophysical Journal Supplement Series, 30, 273-306.

Abt, H. A. and D. W. Willmarth. (1999). "Binaries in the Praesepe and Coma Star Clusters and Their Implications for Binary Evolution," Astrophysical Journal, 521, 682-690.

Andersen, J. (1991). "Accurate Masses and Radii of Normal Stars," Astronomy and Astrophysics Review, 3, 91-126.

Armstrong, J. T., D. Mozurkewich, L. J. Rickard, D. J. Hutter, J. A. Benson, P. F. Powers, N. M. Elias II, C. A. Hummel, K. J. Johnston, D. F. Buscher, J. H. Clark III, L. Ha, L.-C. Ling, N. M. White and R. S. Simon. (1998). "The Navy Prototype Optical Interferometer," Astrophysical Journal, 496, 550-571.

Clarke, C. J. (1999). "Fragmentation of Cold Slabs: Application to the Formation of Clusters," Monthly Notices of the Royal Astronomical Society, 307, 328-336.

Duchêne, G. (1999). "Binary Fraction in Low-Mass Star Forming Regions: A Reexamination of the Possible Excess and Implications," Astronomy and Astrophysics, 341, 547-552.

Duquennoy, A. and M. Mayor. (1991). "Multiplicity Among Solar-Type Stars in the Solar Neighborhood," Astronomy and Astrophysics, 248, 485-524.

ESA. (1997). The Hipparcos and Tycho Catalogues, European Space Agency SP 1200.

Franz, O. G., L. H. Wasserman, E. Nelan, M. G. Lattanzi, B. Bucciarelli, and L. G. Taff. (1992). "Binary Star Observations with the Hubble Space Telescope Fine Guidance Sensors. II. Bright Hyades," Astronomical Journal, 103, 190-196.

Ghez, A. M., G. Neugebauer, and K. Matthews. (1993). "The Multiplicity of T Tauri Stars in the Star Forming Regions Taurus-Auriga and Ophiuchus-Scorpius: A 2.2 Micron Speckle Imaging Survey," Astronomical Journal, 106, 2005-2023.

Ghez, A. M., R. J. White and M. Simon. (1997). "High Spatial Resolution Imaging of Pre-Main-Sequence Binary Stars: Resolving the Relationship Between Disks and Close Companions," Astrophysical Journal, 490, 353-367.

Hartkopf, W. I., B. D. Mason, H. A. McAlister, N. H. Turner and D. J. Barry. (1996). "ICCD Speckle Observations of Binary Stars. XIII. Measurements During 1989—1994 From the Cerro Tololo 4-m Telescope," Astronomical Journal, 111}, 936-945.

Hartkopf, W. I., H. A. McAlister and B. D. Mason. (2000). Third Catalog of Interferometric Measurements of Binary Stars, CHARA Contribution No. 4.

Henry, T. J. and D. W. McCarthy, Jr. (1993). "The Mass-Luminosity Relation for Stars of Mass 1.0 to 0.08 M⊙" Astronomical Journal, 106, 773-789.

Henry, T. J., O. G. Franz, L. H. Wasserman, G. F. Benedict, P. J. Shelus, P. A. Ianna, J. D. Kirkpatrick and D. W. McCarthy, Jr. (1999). "The Optical Mass-Luminosity Relation at the End of the Main Sequence (0.08—0.20 M⊙)," Astrophysical Journal, 512, 864-873.

Horch, E., Z Ninkov, W. F. van Altena, R. D. Meyer, T. M. Girard and J. G. Timothy. (1999). "Speckle Observations of Binary Stars with the WIYN Telescope. I. Measures During 1997," Astronomical Journal, 117, 548-561.

Horch, E., Z. Ninkov and O. G. Franz. (2001). "CCD Speckle Observations of Binary Stars from the Southern Hemisphere. III. Differential Photometry," Astronomical Journal, 121, 1583-1596.

Horch, E. P., Z. Ninkov and W. F. van Altena. (1998). "New Low-Noise High-Quantum-Efficiency Speckle Imaging System," Proc. SPIE, 3355, 777-785.

Hummel, C. A., D. Mozurkewich, J. T. Armstrong, A. R. Hajian, N. M. Elias II and D. J. Hutter (1998). "Navy Prototype Optical Interferometer Observations of the Double Stars Mizar A and Matar," Astronomical Journal, 116 2536-2548.

Hummel, C. A., J.-M. Carquillat, N. Ginestet, R. F. Griffin, A. F. Boden, A. R. Hajian, D. Mozurkewich, and T. E. Nordgren. (2001). "Orbital and Stellar Parameters of omicron Leonis from Spectroscopy and Interferometry," Astronomical Journal, 121, 1623-1635.

Klückers, V. A., M. G. Edmunds, R. H. Morris and N. Wooder. (1997). "Reality and the Speckle Imaging of Stellar Surfaces - II. The Asymmetry of alpha Orionis," Monthly Notices of the Royal Astronomical Society, 284, 711-716.

Lebreton, Y. (2000). "Stellar Structure and Evolution: Deductions from Hipparcos," Annual Reviews of Astronomy and Astrophysics, 38, 35-77.

Leinert, Ch., H. Zinnecker, N. Weitzel, J. Christou, S. T. Ridgway, R. Jameson, and M. Haas. (1993). "A Systematic Search for Young Binaries in Taurus," Astronomy and Astrophysics, 278, 129-149.

Lydon, T. J., P. A. Fox and S. Sofia. (1993). "A Formulation of Convection for Stellar structure and evolution calculations without the mixing-length theory approximations. II - Application to alpha Centauri A and B," Astrophysical Journal, 413, 390-400.

Martin, C., F. Mignard, W. I. Hartkopf and H. A. McAlister. (1998). "Mass Determinations of Astrometric Binaries with Hipparcos. III. New Results for 28 Systems," Astronomy and Astrophysics Supplement Series, 133, 149-162.

Mason, B. D., W. I. Hartkopf, H. A. McAlister and J. R. Sowell. (1993). "ICCD Speckle Observations of Binary Stars. IX. A Duplicity Survey of the Pleiades, Praesepe, and IC 4665 Clusters." Astronomical Journal, 106, 637-641.

Mason, B. D., T. J. Henry, W. I. Hartkopf, T. ten Brummelaar and D. R. Söderblom. (1998). "A Multiplicity Survey of Chromospherically Active and Inactive Stars," Astronomical Journal, 116, 2975-2983.

Mason, B. D., G. G. Douglass, W. I. Hartkopf. (1999). "Binary Star Orbits from Speckle Interferometry. I. Improved Orbital Elements of 22 Visual Systems," Astronomical Journal, 117, 1023-1036.

McAlister, H. A. (1985). "High Angular Resolution Measurements of Stellar Properties," Annual Reviews in Astronomy and Astrophysics, 23, 59-87.

McDonald, J. M. and C. J. Clarke. (1993). "Dynamical Biasing in Binary Star Formation—Implications for Brown Dwarfs in Binaries," Monthly Notices of the Royal Astronomical Society, 262, 800-804.

Mignard, F., S. Söderhjelm, H. Bernstein, R. Pannunzio, J. Kovalevsky, M. Froeschlé, J. L. Falin, L. Lindegren, C. Martin, M. Badiali, D. Cardini, A. Emanuele, A. Spagna, P. L. Bernacca, L. Borriello, and G. Prezioso. (1995). "Astrometry of Double Stars with Hipparcos," Astronomy and Astrophysics, 304, 94-104.

Patience, J., A. M. Ghez, I. N. Reid, A. J. Weinberger and K. Matthews. (1998). "The Multiplicity of the Hyades and Its Implications for Binary Star Formation and Evolution," Astronomical Journal, 115, 1972-1988.

Pourbaix, D., C. Neuforge-Verheecke and A. Noels. (1999). "Revised Masses of alpha Centauri," Astronomy and Astrophysics, 344, 172-176.

Sams, B. J. (1991). "The Effect of Microchannel Plate Gain Depression on PAPA Photon Counting Cameras," Review of Scientific Instruments, 62, 595-599.

Söderhjelm, S. (1999). "Visual Binary Orbits and Masses Post Hipparcos," Astronomy and Astrophysics, 341, 121-140.

ten Brummelaar, T., B. D. Mason, H. A. McAlister, L. C. Roberts, Jr., N. H. Turner, W. I. Hartkopf and W. G. Bagnuolo, Jr. (2000). "Binary Star Differential Photometry Using the Adaptive Optics System at Mount Wilson Observatory," Astronomical Journal, 119, 2403-2414.

Tyler, D. W. and Matson, C. L. (1993). "Speckle Imaging Detector Optimization and Comparison," Optical Engineering, 32, 864-869.

Udry, S., M. Mayor, D. W. Latham, R. P. Stefanik, G. Torres, T. Mazeh, D. Goldberg, J. Andersen and B. Nordstrom. (1998). "A Survey for Spectroscopic Binaries in a Large Sample of G Dwarfs," in The Tenth Cambridge Workshop on Cool Stars, Stellar Systems and the Sun, ed. R. A. Donahue and J. A. Bookbinder, ASP Conference Series, Vol 154, 2148-2157.

Worley, C. E. and G. G. Douglass. (1997). "The Washington Double Star Catalog (WDS, 1996.0)," Astronomy and Astrophysics Supplement Series, 125, 523.

[1] See http://www.usno.navy.mil/FAME

[2] See http://sim.jpl.nasa.gov

[3] See http://astro.estec.esa.nl/GAIA

[4] See also http://ad.usno.navy.mil/wds/wds.html\

Chapter 10

Imaging Exoplanets

The Role of Small Telescopes

B. R. Oppenheimer
American Museum of Natural History
New York, New York USA

A. Sivaramakrishnan, R. B. Makidon
Space Telescope Science Institute
Baltimore, Maryland USA

Abstract: The principal difficulties in directly imaging extremely faint objects in orbit about nearby stars stem not from the sensitivity or collecting area of telescopes, but rather from the fact that the central star's light drowns out the signal of the faint companion. We have conducted extensive simulations of very high-order adaptive optics (AO) systems with coronagraphic imagers on relatively small telescopes. High-order AO and optimized coronagraphs enable small (< 4m) telescopes to play a critical role in the burgeoning effort to image exoplanets. The key to understanding the benefits of outfitting a 2 to 4-m telescope with extremely high-order AO systems lies in the nature of the AO correction. An AO system with N_{act} actuators projected across the linear diameter, D, of the telescope's pupil can correct the point spread function (PSF) of a star on the optical axis within the angle $\theta_{AO} = N_{act}\ \lambda/2D$. For a system with $N_{act} = 10$, the AO system corrects the PSF within $5\lambda/D$, or five times the diffraction limit. As N_{act} is increased, successively improved suppression of the seeing halo results, opening a large, as yet unexplored, part of the mass-separation parameter space relevant to faint companion science. For example, an optimized coronagraph behind a 941 element ($N_{act} = 34$) AO system on a 3.6-m telescope is capable of detecting objects >14 magnitudes fainter than the central star at separations between 0.2 and 1.5 arcsec in the H-band. Such a system will also solve or quantify many of the problems that exoplanet imaging projects must face in the near future.

Key words: extrasolar planets, adaptive optics, coronagraphy, brown dwarf, point spread functions

T.D. Oswalt (ed.), The Future of Small Telescopes in the New Millennium, Vol. III, 155-171.

1. INTRODUCTION

In 1994 Angel published a startling set of calculations. He showed, convincingly, for the first time, that the direct imaging and even low-order spectroscopic study of planets orbiting stars other than the Sun (a.k.a. "exoplanets") is not only possible using ground-based telescopes, but also feasible within a decade or two. If Angel's (1994) calculations are shown to be correct, obtaining such images would represent the solution of perhaps one of the longest standing problems in modern astronomy. Indeed, Immanuel Kant (1755) devoted an entire section of his *Allgemeine Naturgeschichte und Theorie des Himmels* to what such planets might be like and who would inhabit them, based on Newtonian physics. In 1733 Alexander Pope, musing about "worlds unnumbered" wrote the following in "An Essay on Man" (a poem which, in its entirety, provides perhaps the best justification for all academic and basic research).

> He, who through vast immensity can pierce,
> See worlds on worlds compose one universe,
> Observe how system into system runs,
> What other planets circle other suns,
> What varied Being peoples every star,
> May tell why Heaven has made us as we are.

Although the empirical methods of science could not until recently provide evidence for the existence of Pope's "other planets," they did not restrain Kant, nor many generations of future scientists, novelists and the interested public, from speculating. The idea of exoplanets is older than empiricism, but the scientific evidence for them is new, and the detailed study of their physics, chemistry and geology has not begun.

Shortly after Angel's paper appeared, the indirect discovery of a body, roughly the mass of Jupiter orbiting a bright, visible star was announced (Mayor and Queloz 1995). The following week saw the publication of the first images and the Jupiter-like spectrum of a substellar companion of a nearby star (Nakajima et al. 1995; Oppenheimer et al. 1995). To date, over 100 extrasolar planets and a handful of brown dwarf companions of other stars have been found through the radial velocity method, which detects the reflex motion of the star due to the gravitational influence of the orbiting body (Marcy and Butler 2000). Almost two-dozen spectra of free-floating, cool brown dwarfs have been obtained (e.g. Burgasser et al. 2001). With the benefit of this knowledge, the problem of direct imaging of exoplanets has been greatly refined. Herein we set out to specify the problem anew and examine solutions, with particular attention to the already fruitful adaptive optics (AO) coronagraphy technique.

Telescopes smaller than 4-m in diameter with AO and coronagraphs play a crucial role in the ultimate goal of imaging exoplanets because they fulfil a necessary set of precursor experiments, not only to establish and test the needed technologies for larger scale space-based projects, but also to begin the systematic investigation of the physics of faint companions.

2. THE MASS-SEPARATION PARAMETER SPACE

The parameter space relevant to the search for objects orbiting nearby stars is defined by the mass of the object, M, and its orbital separation, a. This seems simple to comprehend, but the sensitivity of a direct observing campaign is not uniform in the physically fundamental M-a parameter space, primarily because direct techniques are sensitive to the observed quantities brightness and angular separation. For stellar companions of stars within a given volume-limited sample, brightness can be converted into M and the distribution of distances in the sample determines the coverage in a. However, in detecting substellar companions, those below the hydrogen burning limit, the conversion between the companion's brightness and M is not straight-forward. Indeed, brown dwarfs and planets (the principal objects presumed to exist in the parameter space of interest) cool with time. The cooling introduces an additional parameter, the age. In a volume-limited sample of stars, the ages of the stars are not, in general, measurable.

In the case of the indirect searches the sensitivity to mass is uninfluenced by age, but the parameter space is also explored in a non-uniform manner because of the large phase space of potential orbits defined by orbital period, mass and eccentricity, and also because the timing of observations in these surveys is neither uniform nor complete. For example, a survey that observes monthly has no sensitivity to orbital periods that are an integer multiple of one month. (Unfortunately, this incompleteness and the corresponding corrections to the statistical interpretation of the existing radial velocity surveys has still not been properly assessed, although the technique has been explored by Nelson and Angel 1998.) Thus, the mass-separation parameter space for the entire sample of stars in radial velocity surveys is not uniformly sampled, although for individual stars, large parts of the M-a plane may be well-explored.

Figure 1 displays the M-a parameter space for substellar values of M and solar system scales for a. The regions that have been probed by the indirect and direct techniques to-date are also indicated. Figure 1 clearly shows that neither indirect nor direct techniques have covered even the majority of the M-a parameter space in the regime of substellar objects ($M < 0.08\ M_{\odot}$) and solar system scales ($a < 50$ AU). For example, an analog of Jupiter is barely detectable and a Saturn analog is completely inaccessible now. Furthermore, comparison between the direct and indirect observing methods is difficult because the overlap in parameter space is only now beginning to exist.

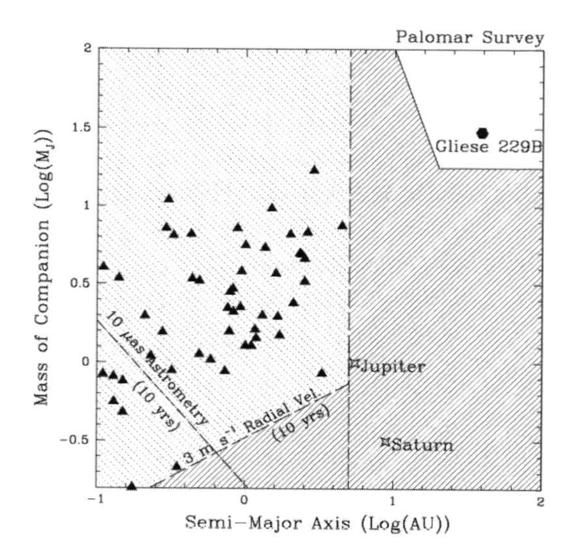

Figure 1. Mass, M, versus separation, a, for substellar companions and solar system scales, shown on a logarithmic scale in both axes. Mass is in units of the mass of Jupiter, M_J, and separation is in units of astronomical units (AU). The shaded regions indicate areas probed by surveys to date (see text). The region in the upper right corner, surveyed by Oppenheimer et al. (2001) using a 1.5-m telescope, is the only complete direct imaging survey to enter this parameter space. On the left, the lightly shaded region has been non-uniformly probed by the 3-m s^{-1} radial velocity searches. The dash-dot line indicates the region accessible to an indirect astrometric search lasting 10 years. To date, no survey of any kind has probed the darkly shaded region, although a few observations of individual stars have probed parts of it.

Although the regions of the *M-a* plane surveyed by the indirect techniques have not been uniformly sampled, tremendous progress has been made (e.g. Marcy and Butler 2000 and references therein). Within the next decade serious treatment of the general physics of companions around G, K and F main sequence stars will be possible with the results from the indirect surveys alone. Ideally, a complete study of this parameter space should cover all relevant ranges of M and a, for all types of primary stars. Furthermore, the true nature of the companions themselves can only be determined through spectroscopic measurements. For these reasons, the techniques must be improved such that they can reach ranges in M and a comparable to those of the indirect techniques. This can be achieved with high-order AO and coronagraphy on ground-based telescopes. Other authors have made similar suggestions in the past (See for example, Angel 1994 or Sandler and Angel 1997). However, what has not hitherto been widely appreciated is that a large part of the *M-a* parameter space is accessible to AO coronagraphs on telescopes smaller than 4-m.

The first consideration in support of this assertion is summarized in Table 1, which shows the required angular resolution of the imaging system used.

Considering only the spatial resolution needed, a 4-m telescope operating at the diffraction limit at a wavelength, $\lambda = 1\ \mu m$ (achievable with AO already), would be capable of resolving physical distances as small as 1 AU around some 2000 stars known within 20 pc. This wavelength is chosen because both observations of brown dwarfs and models of exoplanets show that these cool objects peak in their spectral energy distributions near 1 μm (e.g. Oppenheimer et al. 2000b).

Table 1.
Telescope Diffraction Limit Requirements For Direct Imaging Surveys

Distance	5 pc	10 pc	15 pc	30 pc
Max. Ang. Separation: 1 AU orbit	200 mas	100 mas	67 mas	33.3 mas
Required Telescope Diameter for observations at $\lambda = 1\ \mu m$	1.05-m	2.1-m	3.1-m	6.2-m
J-band Limiting Mag. (5σ, 1000s)	19.5	21	21.8	23.4
Approx. No. Stars Known	64	325	877	3508

Telescope diameter is determined by the angular resolution required for 1 AU physical resolution at the given distance. The *J*-band limiting mag. is an approximation for field point-source detection based on existing state-of-the-art instruments without AO for the telescope aperture given.

However, challengingly small angular separation is not the only difficulty facing a would-be planet hunter. Let us also briefly examine the predicted brightness of a Jupiter sized companion. By reflected light, Jupiter at 5 AU from the Sun viewed from outside the solar system is approximately 18 magnitudes fainter than the Sun (Woolf and Angel 1998). For the purposes of argument, let us assume that a clever optical technique can completely remove the starlight from an exoplanetary system, so that the only object left in our hypothetical image is that of the Jupiter analog. The brightness of the exo-Jupiter is determined solely by the brightness of the parent star. Table 1 shows that a 3-m telescope is capable of detecting point sources without adaptive optics that are as faint as 22 mag. in *J*-band. Thus, stars brighter than $J = 4$ could be searched with reasonable exposure times for gas giant planets.

These arguments serve to show simply that the telescope aperture is not, *a priori*, the limiting factor in the direct imaging of exoplanets. Indeed, the principal problem is finding that "clever optical technique" mentioned above that can remove all or most of the central star's light. The most important element limiting the direct detection of planets is the point spread function (PSF) of the telescope optics and the atmosphere. The light of the primary star, in practical situations, is predominantly outside of the diffraction limited Airy pattern. Indeed, typical "seeing" results in images on the order of 5 to 80 times wider than the angular separations listed in Table 1, with residual light levels extending many arcseconds away from the star centroid

(See, for example King 1971 or Racine 1996). Adaptive optics can control the seeing and coronagraphy can reduce the amount of remaining light from the central star. To understand the utility and efficacy of these techniques, we must first understand the various components of the PSF, particularly with an adaptive optics system in operation. As a figure of merit in the following discussion, we rely heavily upon the Strehl ratio, defined as the ratio of the peak intensity of an actual PSF to that of an ideal, perfect, diffraction-limited PSF.

3. LIGHT: THE FUNDAMENTAL LIMIT

In any signal processing problem, the detection of a signal is dependent on the noise present. In the case we explore here, the signal is light from an exosolar system and the processor is the telescope, AO system, science camera and software. At the location of a faint planet on the science detector, there are several sources of noise, including detector read noise, sky background and any other source of contaminating light. When a bright star is present, even for the shortest duration exposures, the light of the star is, by far, the dominant source of noise impeding the detection of the planet. In order to detect that planet, the random fluctuations in the wings of the star's PSF at the location of the planet must be *at least* a few times smaller than the signal from the planet. One proposed plan of attack on this problem is to marginally reduce the starlight with AO and then simply increase the telescope aperture D to the point where integrations yielding $\gamma_{planet}/\gamma_{star} > 3$ are practical. (Here γ_{planet} is the number of photons detected from the planet and γ_{star} is the number of photons from the star that are superimposed on the location of the planet.) In fact, it is more efficient, at this point in time, to devote astronomical resources to the reduction of γ_{star}. Indeed, this is a cheaper, possibly faster route to a cleaner, less contaminated portrait of an exosolar system. Note that as D increases, telescope cost increases approximately as D^2 to D^4 (Schmidt-Kaler 1997), and the necessary AO system complexity and cost also rise as D^2.

Let us examine the PSF assuming that the telescope optics are diffraction limited (e.g. $\lambda/10$ at .65 μm) and free of ghost reflections or light sources exterior to the field of view. In this case, the PSF is dominated by structure due to turbulence in the atmosphere. Microfluctuations in the air temperature throughout the atmosphere above an observatory translate into microfluctuations in the air's index of refraction. These result in the random disruption of the phase of the planar wave front of light from the distant star. The turbulence is blown past the telescope aperture by winds with speeds on the order of 10-m s^{-1}. Substantial work on the structure and theory of this turbulence has provided useful guidelines on how to correct it, and the body of knowledge and the set of AO instruments that do correct it, at least partially, is growing rapidly. We do not intend to review this material here. Instead, the goal is to provide an understanding of the next few steps necessary for the nearly perfect AO correction that permits planet detection, or at least access to the unprobed regions of the *M-a* parameter space.

3.1 How to Reach 95% Strehl Ratios

In a real operating AO system the PSF after correction is characterized by a sharp, diffraction limited core, sometimes with bright rings visible, along with a broad "seeing halo" which corresponds in width and shape to the uncorrected image. Typical Strehl ratios achieved are near .4 to .7. It is important to appreciate that the sensitivity to faint companions is an extremely strong function of the Strehl ratio (See also Section 4). Once the Strehl ratio exceeds 90%, faint companion detection efficiency improves hyperbolically, such that an improvement from 96 to 97% results in almost an order of magnitude increase in sensitivity. There are a number of effects that prevent existing AO systems from producing PSFs with Strehl ratios above 70%:

- Wave front fitting error, due to finite spatial sampling of the wave front and a finite number of correcting elements whose arrangement does not span all scales of turbulence
- The time-lag error based on the control loop's update frequency and the wind speed during observations
- Wave front sensing error due to a finite number of photons available from the guide star
- Residual alignment error and PSF calibration, i.e. it is difficult to establish the set of control parameters used to drive the control loop to produce a perfect PSF

The fourth issue itemized above can be solved without any major innovation in technology. Residual alignment error can be addressed through additional, lower bandwidth control loops that actively maintain precise alignment throughout the system without incurring extreme costs. PSF calibration, a stumbling block for many existing AO systems, is ultimately a solvable problem using bright calibration sources that permit software, such as phase diversity solutions, to detect extremely low-level aberrations in the final AO system PSF. This is done without the need for telescope time allocated at night.

The first two items in the list above can be addressed essentially by increasing the number of correcting elements in the AO system and speeding up the duty cycle. However this cannot be done arbitrarily. Item 3 constrains the total number of actuators and the update frequency. Table 2 shows a calculation of the reduction of Strehl ratio as a function of the number of actuators placed in a grid on a 3.6-m telescope, with a 4th magnitude star used as a guide star. The system efficiency is assumed to be 20%. N_{act} is the linear number of actuators placed across the telescope pupil, while N_{tot} is the total number of actuators filling the pupil. The update frequency is assumed to be 5 kHz. Clearly, increasing N_{act} beyond 64 results in Strehl degradations that are disadvantageous. The numbers in Table 2, which should be treated as theoretical limits given the system throughput, can only be improved with superior wave front sensing technology. They serve as a useful benchmark for this discussion.

Table 2.
Degradation of Strehl Ratio Due to Number of Photons Available in Each Subaperture
(4th magnitude guide star, 5-kHz update frequency, 20% efficiency, 3.6-m telescope assumed)

N_{act}	4	8	16	34	64	128
N_{tot}	12	50	201	907	3216	12867
Strehl "hit" at 0.8 μm	0.0%	0.0%	0.1%	0.4%	1.5%	6.0%
Strehl "hit" at 1.6 μm	0.0%	0.0%	0.0%	0.1%	0.4%	1.5%

Addressing the issues in this section with a high order AO system can bring the PSF to a Strehl ratio near 95%. We demonstrate this claim in the following section detailing our simulations that include all of the effects itemized above. Then, in Section 3.4, we describe further problems and improvements needed to achieve Strehl ratios above 95%, the regime where coronagraphy and nulling provide the largest gains in tandem with AO.

3.2 Image Improvement by AO

AO PSF improvement is manifest only within a maximum radial distance, θ_{AO}, from the center of the PSF (Sivaramakrishnan et al. 2001). This distance is determined by the linear actuator density, the telescope aperture, D, and the observing wavelength, λ. To understand this intuitively, one can think of AO correction as a high pass filter acting on spatial phase variations of incoming wave fronts. The AO system's spatial frequency cutoff is $k_{AO} = N_{act}/2D$, where N_{act} is the number of deformable mirror actuators projected across the telescope's primary mirror. Since the electric field at the image plane is the Fourier transform of the field in the pupil plane, k_{AO} translates to an angle $\theta_{AO} = N_{act}\lambda/2D$ on the sky. Thus AO only improves the point spread function within a radius θ_{AO}.

Furthermore, even a perfectly corrected image from an AO system has bright rings because the point spread function is the diffraction pattern generated by the telescope's aperture. These bright rings swamp all faint structure around a central on-axis source. (For clarity, the Airy pattern is specific to an unobscured circular pupil and is not, in general, the shape of a telescope's PSF.)

3.3 Halo Suppression from High-Order AO

For low-order AO systems ($N_{act} \sim 10$), θ_{AO} is close to the angular size of the third or fourth ring ($\sim 5\,\lambda / D$) of the perfect, diffraction-limited PSF. As a result, a diffraction limited core is seen in the corrected images, but the bright "seeing halo" remains mostly unaffected.

High-order AO (where $N_{act} > 10$ or $\theta_{AO} > 5\,\lambda / D$) suppresses the seeing halo, which permits an increased dynamic range in the final image. The amount of suppression can be quantified with the following simplistic expression.

$$\eta = \frac{S_{AO}}{S_{see}}\left(1 - S_{AO}\right)^{-1}$$

Here, η is the contrast enhancement provided by the AO correction. S_{AO} and S_{see} are the Strehl ratios of the AO-corrected image and the image made through natural seeing conditions, respectively. The first factor in this expression is the contrast improvement due to sharper images (i.e. the increase in the peaks of the point source images). The second factor represents the contrast improvement due to the removal of light from the seeing halo. In other words, not only is the faint object's PSF more sharply peaked, but the background of light from the central star is reduced. Both effects lead to detection gains.

For existing AO systems such as those at Palomar and Lick, η is approximately 30 to 40. Figure 2 compares the expected PSFs of a 3.6-m telescope with N_{act} = 12 and 34, at an observing wavelength of $\lambda = 1.6\ \mu$m. The importance of the increased number of actuators is apparent. For N_{act} = 34, the PSF has a profoundly darker region within 1.5 arcsec as compared to the lower-order AO system. In this case, η is approximately 500. For N_{act} = 12, close in performance to the Palomar AO system, η is about 45. This is a direct consequence of correcting the higher frequency components of the wave front. This clearly shows that increasing N_{act} provides tremendous gains in the search for faint objects in orbit about nearby stars (details of the simulation techniques are described in Sivaramakrishnan et al. 2001).

These simulations include all of the effects described in Section 3.1. Figure 3 shows radial profiles of similarly simulated PSFs on a 3.6-m telescope subjected to median seeing conditions at a good site, such as Mauna

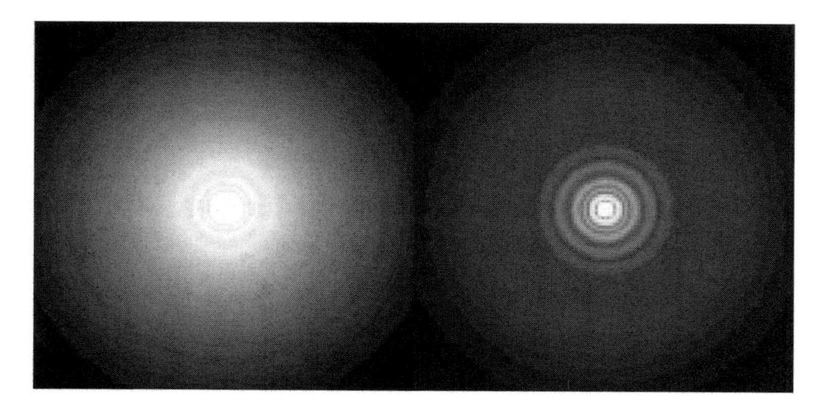

Figure 2. AO PSF simulations for N_{act} = 12 (left) and 34 (right) in H-band using 1000 independent atmospheric phase screen realizations with cell size one 15th of the telescope diameter (3.6-m). Phase screens are scaled to 40 wavelengths across the H-band and passed through a simulated AO system to demonstrate broad-band performance. The effect of increasing N_{act} is manifest in the suppression of the halo in the right hand image within the radius 1.5 arcsec. (See Sivaramakrishnan et al. 2001 for further details.) The Strehl ratios are 0.69 and 0.89 for the left and right PSFs respectively.

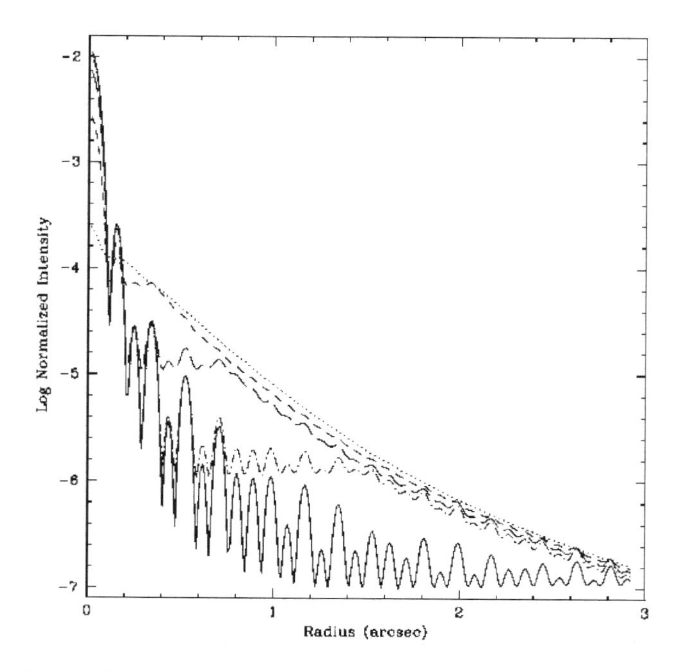

Figure 3. Radial profiles of simulated AO PSFs for a 3.6-m telescope with N_{act} = 4, 8, 16, 34 and 64 (from top to bottom at the 1 arcsecond radius) with median Mauna Kea seeing in H-band (1.6 μm). The Strehl ratios are 0.05, 0.23, 0.63, 0.87 and 0.96 respectively. Note the superior halo clearing as N_{act} increases. θ_{AO} is the radius of the "shoulder" in the PSF.

Kea. Radial profiles are shown for each of the values of N_{act} used in Table 2. Beyond N_{act} = 64, the simulations become unphysical because they do not include additional effects which we describe below.

3.4 Going Beyond 95% Strehl Ratio

Once the effects described in Section 3.1 are accounted for by a sufficiently high-order high-speed AO system (N_{act} = 64 at 5-kHz for a 3.6-m telescope in Hawaii), Strehl ratios around 95% can be achieved at the 1μm wavelength of interest for faint companion studies. However, further improvements cannot be made simply by throwing a larger number of faster actuators at the problem. Indeed, we showed above that for more than 64 actuators across the pupil, there are not enough photons available to permit accurate AO correction (Table 2). Furthermore, additional natural and instrumental effects become important when Strehl ratios above 95% are needed. These effects include speckle noise, atmospheric scintillation, differential refraction and dust or microroughness in the optics.

Speckle noise is an effect that amplifies the noise level in the halo beyond the simple expression γ_{star}. Speckle noise is due to the persistence of remnant speckles in an AO image. Bright spots approximately the size of the diffraction limit wander about. In long exposures these average into the seeing halo, but the accumulation of photons in any given pixel is not governed by simple Poisson statistics because of the behavior of these speckles. A number of investigators have researched the effects of speckle noise. Pessimistic views (Sandler and Angel 1997, Racine 1999) suggest that the effect requires image integration times that must be modified by a factor related to the effective speckle lifetime (which is on the order of milliseconds). Other studies show that the spatial distribution of speckles is actually strongly correlated with the structure of the PSF, such that the speckles tend to exist in bright parts of the PSF (such as the bright rings surrounding the core) far more frequently than elsewhere (Bloemhof et al. 2001). Sivaramakrishnan et al. (2001) show that such "first order" speckle patterns necessarily form a perturbation (to the perfectly corrected PSF) which is antisymmetric about its center. In any case the principal effect is that integration times must be somewhat longer than those expected from Poisson statistics. This is not particularly prohibitive, and speckle noise loses its significance when a coronagraph or nulling technique is used because the speckles themselves are also suppressed.

Atmospheric scintillation is a low level effect investigated most thoroughly by Dravins (1997a, 1997b, 1998). The atmosphere's turbulence affects not only the phase but also the amplitude of the star's wave front at a level of about 1 part in 10^6. Scintillation reduces the Strehl ratio by about 3 to 4% (e.g. Angel 1994) at a relatively good observing site. In principle scintillation can be corrected by an additional adaptive element and a second wave front sensor. This has not been investigated seriously yet, but it must be in the near future, when AO systems approach the 90 to 95% Strehl ratio regime.

Differential refraction is an effect that probably becomes important at Strehl ratios above 97 or 98% and is due to a very weak wavelength dependence of the index of refraction of air. Thus, in an AO system where the optical path difference is corrected without regard for wavelength dependence, image degradation will occur when broadband filters are used. Differential refraction probably affects the PSF at the level of one part in about 10^7 or 10^8, although it has also not been properly investigated (See Nakajima 1994). This effect is also not included in our simulations.

Furthermore, dust in the optical system or microscopic scratches and pits in the optics can scatter light over the entire field of view. Fortunately these effects result in a very low surface brightness, essentially uniform background. However, proper understanding of these effects is also needed since they represent an additional source of noise.

Finally, a few authors have raised the possibility that natural Zodiacal light in an exosolar system may prevent exoplanet imaging (e.g. Woolf and Angel 1998, Backman et al. 1998). However, this, too, is largely an unknown aspect of the problem. The amount of Zodiacal dust in our own system is not well quantified and comprehensive surveys of such optical dust disks around other stars are only possible with the techniques described in this paper.

Each of these effects absolutely must be assessed theoretically and measured observationally if very high Strehl ratios are ever going to be seen, especially from the ground. The precise contribution of these effects is unknown, but present indications are that they must be understood and controlled for imaging planets smaller than Jupiter around nearby stars. To this end, we discuss the use of a coronagraph behind the sort of high order AO system described above, capable of a 95% Strehl ratio in the near infrared wavelengths. Such a system, which would push current AO technology to its limit (as described in Section 3.1 to 3.3), will determine exactly how to go about solving the problems raised in this section.

4. CONTRAST ENHANCEMENT BY CORONAGRAPHY

To demonstrate that much can be achieved immediately with systems approaching 90 to 95% Strehl ratios, we have simulated coronagraphic PSFs for N_{act} = 12 and 34 on a 3.6-m telescope observing in the *H*-band (Sivaramakrishnan et al. 2001). These images, shown in Figure 4, directly complement those in Figure 2. The N_{act}= 34 case shows that halo suppression and optimized coronagraphy reduce the starlight in the 0.2 to 1.5 arcsec range by a factor of more than 30, compared to the N_{act}= 12 case. These simulations include all of the effects discussed in Section 3 except for those in Section 3.4.

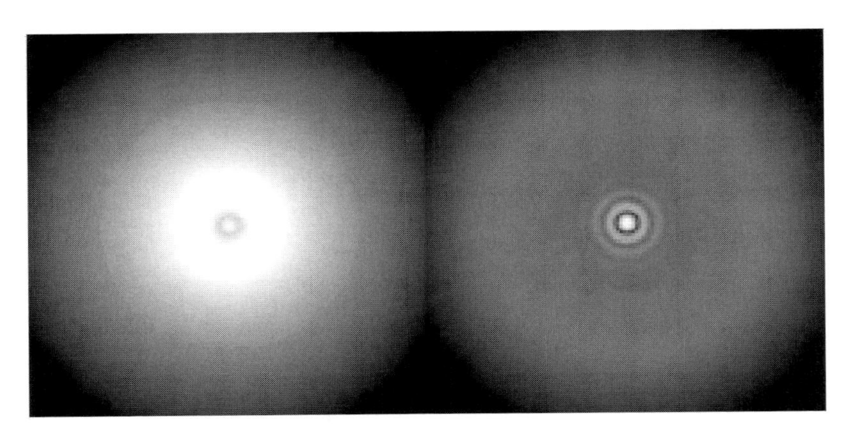

Figure 4. Simulated PSFs similar to those in Figure 2, but with coronagraphic optics inserted.

Using these simulations we can compute the faintest companion detectable in a long exposure image taken with such a system (accounting for detector characteristics, optical quality, positioning and manufacturing errors of 100 nm and thermal background). We compute this by considering the dynamic range, which we define as the difference in magnitudes between the central star and the faintest detectable point source at a given radius from the star. Figure 5 shows the dynamic range of a 3.6-m telescope fitted with an AO system with N_{act} = 34 and an optimized coronagraph with active optics alignment and an additional high-speed tip-tilt system for image stability. Such a system is sensitive, in only 1000-s, to companions more than 14 magnitudes fainter than the primary star.

One can construct a hypothetical survey with such an instrument imaging 300 nearby stars. Using the average ages of these stars as well as their distribution in distance from Earth and their magnitudes, we can compute the approximate region of the *M*-*a* parameter space that such a survey would probe. Figure 6 shows this region, which represents an almost 100-fold increase in the parameter space accessible to direct imaging techniques. We should be careful to remark that we have been conservative here, by only conducting these calculations for the N_{act}= 34 case. The most important conclusion, however, is that a 3.6-m telescope outfitted with a high order AO

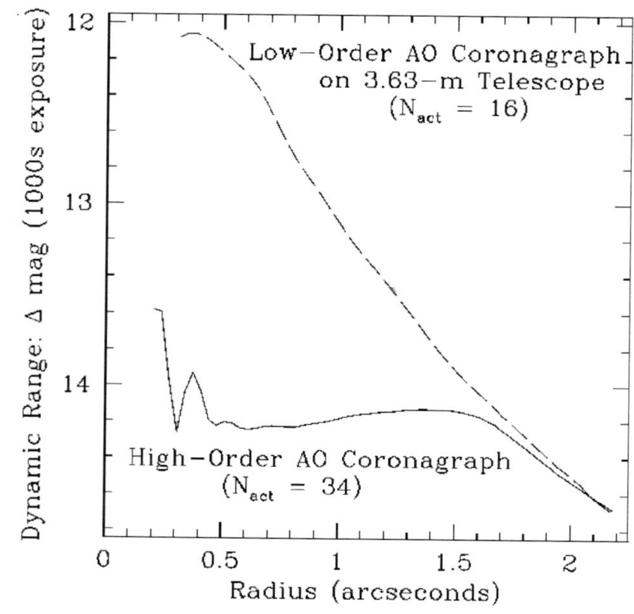

Figure 5. Dynamic range for coronagraphs behind AO systems with $N_{act} = 12$ ("low-order") and $N_{act} = 34$ ("high-order"). We have evaluated the accuracy of the simulations by comparing them directly with real Palomar AO Coronagraph observations (Sivaramakrishnan et al. 2001, Oppenheimer et al. 2000a).

system and an optimized coronagraph will, without doubt, permit the synergism of the indirect and direct techniques for faint companion detection, explore a virtually untouched region of the *M-a* parameter space, and provide extremely important quantitative and empirical direction to the next few steps toward imaging exoplanetary systems.

Furthermore, a number of other techniques might be exploited to improve on Figure 6. Recent papers have suggested that the PSF can be "engineered" in order to provide regions that are particularly dark, i.e. small holes in the PSF where one could search for even fainter companions (e.g. Malbet et al. 1995). A system such as the one discussed here would be capable of testing such techniques and exploiting them for the first time.

Sivaramakrishnan et al. (2001) demonstrate that Lyot coronagraphs (Lyot 1939) are most effective with occulting spots that are at least $4\lambda/D$ in diameter and when Strehl ratios are 80% or higher (except in the case of solar Lyot coronagraphs). (See Sivaramakrishnan et al. 2001 for a detailed description of the principles of Lyot coronagraphy.) When the Strehl ratio exceeds 95%, phase mask coronagraphs (Roddier and Roddier 1997, Rouan et al. 2000, Riaud et al. 2001) enable image core suppression and may be superior to Lyot coronagraphs.

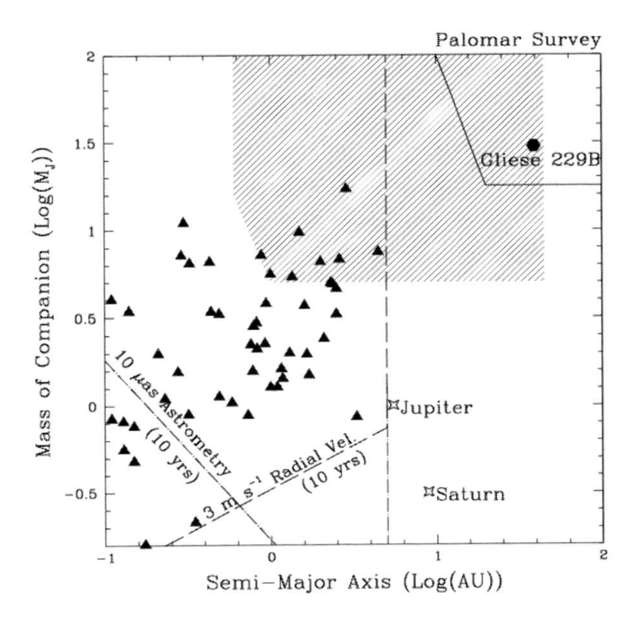

Figure 6. Same as Figure 1, but the shaded region indicates the parameter space accessible to the AO coronagraph and survey described in Section 4.

5. SEVEN REASONS TO PURSUE HIGH-ORDER AO ON 3 TO 4-m TELESCOPES

In conclusion and summary, there is substantial evidence that small telescopes can and will play a vital role in the mounting effort to image exoplanets. We have shown that the scientific gains to be gleaned from a high-order AO system with a coronagraph on a 3.6-m telescope are tremendous and will help solve many of the issues that this effort will face in the next two decades. We list of seven principal reasons why a small aperture telescope should be fitted with a high order AO system and an optimized, diffraction-limited coronagraph:

- AO PSF calibration to extremely high precision will be solved.
- The true extent and importance of speckle noise, atmospheric scintillation and differential refraction will be measured.
- Innovative coronagraphic or nulling techniques will be evaluated, tested and refined.
- "PSF engineering" will be attempted for the first time on an astronomical telescope.

- Experience with extremely high-order AO will be acquired and some of the control issues surrounding these complex systems will necessarily be used to build the 30 to 100-m telescopes currently on the drawing board.
- A survey of Zodiacal light around other main sequence stars will be undertaken, determining whether this is an important concern for the exoplanet effort and probing the physics of exo-Zodiacal dust disks.
- A 100-fold increase in the *M-a* parameter space accessible to direct imaging will result in the direct imaging of brown dwarfs and hot super-Jupiter planets around stars within 30 pc. These objects will be prime targets of opportunity for coronagraphic spectrographs on larger telescopes.

6. REFERENCES

Angel, J. R. P. 1994, "Groundbased Imaging of Extrasolar Planets Using Adaptive Optics," Nature 368, 203

Backman, D.E., Caroff, L.J., Sandford, S.A., and Wooden, D.H. 1998, "Exozodiacal Dust Workshop, Conference Proceedings," NASA/CP-1998-10155.

Bloemhof, E. E., Troy, M., Dekany, R. G., Oppenheimer, B. R. 2001, "Behavior of Remnant Speckles in an Adaptively Corrected Imaging System," Astrophys. J. Lett. 558, L71

Burgasser, A. J., Kirkpatrick, J. D., Brown, M. E., Reid, I. N., Burrows, A., Liebert, J., Matthews, K., Gizis, J. E., Dahn, C. C., Monet, D. G., Cutri, R. M., Skrutskie, M. F. 2001, "The Spectra of T Dwarfs I: Near-Infrared Data and Spectral Classification," Astrophys. J. 563, in press

Dravins, D., Lindegren, L., Mezey, E., Young, A. T. 1997a, "Atmospheric Intensity Scintillation of Stars. I. Statistical Distributions and Temporal Properties," Pub. Astron. Soc. Pac. 109, 173

Dravins, D., Lindegren, L., Mezey, E., Young, A. T. 1997b, "Atmospheric Intensity Scintillation of Stars. II. Dependence on Wavelength," Pub. Astron. Soc. Pac. 109, 725

Dravins, D., Lindegren, L., Mezey, E., Young, A. T. 1998, "Atmospheric Intensity Scintillation of Stars. III. Effects for Different Telescope Apertures," Pub. Astron. Soc. Pac. 110, 610

Kant, I. 1755 in Allgemeine Naturgeschichte und Theorie des Himmels, Section 3, public domain, English translation by I. Johnston: http://www.mala.bc.ca/~johnstoi/kant1.htm

King, I. R. 1971, "The Profile of a Star Image," Pub. Astron. Soc. Pac. 83, 199

Lyot, M. B. 1939, "Study of the Solar Corona and Prominences Without Eclipses," Mon. Not. Royal Astron. Soc. 99, 578

Malbet, F., Liu, D. T., Yu, J. W., Shao, M. 1995, "Space Adaptive Optics Coronagraphy," in Space Telescopes and Instruments, P. Y. Bely; J. B. Breckinridge, eds., Proceedings of SPIE, Vol. 2478, 230

Marcy, G. W., and Butler, R. P. 2000, "Planets Orbiting Other Suns," Pub. Astron. Soc. Pac. 112, 137

Mayor, M., and Queloz, D. 1995, "A Jupiter-Mass Companion to a Solar-Type Star," Nature 378, 355

Nakajima, T. 1994, "Planet Detectability by an Adaptive Optics Stellar Coronagraph," Astrophys. J. 425, 348

Nakajima, T., Oppenheimer, B. R., Kulkarni, S. R., Golimowski, D. A., Matthews, K., and Durrance, S. T. 1995, "Discovery of a Cool Brown Dwarf," Nature 378, 463
Nelson, A. F. and Angel, J. R. P. 1998, "The Range of Masses and Periods Explored by Radial Velocity Searches for Planetary Companions," Astrophys. J. 500, 490
Oppenheimer, B. R., Kulkarni, S. R., Matthews, K., Nakajima, T. 1995, "Near IR Spectrum of the Cool Brown Dwarf GL 229B," Science 270, 1478
Oppenheimer, B. R., Dekany, R. G., Hayward, T. L., Brandl, B., Troy, M., and Bloemhof, E. E. 2000a, "Companion Detection Limits with Adaptive Optics Coronagraphy" in Adaptive Optical Systems Technology, P. L. Wizinowich, ed., Proceedings of SPIE, Vol. 4007, 899
Oppenheimer, B. R., Kulkarni, S. R. and Stauffer, J. R. 2000b in Protostars and Planets IV , V. Mannings, A. Boss and S. Russell, eds. (Tucson: University of Arizona Press)
Oppenheimer, B., R., Golimowski, D. A., Kulkarni, S. R., Matthews, K., Nakajima, T. 2001, "Coronagraphic Survey for Companions of Stars within 8 Parsecs," Astron. J. 121, 2189
Pope, A. 1733, An Essay on Man, public domain
Racine, R. 1996, "The Telescopic Point-Spread Function," Pub. Astron. Soc. Pac. 108, 699
Racine, R., Walker, G. A. H., Nadeau, D., Doyon, R., and Marois, C. 1999, "Speckle Noise and the Detection of Faint Companions," Pub. Astron. Soc. Pac. 111, 587
Riaud, P., Boccaletti, A., Rouan, D., Lemarquis, F., Labeyrie, A. 2001, "The Four-Quadrant Phase Mask Coronograph. II. Simulations," Pub. Astron. Soc. Pac. in press.
Roddier, F., and Roddier, C. 1997, "Stellar Coronagraph with Phase Mask," Pub. Astron. Soc. Pac. 109, 815
Rouan, D., Riaud, P., Boccaletti, A., Clénet, Y., and Labeyrie, A. 2000, "The Four-Quadrant Phase-Mask Coronagraph. I. Principle," Pub. Astron. Soc. Pac. 112, 1479
Sandler, D. G., and Angel, J. R. 1997, "Direct Imaging of Extra-Solar Planets from the Ground Using Adaptive Optics," in Optical Telescopes of Today and Tomorrow, Arne L. Ardeberg, ed., Proceedings of SPIE, Vol. 2871, 842
Schmidt-Kaler, T. 1997, "Telescope Costs and Cost Reduction," in Optical Telescopes of Today and Tomorrow, A. L. Ardeberg, ed. Proceedings of SPIE Vol. 2871, 635
Sivaramakrishnan, A., Koresko, C. D., Makidon, R. B., Berkefeld, T., and Kuchner, M. J. 2001, "Coronagraphy with High-Order Adaptive Optics," Astrophys. J. 552, 397
Sivaramakrishnan, A., Lloyd, J. P., Hodge, P.E., and Macintosh, B. (2002), Astrophys. J. Lett. 581, L59.
Woolf, N. and Angel, J. R. P., 1998, "Astronomical Searches for Earth-Like Planets and Signs of Life," Ann. Rev. Astron. Astrophys. 36, 507

Chapter 11

The Detection of Extrasolar Planets via the Transit Method

Charles H. McGruder, III
Western Kentucky University
Bowling Green, Kentucky USA

Steve B. Howell and Mark E. Everett
Planetary Science Institute
Tucson, Arizona USA

Abstract: Small telescopes (1- to 2-m), using the transit (or photometric) method for the detection of extrasolar planets, can be employed in dense stellar fields to discover unknown extrasolar planets—not only Jupiter-sized, but also Neptune-sized gas giants, and possibly Earth-sized planets as well. This approach overcomes the current limitations of the radial velocity method. It enables the detection of extrasolar planets around faint ($V > 11^m$) distant stars ($d > 100$ pc), at earlier spectral types (earlier than late F) than the radial velocity method.

Key words: extrasolar planets, photometry, transit method, photometric precision, photometric techniques

1. INTRODUCTION

The radial velocity method has proven to be an extremely powerful method for the detection of low mass companions (LMC) revolving around normal stars (Mayor and Queloz, 1995; Marcy and Butler, 1996). LMCs are stars of spectral M or later as well as brown dwarfs and extrasolar planets. This approach involves the measurement of the small periodic changes in the radial velocity of the parent star caused by the gravitational attraction of a revolving LMC. However, it only yields orbital information and the minimum mass, m, of the LMC. The measured minimum mass is $m = M \sin i$,

T.D. Oswalt (ed.), The Future of Small Telescopes in the New Millennium, Vol. III, 173-187.

where M is the actual mass of the LMC and i, the angle of inclination (between an observer's line-of-sight to the parent star and the normal to the orbital plane of the LMC). It is the projected mass the LMC would have, if its orbital plane is in the line-of-sight (i = 90°). Since neither the actual companion mass nor the orbital inclination are measured in this method there is an inherent ambiguity in classifying an LMC as an extrasolar planet.

Approximately 70 LMCs have been classified as extrasolar planets.[1] All have been discovered with the radial velocity method. No other method to date has lead to a single discovery of an extrasolar planet. However, the classifications are based on the measured minimum masses, m, and not on the actual masses, M. If m is small it is assumed that M is also small. For a randomly oriented sample of orbital planes this assumption would be valid in the vast majority of cases. However, the actual mass of an LMC is given by: $M=m/\sin i$. If i is close to zero then it follows from this equation that M can be large even if the m is small. Thus, the combination of small i and small m leads to a misclassification — non-extrasolar planets are erroneously classified as extrasolar planets. Consequently, the assumption of a randomly orientated sample is crucial for the interpretation of the LMCs as extrasolar planets.

Recently Han et al. (2001) have presented preliminary evidence that a selection of LMCs chosen on the basis of small-amplitude periodic radial velocity variations do in fact have predominantly small angles of inclination and are not randomly orientated. Thus, it may be that many of the LMCs classified as extrasolar planets are really not extrasolar planets at all.

Another approach is required to confirm which LMCs are really extrasolar planets. We describe in this chapter the transit method, which has given us the first confirmation that an LMC is actually an extrasolar planet. If an LMC moves in front of a star relative to a terrestrial observer, it will prevent a portion of the stellar light from reaching the observer. This portion or decrease in stellar brightness to the first order is given simply by ratio of the cross sectional areas of the star and the LMC assuming the LMC contributes negligibly to the stellar brightness. If this occurs not only the radius of the LMC but also the orbital inclination, i, and actual mass, M, (assuming that the minimum mass is known from radial velocity observations) can be derived. Following the discovery of an LMC associated with HD 209458 from radial velocity measurements, Charbonneau et al. (2000) and Henry et al. (2000) detected transits. They concluded that the LMC possesses a planetary mass of $.63M_{Jup}$. Thus, they proved conclusively that an extrasolar gas giant planet in close orbit exists around a normal star.

The search for extrasolar planets is clearly frontline research. The ongoing search for extrasolar planets is one the most exciting scientific endeavors not only to astronomers, but also to the public at large. We will demonstrate in the following how small telescopes using the transit method of extrasolar planet detection can be employed to discover unknown extrasolar planets – not only Jupiter-sized gas giants but even Earth-sized planets.

2. THE TRANSIT METHOD

2.1 Background

Consider a planet, which revolves around its parent star. The Transit (or Photometric) Method for the detection of extrasolar planets involves the measurement of the decrease in the stellar brightness of the parent star that is caused by a planet crossing (transiting) in front of the parent star. The decrease, ΔI, in brightness, I, of the star is given simply to the first order by the ratio of the cross-sectional areas of the planetary, R_p^2, and stellar disks, R_s^2:

$$\frac{\Delta I}{I} = \frac{R_p^2}{R_s^2} \tag{1.1}$$

However, the detailed shape of the transit curve, that is $\Delta I(time)$, is determined by five quantities:

- Planetary radius
- Stellar radius
- Stellar mass
- Orbital Inclination
- Limb Darkening

The best fits of the values for these parameters gives the observed shape of the transit curve and allows the determination of the completely unknown quantities – planetary radius and orbital inclination. This method differs from the microlensing method of detecting extrasolar planets, in which a foreground planet transits a background star (generally not the parent star that the planet revolves around) causing an increase in stellar brightness due to lensing. Microlensing events are not repeatable, whereas in the photometric method a transit occurs once every orbital period.

2.2 Comparison between the radial velocity and the transit method

All extrasolar planet discoveries so far have been made via the radial velocity method. This method leads to orbital information and minimum masses only. In order to determine planetary structure we must know their radii, which cannot be determined by the radial velocity method. In contrast equation 1.1 makes clear that the transit method gives us the planetary radius and in addition orbital information (orbital inclination, period and time at the center of the transit).

The radial velocity method requires high-precision spectroscopic studies of minute changes in a star's radial velocity, which take months to years of work for each star. And only a few of the oversubscribed world's largest telescopes that are equipped with spectrographs of sufficient stability and precision can perform the required work. The transit method can be employed using small telescopes that are equipped with high quality detectors.

The radial velocity method has lead to the discoveries of extrasolar planets associated only with bright stars (V <~ 10) and thus relatively nearby stars (d <~ 50 pc). The transit method with a meter-class telescope will be able to see Jupiter-sized transits down to stellar magnitudes of V <~ 19 (Everett and Howell, 2001), consequently for sun-like stars out to d <~ 8 kpc, if absorption is not significant. We can achieve this because the basic formula of transit method (equation 1.1) is independent of the magnitude of the parent star.

The radial velocity method is currently not at all capable of finding Neptune-sized and Earth-sized planets, whereas the transit method can. The radial velocity method has lead to the discoveries of extrasolar planets only around late type stars (from spectral types late F to M). In contrast the transit method may lead to discoveries of extrasolar planets around A and early type F stars. Table 1 summarizes a comparison between the radial velocity and transit methods.

Table 1.
Comparison of Radial Velocity and Transit Methods

Property	Radial Velocity	Transit
Period	Yes	Yes
Eccentricity	Yes	Yes
Semi-Major Axis	Yes	Yes
Minimum Mass	Yes	No
Orbital Inclination	No	Yes
Planetary Radius	No	Yes
Planets around F, G, K, M stars	Yes	Yes
Planets around early F	No	Yes
Planets around A stars	No	Yes
Planets around B stars	No	Yes
Planets around faint stars ($m > 11$)	No	Yes
Planets around distant stars ($d > 100$ pc)	No	Yes
Earth-sized planets	No	Yes

2.3 Application

There are two major difficulties in the application of the transit method for the discovery of extrasolar planets. First, the attainment of high photometric precision (<~ 1%) and secondly the fact that the probability that a planet will transit is small (<~ 10% for "hot Jupiters" or 0.47% for a cold Jupiter in a Jupiter-size solar system orbit) because the plane of the orbit may not be in the line of sight. Below we will show how the first difficulty can be overcome. There are two approaches to overcoming the second difficulty. First, the monitoring of eclipsing binary stars, whose orbital inclinations are $i \approx 90°$ (TEP[2], Deeg et al. 1998). The second approach involves monitoring a large number of stars. There are two ways to accomplish this. One way is to observe bright stars with a small aperture telescope with a large field of view (VULCAN[3], Borucki et al. 1999; STARE[4], Brown and Charbonneau 2000, PSST[5]). Another possibility is to monitor faint stars in dense stellar fields (ASP[6], Everett and Howell 2001; STARBASE[7], McGruder et al. 2001). This chapter focuses on this later possibility.

3. DETECTABLE PLANETS

In this section we discuss what types of planets can be detected with the Transit Method emphasizing the photometric precision that is required to detect them. Our discussion is based on Figure 1. In the next section we discuss how to achieve the required precision.

Figure 1 is a plot of main sequence spectral type vs. the logarithm of the change in magnitude, Δm, caused by a transiting planet. From equation 1.1 the differential magnitude (transit depth), Δm, between the stellar brightness, I, without the planet and its brightness with the planet transiting it, $I - \Delta I$, is given by

$$\Delta m = -2.5 \log_{10}\left(\frac{I - \Delta I}{I}\right) \tag{1.2}$$

In Figure 1 we consider four types of planets – hot Jupiter ($1.6R_J$), cold Jupiter (R_J), Neptune ($.34R_J$), and Earth ($.09R_J$). We have chosen to use $1.6R_J$ for hot Jupiters because this value agrees with the only radius of a hot Jupiter that has been measured (Henry et al., 2000; Jha et al., 2000). However, theory (Guillot et al., 1996) indicates that R for hot Jupiters can go up to just shy of $3R_J$. Thus our predictions may be conservative.

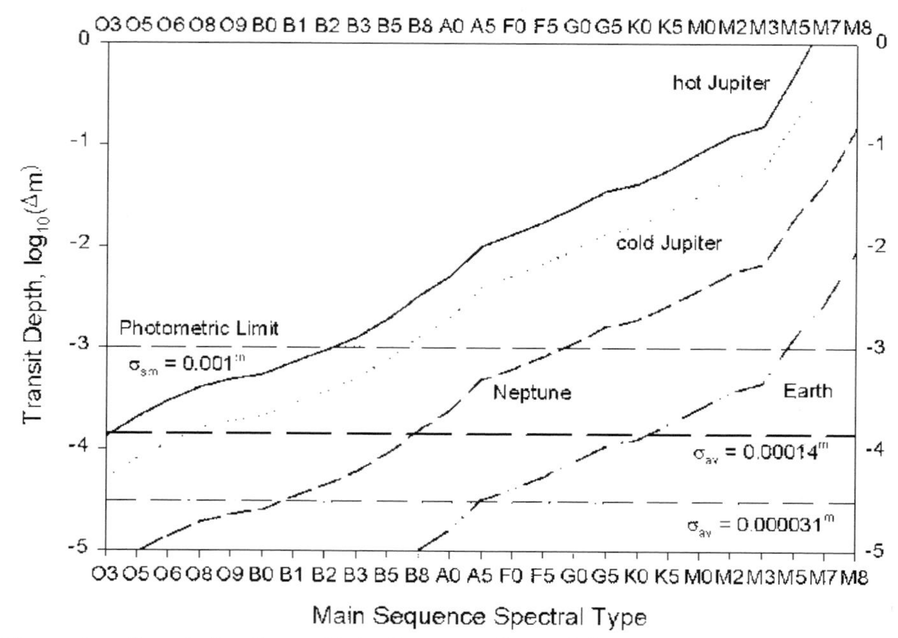

Figure 1. Transit Depth vs. Main Sequence Spectral Type. Three photometric limits (horizontal lines) are shown – precision for a single measurement ($\sigma_{sm} = 0.001^m$), an average over all measurements made in a single transit window ($\sigma_{av} = 0.00014^m$) and an average over all transit windows during an entire observing season ($\sigma_{av} = 0.000031^m$).

3.1 Photometric Limits

It is clear from Figure 1 that the minimum photometric precision for a single measurement, *sm*, required to detect earth-sized planets is ~3 = $\log_{10}(\Delta m)$ or $\sigma_{sm} = 0.001^{m}$ (0.1%). This limit is shown in Figure 1. Planetary detections with precisions of ~σ_{sm} are not believable detections (normally one would prefer ~$3\sigma_{sm}$ or greater). However, in the photometric method higher precisions are achievable through time averaging. Consequently, smaller planets or planets at earlier spectral types can be detected. This higher precision is achieved however at the loss of time resolution. We will consider two extreme cases.

First we consider averaging all of the points of a single transit curve. This means no time resolution at all. That is, the actual transit curve cannot be determined, only whether a transit exists or not. Thus, the stellar mass, orbital inclination and limb darkening will not be able to be used in the determination of the planetary radius. Instead, the planetary radius must be calculated from equation 1.1. In order to compute a photometric limit corresponding to this case, we consider the 1.3-m Robotically Controlled Telescope (RCT) at Kitt Peak National Observatory. For a single measurement it is expected to achieve, $\sigma_{sm} = 0.001^m$ (McGruder et al. 2000). For an exposure time of 2 minutes with a readout time of 1.5 minutes we have a sampling time is 3.5 minutes. A "hot Jupiter" requires about 3 hours to transit. Thus, 51 measurements take place in the 3 hours. The precision achieved is $.001/\sqrt{(51)} = .00014^m$. Figure 1 also shows this photometric limit, 140 μmags. It will be necessary to compare the results of many periods for a confirmation.

We consider a second extreme case. Averaging over an entire observing season with continuous telescopic observations (via a worldwide network of telescopes). This scenario would mean that there would be no information on the transit curve only on detection of the planet. If we take the observing period to be three months (90 days), a 51 Pegasi-like planet with a period of 4.2 days, and again as above a transit duration of 3 hour and a RCT sampling time of 3.5 minutes, then there are 21 transits in the observing period or 1071 observations in the transit widows. Averaging gives a precision of: $.001/\sqrt{(1071)} = .00003^m$. Figure 1 also contains this photometric limit, 30.6 μmags. The necessary confirmation must be achieved by comparing the results over a number of years.

3.2 Jupiters

Figure 1 makes abundantly clear one of the major advantages of the transit method – transits can be detected over a much wider range of spectral class than is possible with the radial velocity method. Transits can be found for B and A stars as well as early F. Radial velocity searches can detect only planets associated with late F down to M stars. Planets around such stars can also be detected by the transit method as well.

3.3 Neptunes

It is apparent from Figure 1 that the Transit Method is capable of detecting Neptune-sized planets revolving around late F as well as K and M stars. The radial velocity method cannot achieve this either. Thus, since Neptune is the smallest of the gas giants of the solar system, the photometric approach can detect all types of solar system gas giants. If one averages over an entire transit window Neptune-sized planets can be found around B stars.

Based on radius alone Jupiter-sized planets cannot clearly be differentiated from brown dwarfs or very late type stars. On the contrary Neptune-sized bodies can be unambiguously classified as planets. Thus, discovering a Neptune-sized body means directly confirming the existence of an extrasolar planet.

3.4 Earths

Most importantly Figure 1 makes clear that earth-sized planets can in principle be detected via the photometric method unlike the radial velocity method. In fact, apart from unrepeatable microlensing events the transit method is the only ground-based method capable of detecting earth-sized planets. However, for single measurements with an RCT-like setup this is only true for late M stars. And there are two caveats: Firstly, there may be relatively few late type M stars found down to the limiting magnitude of the exposures employed. Secondly, these stars exhibit intrinsic photometric variability and we must learn to differentiate between their stellar variability and that caused by a planetary transit. However, by averaging over an entire transit window earth-sized planets can be detected around K stars and they can even be seen around A type stars if averaging is done over an entire observing season.

4. PHOTOMETRIC PRECISION

In this section we investigate the photometric precision, which is attainable by ground-based telescopes. It is clear that the maximum precision obtainable by a single measurement will be proportional to the number of photons collected. The higher the number of photons, N, the greater the precision that can be achieved. Thus, for a given number of photons collected in a single measurement there is a maximum achievable precision. The exact relationship is derived from the Poisson distribution (Howell, 1993):

$$\sigma_{sm} = \frac{1.086}{\sqrt{N}} \tag{1.3}$$

In actual observations there are a number of factors, which prevent observers from achieving the precision given by the above equation. First, there are the factors, which describe the quality of the telescope - guiding, tracking, optics, etc. One important non-telescopic factor, which can determine the precision, is scintillation.

Even though high photometric precision ($\sigma_{sm} \leq 1\%$ i.e. $\sigma_{sm} \leq .01^m$) with CCDs is theoretically possible, it has proven to be difficult to attain in practice. However, a few observers have achieved this milestone. Employing differential photometric techniques the best consistent ground-based precisions obtained with CCD imagers per few minute sampling are in the range, $0.002^m \leq \sigma_{sm} \leq 0.006^m$ (Howell et al., 1988; Honeycutt 1992; Gilliland et al., 1993). A transit of a Jupiter-like planet in front of a sun-like star produces a 1% decrease in the stellar brightness, and is therefore easily detectable using ground-based telescopes.

The prerequisite for the detection of Neptune and Earth-sized extrasolar planets via the transit method is however the attainment of ultra-high photometric precision ($\sigma_{sm} \sim 0.1\%$ i.e. $\sigma_{sm} \sim 0.001^m$) for single observations with CCD imagers.

4.1 Ultra-High Precision

Many astronomers believe that ultra-high precision CCD photometry is not attainable with ground-based telescopes. With an exposure time of 3 minutes and a specific set of observing conditions equation 1.3 yields a theoretical limit of photometric precision for a single measurement of 0.002^m. Everett and Howell (2001) using the NOAO 0.9-m telescope on Kitt Peak with the wide-field MOSAIC CCD Camera obtained precisions as high as 0.002^m for bright stars ($V \sim 14$). Thus, for the brightest stars they reached the theoretical limit of precision and demonstrated that ultra-high precision ground-based CCD photometry is indeed attainable.

The attainment of the theoretical limit of photometric precision means that essentially all sources of systematic error have been eliminated. Thus, random noise sources can be reduced to low levels. This is achieved through careful application of differential photometry (employment of local ensemble differencing on the images), use of a well-sampled Point Spread Function (PSF), limiting the bandpass of the observations to avoid differential refraction and other color effects as well as intrapixel quantum efficiency issues, use of high-quality hardware, and proper error assignment to each data point (not an approximate value assigned to some mean datum).

5. PROBABILITY OF PLANET DETECTION

The number of stars, n, with an observable planetary transit at any given instant is (Giampapa 1995):

$$n = \frac{f_p N}{\pi}\left(\frac{R_s}{a}\right)^2 \tag{1.4}$$

where R_s is the stellar radius, f_p the fraction of stars with at least one planet, N the number of stars in the sample, and a is the orbital radius of a planet. Equation 1.4 assumes circular planetary orbits. We will use it to estimate the probability of planet detection.

According to Cumming et al. (1999) approximately 3% of solar type stars have giant planets within 0.1 AU. Perhaps 60% of the stars in a field are solar type (late F down to M). Thus about 18 in every 1000 stars will be solar type with a planet.

Clearly, a successful photometric detection is only possible if the planet is in the line of sight (Borucki and Summers 1984). Given that a star possesses a planet, then the probability that the planet produces an observable transit is proportional to the radius of the star and inversely proportional to the radius of the planet's orbit (Koch and Borucki 1996). For a randomly oriented sample of target systems, the probability of a transit is only 0.47% for a Jupiter-size planet in a Jupiter-like orbit. But "hot Jupiters" have a relatively high probability (~5%, $R_s/a = R_{sun}/.1 = 0.05$) of producing a transit because they are very close to their parent stars. Thus, about 1 in 1000 stars will have an observable transit.

Now given that a star possesses an observable transit, what is the probability at any given instant of actually seeing the planet in transit? This probability is simply the ratio of transit duration to the orbital period. For a circular orbit this is $R_s/\pi a$. For a sun-like star this probability is 0.016. It follows that we have to observe about 100,000 stars at any given instant just to find 1-2 stars with planets that are transiting.

We conclude that in order to find an appreciable number of planets we need to observe a large number of stars *continuously*.

6. TELESCOPE NETWORK

The above section makes it clear that we need to observe dense star fields continuously in order to maximize the number of planetary detections. Obviously, a single telescope cannot observe continuously because stars cannot be seen in the daytime. What is required is a network of telescopes.

6.1 Full Longitudinal Coverage

There are two major reasons why a worldwide network of telescopes (full longitudinal coverage) is required. First is the day/night cycle and secondly weather related phenomena. Each telescope should limit observations to star fields at 1.4 airmasses (~45° from the zenith) or less. Because of weather and temporal overlap coverage, a redundancy near each longitude is required (but at a different latitude with different weather patterns). Consequently, a network of 4-8 telescopes is needed in each hemisphere. Total coverage of both the northern and southern hemispheres requires 8-16 telescopes. At approximately \$2 million dollars per telescope and an additional \$2 million for each focal plane array and associated computer hardware, the total cost for a network of 4-8 such systems is only \$16-32 million dollars or \$32-64 million for total worldwide coverage. This cost estimate assumes that sites for the telescopes and the associated infrastructure are already in place.

Compare this cost to that of a typical NASA discovery mission (~ $300 million) and one can immediately see the advantages of a ground-based search for extra-solar planets.

6.2 Robotic Telescopes

Because the times for transits of undiscovered extrasolar planets are completely unknown, observations must be conducted 24 hours/day. A manual operated classical observatory would require at least two observers per telescope, costing ~$150,000 in external funding per year per telescope. For a complete worldwide network in both hemispheres this is $1,200,000 to $2,400,000 in external funding each year. It is clear that robotic telescopes are required to eliminate the need for unaffordable manpower.

7. A GROUND-BASED PLANETARY TRANSIT TELESCOPE

Ground-based telescopes are cheaper to build and operate and can be set into operation faster than space missions. Space missions have distinct advantages which set them apart from ground-based telescopes; they have no day-night cycles, no weather, and no atmospheric disturbances such as seeing and color dependent refraction. However, space-based telescopes are expensive and require many years lead time prior to launch and operation as well as having their own suite of issues such as cosmic rays and spacecraft jitter. Searches from space are expected to begin soon with missions such as Kepler[8] and COROT[9], both using a design goal of obtaining photometric precisions near 1-5 x 10^{-5}. This is the value needed to discover an earth-size planet transiting a star like our Sun.

For transit searches, ground-based telescopes do not need to be large in aperture. In equation 1.3 the number of photons, $N \sim D^2$, where D is the diameter of the telescope. Thus we have:

$$\sigma_{sm} \sim \frac{1}{D} \qquad (1.5)$$

Remembering that Everett and Howell (2001) achieve a precision of 0.002^m with a 0.9-m telescope, we can construct Figure 2. It makes clear that large aperture telescopes do not achieve an appreciably higher photometric precision and are therefore not required for planetary searches via the Transit Method. Figure 2 also makes clear that small telescopes (1-2-m) are quite adequate for an extra-solar planet transit search. The largest obstacles are the ability to effectively deal with the lack of perfect on-source cadence, the need for good optical quality across the field of view, and the atmosphere. Additionally, the high photometric precision obtained must be maintained over time periods of months to years for many thousands of stars.

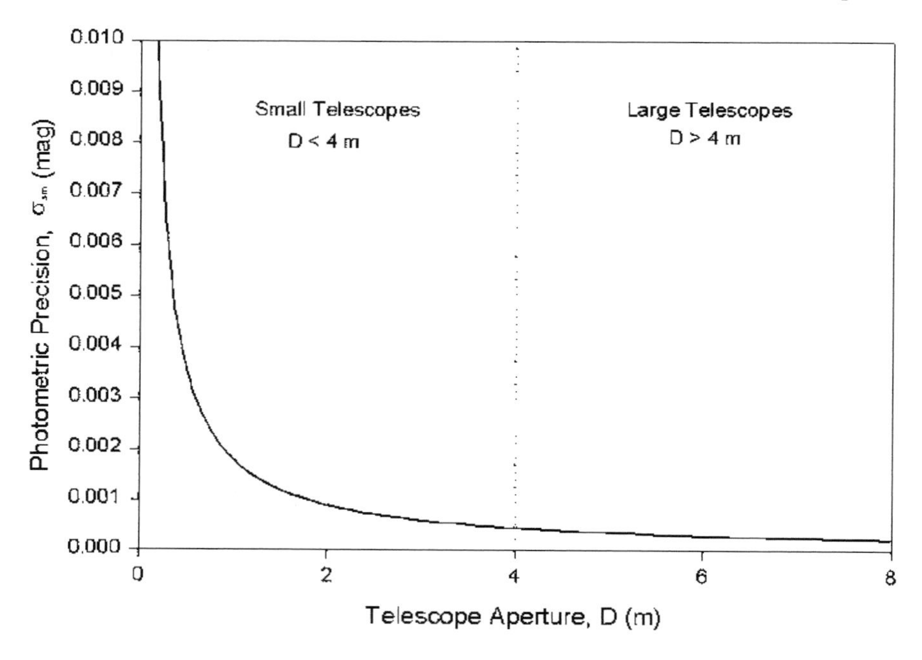

Figure 2. Photometric Precision vs. Telescope Aperture. This plot shows small telescopes are appropriate for extrasolar planet search.

To date, ground-based telescopes have achieved consistent high photometric precisions. This level of precision is, in principle, reachable with any aperture telescope, and in fact has usually been done with 1-m class optical systems. The typical determining factor for the precision available, to first order, is the available number of stellar photons one can detect. With well-behaved telescope + CCD combinations precisions of near 2 (single measurement) to 0.2 mmag (5 hour averages) have been achieved for extra-solar planetary transit time scales. These levels are still a factor of about 10 above that needed for a single exposure transit detection of an Earth-sized planet about a solar-like star (F-K) but through co-addition of frames and detailed computational analysis of the light curves, transits can be found.

Thus, we need to explore what can be done to improve ground-based transit searches in order to make them viable as extra-solar planet finders. What would be an ideal planet finder?

One important issue is that of the statistical nature of finding extra-solar planets orbiting their stars in a (nearly) edge on orbit. While only about 1 in 20 stars are viewed at the proper inclination, and then maybe 2-4% of these have planetary systems, the use of a wide-field imager is needed to allow observations of millions of stars within a specific range of magnitudes over months to years of time. This requirement for a good planet finder simply means a wide-field telescope with a focal plane array of detectors such as

CCDs is critical to the success of a detailed search. We will see below, that the requirements for a large field imager are not simple however, as good stable image quality is demanded.

Now we look deeper into the details as we need to measure very small changes of order 0.01-0.05 mmag. Given the fact that we need to get to levels 10 times those achieved so far, the details of scintillation, airmass effects, differential refraction, seeing, and extinction remain, even for a network of telescopes. Observations at low air mass, sites with good atmospheric conditions (e.g., island mountain tops) and, as we will see below, use of well sampled point spread functions and local differential ensemble measurement technique all combine to essentially bring these atmospheric factors under control. These parameters simply need to be compensated for in some consistent and very precise manner over the long time-scales involved.

On a given night seeing effects, even changes by only 20%, can cause ensemble differential photometric uncertainties to increase by as much as a factor of two. Thus, time of poor seeing may need to be removed from the datasets. Observations at air masses greater than ~2 and observing over a wide field of view during large atmospheric extinction changes are also candidates for removal from the datasets in order to achieve the very precise levels needed. The point spread functions of the stars do not necessarily need to be perfect at all locations within the focal plane, simply they should remain stable and mostly color invariant during the course of the observations. For example, typical off-axis images in ground-based telescopes change in different ways with respect to on-axis images and with respect to each other. These changes occur with changes in air mass, zenith distance, and distance from the optic axis; they do not simply scale linearly with seeing or airmass.

Given the possible need for removal of some of the light curve samples during say a given 24-hour period due to the intricacies mentioned above, the resultant non-uniformly sampled datasets may be very prone to aliasing, especially near the times of interest (hours) to transit hunters. Stars move from east to west within the zone of "good" airmass (above 2.0) over the course of hours and typical seeing and extinction changes are also hour time scale events, thus a power spectrum search for few to ten or so hour long transit events becomes difficult. The number of false positives could be large. The aliasing issues can be managed if redundant and/or continuous observational coverage occurs.

Proper transit detections at very low signal levels are best done if one observes consecutive transits. This eliminates the possibility of period believability issues and aliasing effects. The above examples show that even for a telescope network with 70-80% coverage (i.e. observing efficiency) the number of consecutive transits observable (taking typical earth-like transit times of 5-15 hours and observational sampling) is only about 60% of those available with a continuous program such as could be performed in space.

Issues of ground-based limiting precisions still exist. Does atmospheric scintillation set an ultimate limit on the precisions obtainable? Is there a low frequency noise component due to transit time phenomena that truly limits taking few minute integrations and co-adding to get simply $\sqrt{N}$ increases in S/N? Can one effectively deal with second order differential refraction? So far, the limiting precision obtainable with a ground-based telescope, CCD detector, and the use of local differential ensemble photometric techniques appears not to have been reached (Everett & Howell 2001).

Given the high cost savings of ground-based telescopes relative to space missions, the ability to replace and upgrade instrumentation and computer systems, and the tremendous scientific potential of finding terrestrial size planets, we believe that timely and productive searches for extra-solar planets can be effectively performed from the ground. The ideal system would consist of a 2-m class telescope (to allow for a large enough focal plane array with an unvignetted field of view of reasonable size, a few-to-many square degrees), a focal plane array of CCD detectors with high quantum efficiency, low read noise, and deep pixel wells, a high quality field flattener, an image corrector with color compensation, and an image scale allowing well sampled point spread functions, at least 3-5 pixels across the full width, half maximum value.

The real requirement that must be met for detectors is to have the quantity N x M x W as large as possible for a given integration. N and M are the x and y number of pixels covered by a stellar profile (usually N=M) and W is the well depth of a single CCD pixel. Very good guiding and tracking are essential as well. This last item makes data reduction easier and more importantly, allow for consistent stellar images. A longitudinal distribution of robotic telescopes connected by high-speed networks (to allow for ease of transport of the collected data) as well as the use of filtered (bandpass limiting) observations and a robust data reduction and analysis package round out the requirements. Details of the observational techniques and a well-suited data reduction package are given in Everett and Howell (2001).

8. REFERENCES

Borucki, W., & Summers, A., 1984, Icarus 58, 121

Borucki, W.J. Caldwell, D.A., Koch, D.G., Webster, L.D., Jenkeins, J.M., & Ninkov, Z. 1999, BAAS 195, 24.01

Brown, T. M. & Charbonneau, D. 2000, in Disks, Planetesimals, and Planets ed. F. Garson & T.J. Mahoney (San Francisco: ASP)

Charbonneau, D., Brown, T., Latham, D., and Mayor, M., 2000, ApJ. 529, L45

Cumming, A., Marcy, G., & Butler, R.P. 1999, ApJ 526, 890

Deeg, H. J., et al. 1998, A&A 338. 479

Everett, M. & Howell, S. B. 2001, PASP 113, 1428

Giampapa, M.S., Craine, E.R. & Hott, D.A. 1995, Icarus 118, 199
Gilland, R., et al. 1993, AJ 106, 2441
Guillot, T., Burrows, A., Hubbard, W. B., Lunine, J. I., & Saumon, D. 1996, ApJ 459, L35
Han, I., Black, D.C., & Gatewood, G., 2001, ApJ 548, L57
Henry, G., Marcy, G. W., Butler, R. P. & Vogt, S.S. 2000, ApJ 529, L41
Honeycutt, R. K., 1992, PASP 104, 435
Howell, S. B., Mitchell, K. J., & Warnock, A., III, 1988, AJ 95, 247
Howell, S. B., 1993, in "Stellar Photometry - Current techniques and Future Developments", eds. C. Butler & I. Elliott, IAU Colloquim 136, Cambridge University Press, p. 318
Jha, S. et al. 2000, ApJ 540, L45
Koch, D. & Borucki, W., 1996, A Search For Earth-Sized Planets In Habitable Zones Using Photometry, First International Conf on Circumstellar Habitable Zones, Travis House Pub., 229
Marcy, G. W. & Butler, P. P. 1996, ApJ 464, L147
Mayor M., & Queloz D., 1995, Nature 378, 355
McGruder, III, C. H., Everett, M., Howell, S. B., & Barnaby, D., 2000, in "Instruments, Methods and Missions for Astrobiology III", ed. R. Hoover, Proceedings of SPIE, 4137
McGruder, III, C. H., Everett, M. & Howell, S. B., 2001, in "Small-Telescope Astronomy on Global Scales" eds. B. Paczynski, W. Chen, and C. Lemme, in press, (San Francisco: ASP)

---

[1] http://exoplanets.org/science.html

[2] http://www.iac.es/proyect/tep/tephome.html

[3] http://web99.arc.nasa.gov/~mars/vulcan/

[4] http://www.hao.ucar.edu/public/research/stare/stare.html

[5] Ted Dunham at Lowell Observatory

[6] http://www.psi.edu/~esquerdo/asp/asp.html

[7] http://starbase.wku.edu/telescopes.html

[8] http://www.kepler.arc.nasa.gov/

[9] http://www.obspm.fr/encycl/corot.html

Chapter 12

Big Planets with Small Telescopes

Debra Ann Fischer
University of California, Berkeley
Berkeley, California USA

Abstract: In an era where exciting breakthroughs in science typically emerge with technological advances and bigger pieces of glass, the detection of extrasolar planets stands as a tribute to small telescopes equipped with high resolution spectrographs.

Key words: extrasolar planets

1. INTRODUCTION & HISTORICAL PERSPECTIVE

In 1995, the first extrasolar planet to a main sequence star was detected orbiting the star 51 Pegasi (Mayor & Queloz 1995). Since then, about 80 extrasolar planets[1] (Table 1) have been discovered as companions to solar-type stars (Butler et al. 1997, 1998, 1999, 2000, 2001; Cochran et al. 1997; Delfosse et al. 1998; Fischer et al. 1999, 2001, 2002; Hatzes et al. 2000; Henry et al. 2000; Korzennik et al. 2000; Latham et al. 1989; Marcy & Butler 1996; Marcy et al. 1998, 1999, 2000, 2001a, 2001b; Mayor & Queloz 1995; Mazeh et al. 2000; Naef et al. 2001; Noyes et al. 1997; Queloz et al. 2000; Santos et al. 2001a; Tinney et al. 2001; Udry et al. 2000a, 2000b; Vogt et al. 2000, 2002; Zucker et al. 2001).

While the Keck telescope has discovered many of the known extrasolar planets, the initial discoveries all used small telescopes and these programs continue to contribute to the detection of extrasolar planets. Collectively, these small telescopes now survey about 2500 stars (with some overlap in

[1] See http://cfa-www.harvard.edu/planets, http://obswww.unige.ch/~udry/planet/planet.html or http://exoplanets.org

T.D. Oswalt (ed.), The Future of Small Telescopes in the New Millennium, Vol. III, 189-205.

the samples likely). Ranked by increasing telescope aperture, "small telescope" contributions have emerged from:

- *0.6-m Coudé Auxiliary Telescope (CAT) and the 3-m Shane Telescope at Lick Observatory.* These telescopes survey about 360 stars with a magnitude limit of about V = 6.0 for the CAT and V = 7.5 for the Shane. This project achieves a velocity precision down to a few meters per second and several extrasolar planets have been detected in this program.
- *1-m Leonard Euler Swiss Telescope at La Silla Observatory.* This telescope is being used ~200 nights per year with the Coralie spectrograph to survey a volume-limited sample of 1600 stars in the southern hemisphere with a velocity precision of about 6 ms^{-1}. Several extrasolar planets have been discovered in this program.
- *1.5-m Wyeth Refractor at Oak Ridge Observatory.* This telescope surveys about 1850 G dwarfs with a velocity precision of 500 ms^{-1}. While the goal of this project is not extrasolar planet detection, the first Doppler discovery of a substellar (brown dwarf) companion was made here in 1989.
- *1.5-m Advanced Fiber Optic Echelle (AFOE) telescope at the Whipple Observatory* surveys about 120 stars with a precision of 6-7 ms^{-1} with a magnitude limit of about V = 7. This project has discovered three extrasolar planets and confirmed other detections.
- *1.93-m telescope at Haute-Provence Observatory* with the Elodie spectrograph surveys 320 stars brighter than V = 7.65. The first Jupiter-like extrasolar planet was detected in this survey (orbiting 51 Peg) and several other extrasolar planets have been detected in this program.
- *2-m McDonald Telescope* surveyed about 32 bright stars. Two co-discoveries of extrasolar planets have emerged from this project which will eventually be expanded on the HET.
- *3.9-m Anglo Australian Telescope at the Anglo Australian Observatory* surveys about 200 stars brighter than V = 7.5 with a velocity precision of a few meters per second. Several extrasolar planets have been discovered in this program.

Table 1.
Masses and Orbital Characteristics of Extrasolar Planets[2].
(Stellar masses derived from Hipparcos, metallicity, and stellar evolution)

Star Name	M sin i (M_{JUP})	Period (d)	Semimajor Axis (AU)	Eccen- tricity	K (m/s)
1 HD83443 b	0.35	2.986	0.038	0.00	57.0
2 HD46375	0.25	3.024	0.041	0.02	35.2
3 HD179949	0.93	3.092	0.045	0.00	112.0
4 HD187123	0.54	3.097	0.042	0.01	72.0
5 tau Boo	4.14	3.313	0.047	0.02	474.0
6 BD-103166	0.48	3.487	0.046	0.05	60.6
7 HD75289	0.46	3.508	0.047	0.01	56.0
8 HD209458	0.63	3.524	0.046	0.02	82.0
9 51 Peg	0.46	4.231	0.052	0.01	55.2
10 ups And b	0.68	4.617	0.059	0.02	70.2
11 HD68988	1.90	6.276	0.071	0.14	187.0
12 HD168746	0.24	6.400	0.066	0.00	28.0
13 HD217107	1.29	7.130	0.072	0.14	139.7
14 HD162020	13.73	8.420	0.072	0.28	1813.0
15 HD130322	1.15	10.72	0.092	0.05	115.0
16 HD108147	0.35	10.88	0.098	0.56	37.0
17 HD38529	0.79	14.31	0.129	0.12	53.6
18 55 Cnc	0.93	14.66	0.118	0.03	75.8
19 GJ86	4.23	15.80	0.117	0.04	379.0
20 HD195019	3.55	18.20	0.136	0.02	271.0
21 HD6434	0.48	22.09	0.154	0.30	37.0
22 HD192263	0.81	24.35	0.152	0.22	68.2
23 HD83443 c	0.17	29.83	0.174	0.42	14.0
24 GJ876 c	0.56	30.12	0.130	0.27	81.0
25 rho CrB	0.99	39.81	0.224	0.07	61.3
26 HD74156 b	1.55	51.60	0.276	0.65	108.0
27 HD168443 b	7.64	58.10	0.295	0.53	470.0
28 GJ876 b	1.89	61.02	0.207	0.10	210.0
29 HD121504	0.89	64.62	0.317	0.13	45.0
30 HD178911 B	6.46	71.50	0.326	0.14	343.0
31 HD16141	0.22	75.80	0.351	0.00	10.8
32 HD114762	10.96	84.03	0.351	0.33	615.0
33 HD80606	3.43	111.8	0.438	0.93	414.0
34 70 Vir	7.42	116.7	0.482	0.40	316.2
35 HD52265	1.14	119.0	0.493	0.29	45.4

[2] Dec 29, taken from http://exoplanets.org

Table 1, continued.
Masses and Orbital Characteristics of Extrasolar Planets[2].
(Stellar masses derived from Hipparcos, metallicity, and stellar evolution)

Star Name	M sin i (M_{JUP})	Period (d)	Semimajor Axis (AU)	Eccen-tricity	K (m/s)
36 HD1237	3.45	133.8	0.505	0.51	164.0
37 HD37124	1.13	154.8	0.547	0.31	48.0
38 HD82943 c	0.88	221.6	0.728	0.54	34.0
39 HD8574	2.23	228.8	0.756	0.40	76.0
40 HD169830	2.95	230.4	0.823	0.34	83.0
41 ups And c	2.05	241.3	0.828	0.24	58.0
42 HD12661	2.84	250.5	0.795	0.19	89.1
43 HD89744	7.17	256.0	0.883	0.70	257.0
44 HD202206	14.68	258.9	0.768	0.42	554.0
45 HD134987	1.63	265.0	0.821	0.37	53.7
46 HD17051	2.12	312.0	0.909	0.15	63.0
47 HD92788	3.88	337.0	0.969	0.28	113.0
48 HD142	1.00	337.1	0.980	0.38	29.6
49 HD28185	5.59	385.0	1.000	0.06	168.0
50 HD177830	1.24	391.0	1.10	0.40	34.0
51 HD4203	1.64	406.0	1.09	0.53	51.0
52 HD27442	1.32	415.0	1.16	0.06	32.0
53 HD210277	1.29	436.6	1.12	0.45	39.1
54 HD82943 b	1.63	444.6	1.16	0.41	46.0
55 HD19994	1.66	454.0	1.19	0.20	42.0
56 HD114783	0.99	501.0	1.20	0.10	27.0
57 HD222582	5.18	576.0	1.35	0.71	179.6
58 HD 23079	2.54	627.3	1.48	0.06	56.7
59 HD141937	9.67	658.8	1.48	0.40	247.0
60 HD160691	1.99	743.0	1.65	0.62	54.0
61 HD213240	3.75	759.0	1.60	0.31	91.0
62 16 Cyg B	1.68	796.7	1.69	0.68	50.0
63 HD4208	0.81	829.0	1.69	0.04	18.3
64 HD10697	6.08	1074.0	2.12	0.11	114.0
65 47 UMa b	2.56	1090.5	2.09	0.06	49.7
66 HD190228	5.01	1127.0	2.25	0.43	96.0
67 HD50554	4.49	1296.0	2.36	0.51	94.9
68 ups And d	4.29	1308.5	2.56	0.31	70.4
69 HD33636	7.71	1553.0	2.62	0.39	148.0
70 HD106252	7.10	1722.0	2.77	0.57	150.7
71 HD168443 c	16.96	1770.0	2.87	0.20	289.0
72 HD145675	4.05	1775.0	2.93	0.37	70.4
73 HD 39091	10.37	2115.2	3.34	0.62	196.2
74 HD74156 c	7.46	2300.0	3.47	0.40	121.0
75 eps Eri	0.88	2518.0	3.36	0.60	19.0
76 47 UMa c	0.76	2640.0	3.78	0.00	11.0

2. 51 PEG PLANETS

The detection of a planet orbiting 51 Peg surprised everyone who expected that other solar systems would resemble our own, with rocky planets inhabiting the inner few AU and gas giant planets in much wider orbits. With a Doppler-detected mass, Msin i = 0.75 M_{JUP}, 51 Peg b orbits at a distance of about 0.05 AU with an orbital period just over four days. This star is now thought of as the prototype for 10 other extra-solar planets with orbital periods shorter than 6 days. In addition to short orbital periods, these planets are all in orbits that have been circularized, presumably by tidal interactions with their host star.

The 51 Peg-like planets offer a potential scientific bonus; because of their close proximity to the host stars, about 10% of these planets are expected to have inclinations close enough to edge-on that a planet transit might be observed. True to the expected statistics, HD 209458 is the one case (out of eleven) where photometric dimming of the star was observed at the predicted transit time (Charbonneau et al. 1999, Henry et al. 1999). The transit photometry provided a direct measurement of the planet radius and hence the mean planet density and showed that this Jupiter-mass planet was gaseous rather than solid. Follow-up observations have now detected a spectroscopic signature of sodium in the planet atmosphere (Charbonneau et al. 2001). This first direct observation of an extrasolar planet atmosphere advances planet hunting to the next exciting stage of extrasolar planet characterization.

3. EXOPLANET ORBITAL CHARACTERISTICS

In order to detect an extrasolar planet using the Doppler technique, precise radial velocity measurements must map out Keplerian velocity variations over at least one full orbital period. From the Doppler perspective, the fundamental properties of a planet are not mass and semi-major axis, but the velocity amplitude and the orbital period. The radial velocity amplitude, *K*, is a function of the (star and planet) mass, orbital inclination (*i*), orbital period (*P*) and eccentricity:

$$K = \left(\frac{2\pi G}{P}\right)^{1/3} \frac{M_{PL} \sin i}{\left(M_{STAR} + M_{PL}\right)^{2/3} \left(1 - e^2\right)^{1/2}}$$

This velocity amplitude is modulated over the orbital period; circular orbits exhibit sinusoidal velocity variations, but the shape of the Keplerian curve for noncircular orbits is determined by the orbital eccentricity and omega, the angle between periapse and the line of nodes. In practice, M_{star}

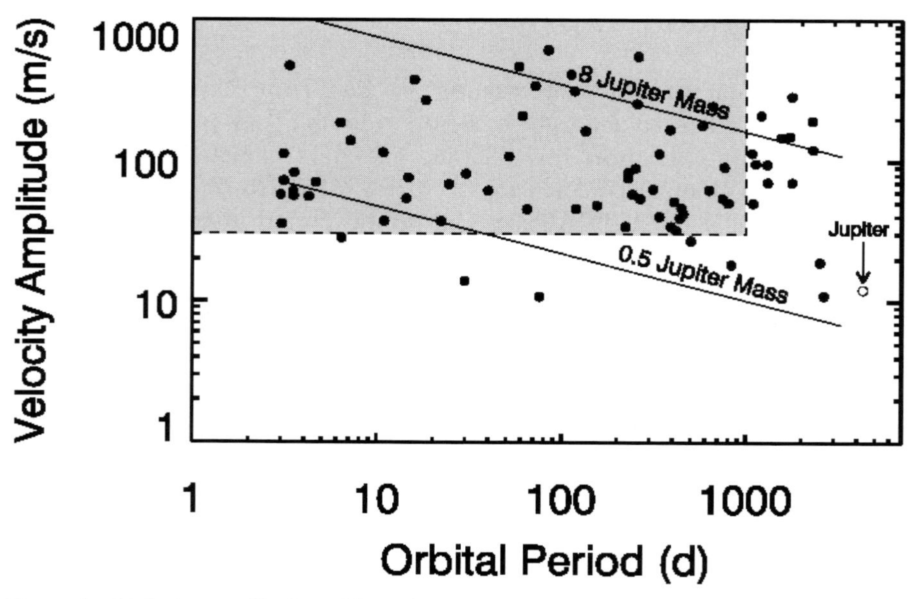

Figure 1. Velocity amplitudes of Doppler-detected extrasolar planets. The shaded region above K=30 m/s and below P=1000d has been well observed for many stars on current planet search surveys. This domain of parameter space should be statistically complete in the next few years.

is assumed from the spectral type of the star and the data are fit using a Marquardt minimization scheme to derive the orbital elements.

The orbital period affects planet detectability in two ways. First, as the above equation shows, longer orbital periods are intrinsically harder to detect because they induce a smaller reflex velocity amplitude in the star. Second, in order to fit a Keplerian model to the data, the velocity observations must extend over at least one full orbital period. So, the maximum detectable orbital period, currently about 7 years, is set by the time baseline of the Doppler observations. As a result, the "typical" orbital period for an extrasolar planet, tracks the timeline of Doppler surveys. In 1997, there was a leap in the number of stars on planet search surveys as software pipelines for Doppler analysis were developed and as people realized that success in the search for extra-solar planets required sifting through large samples. In 1999, most of the extrasolar planets had orbital periods of two years or less. In 2001, orbital periods of a few years were more typical. Apparently most of the ~1 MJUP extrasolar planets in shorter period orbits have already been harvested from current extrasolar planet surveys.

Figure 1 shows the detection parameter space for the Doppler technique populated with the known extrasolar planets listed in Table 1. The detected extrasolar planets tend to cluster in a box in this parameter space demarcated by dashed lines at K= 30 m/s and P= 1000 days (hereafter, the "K_{30}-P_{1000}" domain). While these constitute the planets that we can detect, it is not yet clear whether this type of solar system, with close-in gas giant planets, is more or less common than our own solar system architecture. Observational incompleteness in this domain will continue to shrink and in the next few years, it is likely that these extrasolar planets will constitute a statistically complete sample. It is impressive that we are leaping from the first planet discovery in 1995 to a nearly complete inventory of (K_{30}-P_{1000}) gas giant planets around nearby stars in just under a decade.

Our best estimate, based on sample subsets of a few hundred well-observed stars, is that between 6 – 8 % of solar type stars have extrasolar planets like those found in the K_{30}-P_{1000} parameter space. One of the most intriguing questions yet to be answered is what kind of planetary systems orbit the remaining ~92% of solar type stars. Our solar system is one example of a planetary system that falls outside the K_{30}-P_{1000} domain. Jupiter is the only planet in our solar system that could be detected with state-of-the-art Doppler projects, but our time baseline of observations is not yet long enough to have detected this signpost orbiting other stars. In the next five years our time baseline of observations will be long enough to detect true Jupiter analogues with K~12 m/s and orbital periods of about 12 years. High velocity precision and long-term instrumental stability will be more important than ever as we break out of the K_{30}-P_{1000} box to learn whether our solar system is freakishly unusual or a common by product of star formation.

One important product of the past few years of planet hunting is the mass distribution of detected extrasolar planets. This distribution, shown in Figure 2, is strongly peaked toward the least massive companions *even though we are biased against detecting these low velocity amplitude objects.* The Doppler technique may currently be missing planets with Msin *i* below about 0.2 M_{JUP} since these planets have intrinsically small velocity amplitudes (10 < K < 20 m/s, even in short period orbits). If the mass distribution continues to rise below 0.2 M_{JUP} then the search for these lower mass planets could yield rich scientific returns as velocity precision is pressed below three meters per second in the next few years. Velocity precision of 1 – 2 m/s should enable the detection of companions with Msin *i* down to Neptune masses (Msin *i* ~ 0.06 M_{JUP}) if they exist in close orbits.

Gizis et al. (2001) note that brown dwarfs may be common companions at separations wider than 1000 AU. However, one striking characteristic of the extrasolar planet mass distribution is the paucity of objects with Msin *i* greater than 8 M_{JUP} (where the true mass may be above the deuterium-burning threshold). Despite the fact that higher mass planets are easier to detect, these objects are relatively rare, suggesting truncation of planet formation in the primordial disk at masses above ~8 M_{JUP}. The low observational incompleteness for these companions led Marcy & Butler

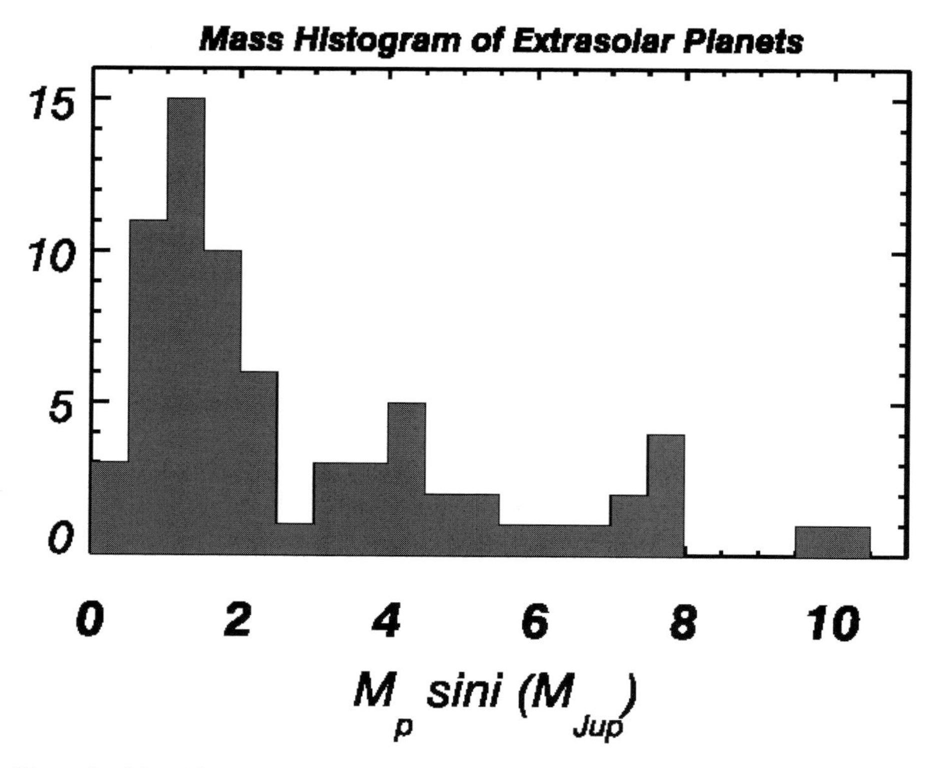

Figure 2. Mass histogram of extrasolar planets. Despite an observational bias that favors massive objects, the gas giant planet distribution is peaked at the lowest detectable masses

(2000) to conclude that fewer than 0.5% of ~500 stars on their Keck planet search survey have brown dwarf companions at separations less than a few AU. Halbwachs et al. (2000) likewise find a minimum in their distribution of companions between 10 – 80 M_{JUP}.

The planets in our solar system orbit on nearly circular tracks. Therefore, one surprising characteristic of the orbits of extrasolar planets is their relatively high orbital eccentricities. In Figure 3 the orbital eccentricity is plotted as a function of orbital period for the known extrasolar planets. One characteristic of this plot has already been noted; planets with orbital periods less than 6 days (i.e., the 51 Peg planets) all reside in circularized orbits. However, for longer orbital periods, the eccentricity appears to be essentially uniform within an envelope that gradually rises from 10 – 100 days and then levels out with a peak value of ~0.7.

The origin of these high eccentricities is presumed to result from dynamical interactions during planet formation stage. Some of the proposed mechanisms are tidal interactions between the proto-planet and the disk (Goldreich and Tremaine 1980, Lin et al., 1996, Bryden et al. 1999), gravitational scattering between growing planetesimals (Rasio & Ford 1996,

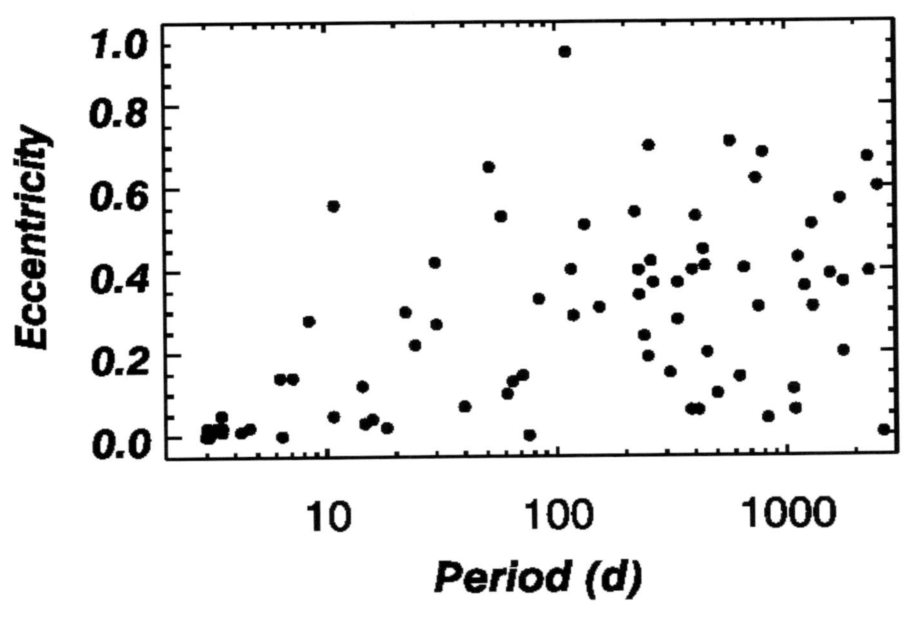

Figure 3. The eccentricity distribution of extrasolar planets. Planets with orbital periods shorter than 10 days are generally in circular orbits. For longer orbital periods, the eccentricities appear to be uniformly distributed below a maximum level that rises from 10 – 100 days to 0.7.

Weidenschilling & Marzari 1996, Levison et al. 1998) and resonant gravitational interactions between planets or planetesimals in the disk (Murray & Holman 1999, Murray et al 2001, Chiang et al 2002). This active dynamical history contrasts with the gentler evolution of solar system planets. The fossil evidence of this history, the eccentricity distribution of extrasolar planets, will offer an important constraint for theoretical models of planet formation.

4. MULTIPLE PLANET SYSTEMS

The first system of planets around a main sequence star, upsilon Andromedae (Butler et al., 1999) has been followed by the publication of additional multiple planet systems (HD 168443, Marcy et al. 2001, Udry et al. 2000; GJ 876, Marcy et al. 2001; 47 UMa, Fischer et al, 2002). There is an observational bias against the detection of multiple systems, so it is notable that a study of twelve planet-bearing stars at Lick with one announced planet and a time baseline of observations longer than 3 years shows that seven out of twelve stars have residual velocity variations

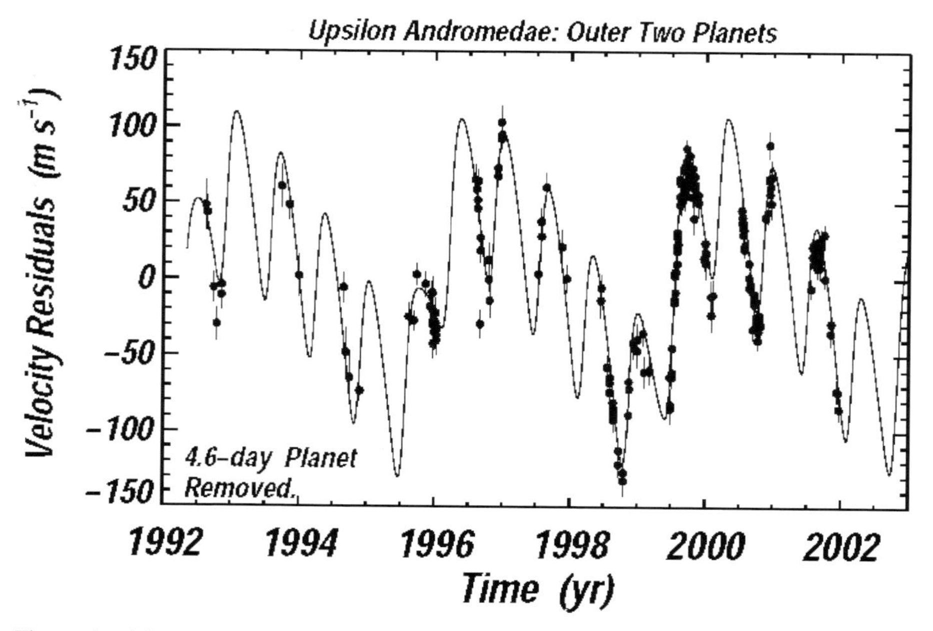

Figure 4. After removing theoretical velocities for a 4.6 day Keplerian orbit, the residual velocities for upsilon Andromedae show evidence of two additional planets.

consistent with an additional gas giant planet (Fischer et al., 2001). The detection of so many trends suggests that multiple planet systems may be common.

Multiple planet systems present a complex velocity signature requiring many high precision observations in order to derive a unique Keplerian model. In Figure 4 theoretical velocities for the inner planet orbiting upsilon Andromedae have been subtracted off in order to make the outer two Keplerian variations visible. This system is well modelled with a superposition of three Keplerians because the dynamical interaction between the planets is small over the time scale of the observations. However, in some cases (GJ 876, HD 168443) there are strong gravitational interactions between the planets and the orbital elements are a function of time. Measurable deviations from a simple superposition of Keplerians can be observed in these systems on relatively short time scales and proper modeling requires dense observational sampling over relatively short time scales and numerical integration of the Newtonian equations of motion.

The star GJ 876 has two gas giant planets: a 0.56 M_{JUP} planet in a 30-day orbit and a 1.89 M_{JUP} planet in a 61-day orbit. The 2:1 resonance orbit of these planets is strongly interacting, making this an extremely challenging detection. The 1.89 M_{JUP} planet was discovered in 1999 and although the Keplerian model left large residual velocities, it took two more years of data (with the Keck telescope for better S/N on this V=9.9 star) to finally model

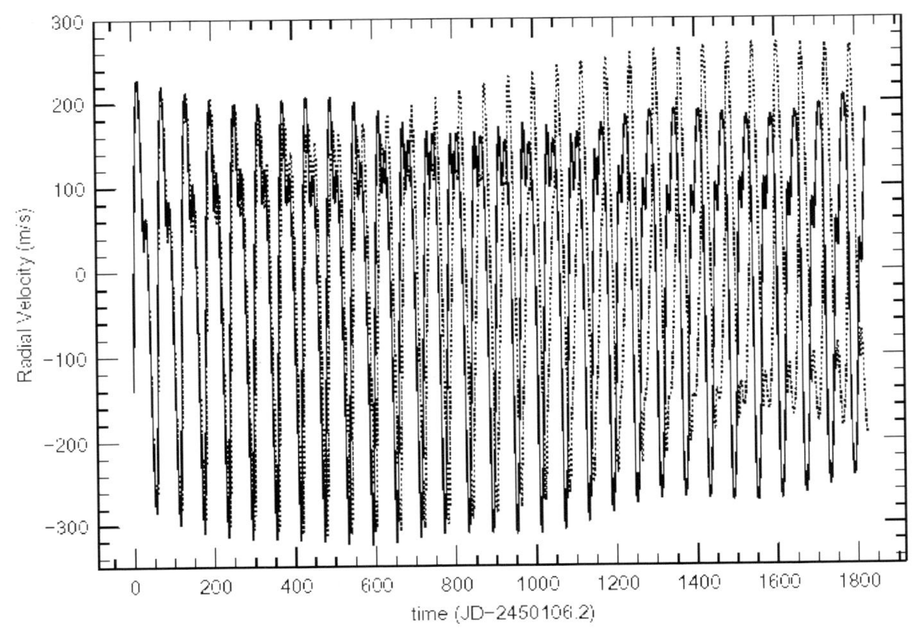

Figure 5. A double Keplerian (solid line) is compared with a numerical integration's model (courtesy of Greg Laughlin) for the 2:1 resonance system orbiting GJ 876. Measurable deviations from the double Keplerian solution are apparent after just a few years.

the second planet. Figure 5 (courtesy of Greg Laughlin) shows theoretical velocities from a superposition of two Keplerians (solid line) compared with the proper numerical integration model (dashed line). These two models diverge significantly after just a few years. Numerical integration models include the full mass of the two planets, so a good fit between the model and the velocities offers the extraordinary opportunity to constrain the orbital inclination (Lee & Peale 2001, Laughlin & Chambers 2001, Rivera & Lissauer 2001).

Dynamical simulations help to illuminate fundamental characteristics of extrasolar planet architecture. In the case of upsilon Andromedae, Lissauer & Rivera (2001) show that if the three detected planets are coplanar and nearly edge-on, then there are only a few resonance windows in the inner 8 AU surrounding this star where an additional low mass planet could reside in a stable orbit. These dynamically stable resonances disappear from their simulations as the assumed orbital inclination increases.

The Lissauer & Rivera simulations place theoretical constraints on the relative inclinations of the three planets, but they also highlight another characteristic: the inner 8 AU surrounding upsilon Andromedae is dynamically "full" of planets. Our own solar system shares this characteristic there are only a few orbital niches in the inner solar system where the existing planets would permit an additional planet to orbit undisturbed. This leads to an intriguing question: when we identify dynamically empty zones around other stars with the Doppler technique, are they really empty, or do they contain planets that are simply below our detection threshold? The identification of architectural characteristics common in extrasolar planet systems will help to identify the best target stars for missions like the Terrestrial Planet Finder (NASA) and Darwin (ESA) which will search for terrestrial planets orbiting in the habitable zones of nearby stars.

5. CHEMICAL COMPOSITION OF PLANET-BEARING STARS

Planet-bearing stars appear to be more metal-rich than field stars without detected planets (Gonzalez 1997, Butler et al. 2000, Santos et al. 2001b). This observation has led to speculation over whether high metallicity is a causal initial condition, or an effect from the accretion of gas-depleted material that pollutes the stellar convective zone.

In support of the causal view, high metallicity provides more raw materials, the dust grains that begin the process of planetesimal formation. The first generations of stars in our galaxy could not have developed planetary systems like our own because they lacked these planet building blocks. So, it seems clear that some threshold metallicity is required for planet formation, but it is not at all clear what that threshold might be.

To test the metallicity correlation, Gilliland et al. (2000) used the Hubble Space Telescope to search for transits in the relatively metal-poor cluster, 47 Tuc ([Fe/H] ~ -0.7). If the stars in this cluster had hosted 51 Peg-like systems similar in occurrence to those found orbiting main sequence field stars, then they should have detected ~17 transits. However, no transits were observed in the sample of ~34,000 stars. The explanation for this result remains somewhat ambiguous because the crowded stellar environment in 47 Tuc is so different from that of field stars on Doppler planet search projects. The low metallicity of this cluster may have prevented planets from forming in the first place. It is also possible that there were more planets destabilizing dynamical interactions in the cluster environment. Or perhaps photo-ionization of protostellar disks by rapidly evolving high mass stars retarded the formation of giant planets in 47 Tuc.

If high metallicity is an initial condition rather than an accretion signature then one might naively guess that planet mass would scale with metallicity. Although any correlation will be softened because we are using Msin *i* rather than the absolute mass, the tendency seems to go in the opposite sense, with the more massive companions having slightly lower metallicity. Grouping the detected extrasolar planets in arbitrary bins of: Msin $i < 3$ M_{JUP} (44 objects), Msin i between 3 - 8 M_{JUP} (20 objects) and Msin $i > 8$ M_{JUP} (7 objects) we find average metallicities of 0.14, 0.11, 0.005 respectively. Lin & Fischer (2002) suggest that the most massive objects could have formed early on when the disk contained more gas. In that case, accretion triggered by dynamical instabilities might have added material that was not yet gas-depleted enough to enrich the metal content of the convective zone.

The hypothesis that enhanced metallicity is a product of accretion (Lin, Bodenheimer & Richardson 1996, Gonzalez 1997, Laughlin 2000, Murray & Chaboyer 2001) fits intriguingly together with the observation that most of the detected extrasolar planets reside in close-in, eccentric orbits. These orbital characteristics may be fossil evidence for disk evolution that was more violent than that experienced by planets in our own solar system. It is easy to imagine that accretion might have been more important in systems sculpted by strong dynamical interactions. But the puzzle pieces of orbital eccentricity, accretion, and increased metallicity do not yet fit neatly together. Santos et al (2001b) do not find evidence for accretion enhancement of metallicity. Similarly, Pinsonneault, DePoy & Coffee (2001) find that if accretion were responsible for the increased metallicity of planet-hosting stars, then earlier type stars with thinner convective envelopes should show dramatically higher metal abundances than are actually measured.

6. FUTURE CONTRIBUTIONS OF SMALL TELESCOPES

New detection abilities can either add planets to the parameter space demarcated in Figure 1 (i.e., already filled with the current crop of extrasolar planets) or they can break new scientific ground by discovering planets outside of that regime.

About 6 – 8% of main sequence stars have gas giant planets with orbital periods shorter than a few years (i.e., in the K_{30}-P_{1000} domain). Based on that statistic, there are probably many detections that can still be made with Doppler projects that have 5 – 10 m/s precision.

The typical stellar sample for one to three meter class telescopes is limited to stars brighter than V = 7.5. High velocity precision also depends on the presence of narrow spectral lines (generally found in slowly rotating F,G,K,M stars) and stable stellar photospheres (i.e., main sequence, rather than K giant stars). There are 3,126 stars listed in the Hipparcos catalogue that are closer than 100 pc, brighter than V = 7.5 and redder than B – V = 0.5. Most of these bright stars have been on Doppler projects for several years now and in the next few years, this sample of nearby, bright dwarfs should yield a total of ~200 extrasolar planets, or roughly an additional 100 planets in the K_{30}-P_{1000} domain.

Once these planets have been gleaned from the currently observed sample of bright stars it will be difficult for small telescopes to tap into the reservoir of ~7000 fainter dwarfs ($7.5 < V < 9.0$) to find additional planets because the velocity precision for these stars will be limited by the poorer photon statistics. While large telescopes can achieve higher velocity precision on fainter stars, surveying 7000 stars, even with large telescopes, is a staggering project. However, at least one compelling scientific goal for a survey of these fainter stars is the detection of new 51 Peg systems which offer a higher probability of a transit detection and the rare opportunity to follow-up with direct observations of the planet atmosphere. Because only one out of a few thousand stars is expected to show a transit by a 51 Peg like planet, the next planet transit will likely come from this sample of ~7000 fainter stars.

One efficient way to carry out a systematic search for 51 Peg systems is with a distributed observing program. Small telescopes could play an important role by collaboratively screening these ~7000 stars with three or four consecutive observations. Even a modest velocity precision of 10 – 15 m/s would be sufficient to identify the short period 51 Peg candidates. This project would be most efficient if the observing list were cooperatively distributed. Although such large collaborations may be politically difficult, they may be the key to survival for planet searches with small telescopes. After identification, these candidates could be followed up by a large telescope (like Keck) to obtain secure orbital parameters and good ephemeris times.

Breaking new scientific ground with the detection of Neptune-mass planets or solar system analogues will demand Doppler precision to be below 3-5 m/s that are not overwhelmed by long-term systematic errors. This requirement represents a dramatic challenge, particularly for small telescopes with photon-limited velocity precision. Small telescopes will need a new strategy to compete in this regime. In addition to improvements in hardware and software, higher S/N will probably be required to increase velocity precision. Since the telescope aperture is fixed, more frequent observations of smaller, brighter samples may be warranted. This requires a serious commitment from participating Observatories to provide generous time allocations to help compensate for small apertures. In the end, an intense observing strategy, invested in a smaller number of stars will only be successful if there are large numbers of extrasolar planets with velocity amplitudes between 5 and 20 m/s.

7. SUMMARY

With the number of Doppler-detected extrasolar planets approaching 100, there are many more questions than answers. A few questions that have been touched on in this chapter are:

- Six to eight percent of stars have gas giant planets within a few AU of the host star. How often do gas giant planets reside at wider separations? (i.e., are true Jupiter analogs typical or rare?)
- What is the semi-major axis distribution of gas giant planets?
- For planets more massive than 0.5 M_{JUP}, the mass distribution is rising toward the Low Mass end. Does the mass distribution of extrasolar planets continue to rise below 0.5 M_{JUP}?
- Do Neptune-mass planets ever reside in 51 Peg-like orbits? Out to potentially detectable separations (~1 AU), what is the semi-major axis distribution of these objects?
- If eccentricity is a signature of orbital migration, are the gas giant planets at wider separations also on eccentric orbits or do they retain primordial circular orbits?
- Can the gas giant planet architecture of some extrasolar systems be sketched out to find dynamically empty gaps where terrestrial planets could reside?
- How does stellar metallicity fit in?

Small telescopes gave birth to this new field of extrasolar planet detection. There are still thousands of stars to be monitored with time series Doppler observations and with adequate allocation of dedicated telescope time, and a collaborative spirit, small telescopes should continue to play an important role.

8. ACKNOWLEDGEMENTS

I would like to thank Geoff Marcy, Greg Laughlin, Doug Lin and Andy Gould for useful conversations while writing this manuscript.

9. REFERENCES

Butler, R. P., Tinney, C. G, Marcy, G. W., Jones, H. R. A., Penny, A. J., Apps, K. 2001. ApJ 555, 410.

Butler, R. P., Vogt, S. S., Marcy, G. W., Fischer, D. A., Henry, G. W., Apps, K., 2000. ApJ 545, 504.

Butler, R. P., Marcy, G. W., Fischer, D. A., Brown, T. M., Contos, A. R., Korzennik, S.G., Nisenson, P. & Noyes, R.W. 1999. ApJ 526, 916.

Butler, R. P., Marcy, G. W., Vogt, S. S. 1998. PASP 110, 1389.
Butler, R. P., Marcy, G. W., Williams, E., Hauser, H., Shirts, P. 1997. ApJL 474, L115.
Bryden, G., Chen, X., Lin, D. N. C., Nelson, R. P., Papaloizou, J. C. B. 1999. ApJ 514, 344.
Cochran, W. D., Hatzes, A. P., Butler, R. P., Marcy, G. W. 1997. ApJ 483, 457.
Charbonneau, D., Brown, T. M., Noyes, R.W., Gilliland, R. L. 2002. Accepted, ApJ.
Charbonneau, D., Brown, T. M., Latham, D. W., Mayor, M. M. 2000. ApJL 529, 45.
Chiang, E. I., Fischer, D., Thommes, E. 2002. Accepted ApJL.
Fischer, D. A., Marcy, G. W., Butler, R. P., Laughlin, G. P., Vogt, S. S. To appear Jan 10, 2002 ApJ .
Fischer, D. A., Marcy, G. W., Butler, R. P., Vogt, S. S., Frink, S. & Apps, K. 2001. ApJ 551, 1107.
Fischer, D. A., Marcy, G. W., Butler, R. P., Vogt, S. S. & Apps, K. 1999. PASP, 111, 50.
Gilliland, R. L., Brown, T. M., Guhathakurta, P., Sarajedini, A., Milone, E. F., Albrow, M. D., Baliber, N. R., Bruntt, H., Burrows, A., Charbonneau, D., Choi, P., Cochran, W. D., Edmonds, P. D., Frandsen, S., Howell, J. H., Lin, D. N. C., Marcy, G. W., Mayor, M., Naef, D., Sigurdsson, S., Stagg, C. R., Vandenberg, D. A., Vogt, S. S., Williams, M. D. 2001. ApJL 545, 47.
Gizis, J. E., Kirkpatrick, J. D., Burgasser, A., Reid, I. N., Monet, D. G., Liebert, J., Wilson, J. C. 2001. ApJL 551, L163.
Goldreich, P., Tremain, S., 1980. ApJ 241, 425.
Gonzalez, G. 1997, MNRAS, 285, 403.
Halbwachs, J. L., Arenou, F., Mayor, M., Udry, S., Queloz, D. 2000. A&A 355, 581.
Hatzes, A. P., Cochran, W. D., McArthur, B., Baliunas, S. L., Walker, G. A. H., Campbell, B., Irwin, A. W., Yang, S., Kürster, M., Endl, M., Els, S., Butler, R. P., Marcy, G. W., 2000. ApJL 544, 145.
Henry, G. W., Marcy, G. W., Butler, R. P., Vogt, S. S., 2000. ApJL 529, L41.
Korzennik, S., Brown, T., Fischer, D., Nisenson, P., Noyes, R. W. 2000. ApJL 533, L147.
Latham, D.W., Mazeh, T., Stefanik, R.P., Mayor, M., Burki, G. 1989. Nature 339, 38.
Laughlin, G. P. 2000. ApJL 545, 1064.
Laughlin, G. P., Chambers, J. 2001. ApJL 551, 109.
Laughlin, G. P., Adams, F. C., 1999. ApJ 526, 881.
Lee, M. H., Peale, S. J. 2002. Accepted for publication ApJ.
Levison, H. F., Lissauer, J. J., Duncan, M. J., 1998. AJ 116, 1998.
Lin, D. N. C., Bodenheimer, P., Richardson, D. C., Nature, 1996. 380, 606.
Lin & Fischer ApJL submitted.
Lissauer, J. J., Rivera, E. J., 2001. ApJL 554, 1141.
Marcy, G. W., Butler, R. P., Fischer, D. A., Vogt, S. S., Lissauer, J. J., Rivera, E. J. 2001b. ApJ 556, 296.
Marcy, G. W., Butler, R. P., Vogt, S. S., Liu, M. C., Laughlin, G., Apps, K., Graham, J. R., Lloyd, J., Luhman, K. L., Jaywardhana, R. 2001a. ApJ 555, 418.
Marcy, G. W., Butler, R. P. 2000. PASP 112, 137.
Marcy, G. W., Butler, R. P., Vogt, S. S. 2000. ApJL 536,L43.
Marcy, G. W., Butler, R. P., Vogt, S. S., Fischer, D. A., Liu, M. C., 1999. ApJ 520, 239.
Marcy, G. W., Butler, R. P., Vogt, S. S., Fischer, D. A., Lissauer, J. 1998. ApJL 505, L147.
Marcy, G. W. & Butler R. P. 1996,.ApJL 464, L151.

Mayor, M. et al. Apr 4, 2001. http://obswww.unige.ch/~udry/planet/new_planet.html
Mayor, M., Queloz, D., 1995, Nature 378, 355.
Mazeh, T., Naef, D., Torres, G., Latham, D. W., Mayor, M., Beuzit, J., Brown, T. M., Buchhave, L., Burnet, M., Carney, B. W., Charbonneau, D., Drukier, G. A., Laird, J. B., Pepe, F., Perrier, C., Queloz, D., Santos, N. C., Sivan, J.-P., Udry, S., Zucker, S. 2000. ApJL 532, 55.
Murray, N., Chaboyer, B., Arras, P., Hansen, B., Noyes, R. W. 2001. ApJ 555, 801.
Murray, N., Holman, M., 1999. Science 283, 1877.
Murray, N., Paskowitz, M., Holman, M., 2001. astro-ph/0104475.
Naef, D., Mayor, M., Pepe, F., Queloz, D., Santos, N. C., Udry, S., Burnet, M., 2001. A&A 375, 205.
Noyes, R. W., Jha, S., Korzennik, S. G., Krockenberger, M., Nisenson, P., Brown, T. M., Kennelly, E. J., Horner, S. D. 1997. ApJL 487, 195.
Pinsonneault, M. H., DePoy, D. L., Coffee, M. 2001. ApJL 556, 59.
Queloz, D., Mayor, M., Weber, L., Blécha, A., Burnet, M., Confino, B.. Naef, D., Pepe, F., Santos, N., Udry, S. 2000. A&A 354, 99.
Rasio, F. A., Ford, E. B., 1996. Science, 274, 954.
Rivera, E. J., Lissauer, J. J. 2001. ApJ 558, 392.
Santos, N. C., Mayor, M., Naef, D., Pepe, F., Queloz, D., Udry, S., Burnet, M. 2001a, "The CORALIE Survey for Southern Extrasolar Planets VI: New Long Period Giant Planets Around HD 28185 and HD 213240", 2001. A&A 379, 999.
Santos, N. C., Israelian, G., Mayor, M. 2001b. A&A 373, 1019.
Tinney, C. G., Butler, R. P., Marcy, G. W., Jones, H. R. A., Penny, A. J., Vogt, S. S., Apps, K. Henry, G. W. 2001. ApJ 551, 507.
Udry, S., Mayor, M., Queloz, D. To appear in: Planetary Systems in the Universe: Observations, Formation and Evolution. ASP Conference Series, ed. A.J. Penny, P. Artymowicz, A.M. Lagrange, and S.S. Russell. 2001b.
Udry, S., Mayor, M., Naef, D., Pepe, F., Queloz, D., Santos, N. C., Burnet, M., Confino, B., Melo, C. 2000a. A&A 356, 590.
Vogt, S. S., Marcy, G. W., Butler, R. P., Apps, K. 2000. ApJ 536, 902.
Vogt, S.S., Butler, R.P., Marcy, G.W., Fischer, D.A., Pourbaix, D., Apps, K., Laughlin, G.P. 2001. ApJ accepted for publication.
Weidenschilling, Mazori, 1996.
Zucker, S., Naef, D., Latham, D. W., Mayor, M., Mazeh, T., Beuzit, J. L., Drukier, G. A., Perrier-Bellet, C., Queloz, D., Sivan, J. P., Torres, G., Udry, S. "A Planet in the Stellar Triple System HD 178911" 2001. Submitted to ApJ.

Chapter 13

Variable Star Research with Small Telescopes

László Szabados
Konkoly Observatory of the Hungarian Academy of Sciences
Budapest, HUNGARY

Abstract: The role of small telescopes is decisive in the research of variable stars. In order to reach a scientific level comparable with that characteristic of large telescopes, the small diameter is usually compensated by concerted long-term observational effort. This review gives examples, not a comprehensive list, of how photometry of variable stars can contribute to progress in astrophysics, and also contains a detailed account on photometry of a particular type of variable stars—Cepheids—with the aim of showing versatility in using brightness measurements obtained with small telescopes.

Key words: astrophysics, binary stars, photometry, stellar evolution, variable stars

1. INTRODUCTION

Variable stars are fundamental objects in several disciplines of astronomy, e.g. in the study of stellar structure and evolution, Galactic structure and cosmic distance scale. Small telescopes are indispensable and widely distributed in observing variable stars. For an observer, stars are external astrophysical laboratories. Photometric variability is a common phenomenon in various stages of stellar evolution, and it is not an exaggeration to state that each star is a variable star.

Therefore it is more appropriate to refer to variability episodes in stars instead of the traditional term "variable stars". Nevertheless, the usage of the old term is supported by the fact that physical processes that normally occur in stars (e.g. convection, stellar wind) result in non-episodic photometric variations of low amplitude. In the 1990's the precision of the photometric measurements reached the millimagnitude level, while in the first decade of the new millennium, a micromagnitude precision is within reach, although from orbiting telescopes, not from the ground yet.

T.D. Oswalt (ed.), The Future of Small Telescopes in the New Millennium, Vol. III, 207-223.

Due to the ever-deepening knowledge about stellar structure and evolution, new phenomena previously unobserved are predicted, and the newly discovered (or to be revealed in the future) subtle phenomena initiate theoretical studies leading to their explanation and better understanding of stellar structure. This means that the interplay between theory and observation is fruitful in both directions. Here I mention only a few typical examples of such progress.

Pulsational calculations showed that helium atmosphere white dwarfs can also oscillate, in addition to the earlier recognized pulsation of the hydrogen atmosphere white dwarfs. Since then pulsation of DB white dwarfs has been discovered, although the number of the known DB-type pulsators is smaller than 10. Pulsation of the subdwarf B-stars was also pointed out theoretically, and the first bona fide sdB pulsators (EC 14026 stars) have been discovered subsequently (Kilkenny et al., 1997). More recently Dorfi and Gautschy (2000) concluded that stars more massive than 20 $M_\odot$ can also pulsate. Discovery of such long period, low amplitude pulsators is a task for the near future. Due to their large mass, these stars are luminous, therefore easy observational targets for small photometric telescopes even if they are not found in the solar neighborhood.

There is a wide variety of examples how observations of variable stars contribute to the development of the theory of stellar structure and evolution. A typical case is the Blazhko effect of the RR Lyrae variables. This effect has been known for a century but the proper explanation for the amplitude and phase modulation is still missing. The two recent concurrent theories are the oblique magnetic rotator model (Shibahashi, 2000) and the resonant mode-interaction model (Van Hoolst, 2000). Each of these two hypotheses has ample observational support from photometric data obtained with small telescopes. The oblique rotator model together with the several-year long cycles also observed in Blazhko-effect modulated RR Lyraes lead to solar cycle analogies occurring in such stars. The other possibility, i.e. the resonant mode interaction is similarly promising. Here a non-radial mode is excited resonantly in the otherwise radially pulsating RR Lyrae star. Continued long datasets will help decide which explanation is the correct one. In any case, detailed observation of the Blazhko effect is the best method to get information about the interior of Population II giant stars.

The specific advantage of the small telescopes is that they are not oversubscribed, therefore long data series can be obtained due to the practically unlimited observational time. In variable star research all phenomena that involve a long time-base are studied with small telescopes.

In what follows, we summarize what kind of studies can be based on extensive photometric data of variable stars highlighting the most promising research directions.

2. TYPICAL NEW OBSERVATIONAL RESULTS

An obvious aim, irrespective of any particular type of variability, is to find as many variable stars as possible. In this way, not only a statistically significant sample can be generated but these discoveries are essential for studying the Galactic distribution of each type of variable stars which bears information on the star formation history in the Milky Way galaxy. Moreover, if all variables become known to a given (desirable) magnitude limit, it is relatively easy to find the most curious objects among them. Then, these strangely behaving stars can be followed regularly or upon necessity. Well-known such objects are the rapidly evolving FG Sagittae (Jurcsik and Montesinos, 1999) and V4334 Sagittarii (Duerbeck et al., 2000).

The complete census of variable stars may result in discovering new types of stellar variability. The most recently introduced new types are the γ Doradus variables (Kaye et al., 1999), the red giants pulsating semi-regularly and with short period, and the eclipsing variables in which the star is occulted by its planet (Kazarovets et al., 2001).

The GDOR stars are found at the red edge of the δ Scuti instability strip in the Hertzsprung-Russell diagram (HR-diagram) and they are pulsating in gravity modes. For asteroseismologists, these modes give information on the deep layers of the star, including the stellar core. Among the millions of frequencies of solar oscillations, not a single g-mode has been identified yet. The first GDOR variable was revealed in 1992. Since that time, a dozen or so variables belonging to this type have been discovered and probably many more such stars are to be found.

The rapidly oscillating M giant stars (their GCVS-designation is SRS) were discovered from the photometric data obtained during the Hipparcos astrometric mission (ESA, 1997). Because of the brightness limit of the Hipparcos project, this means that all known representatives of this type are sufficiently bright. The short ($P < 10$ days) period implies that these stars pulsate in a very high overtone mode (Koen and Laney, 2000). Discovery and subsequent photometry of even fainter SRS stars is possible with small telescopes.

The astrophysical importance of the occultation of a star by its planet is that this transit gives an opportunity to follow the limb darkening across the disk of the star with good resolution. The prototype and the only known representative of such variables is V376 Pegasi (= HD 209458).

The rapidly growing number of new variables will certainly lead to discovery of still unknown types of stellar variability. The unique object 15D in the young open cluster NGC 2264 is such an example. The three magnitude periodic dips in the brightness of the star may be caused by a transit of a feature in the circumstellar disk in front of the star (Kearn and Herbst, 1998). Now the patrol of the whole sky with automated telescopes

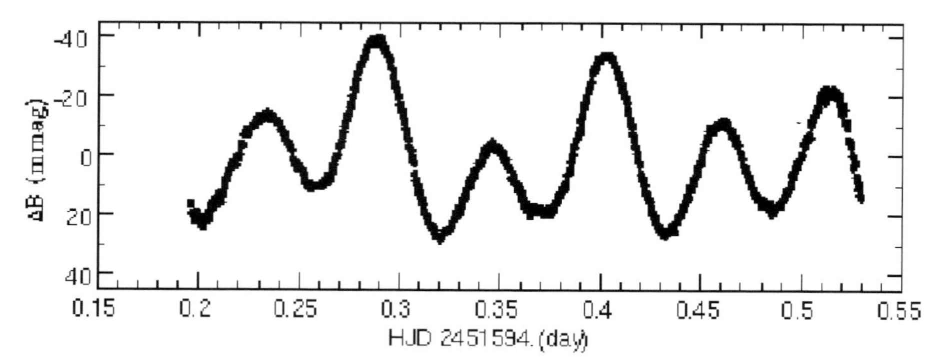

Figure 1. A part of a superb light curve showing the level of precision that can be achieved with ground-based telescopes on a good site. These observations of HD 98851 were obtained by Joshi et al. (2000).

became a routine but the continuous monitoring of individual stars with high photometric precision necessitates dedicated telescopes.

The increasing precision facilitates detection of microvariability in stars and very subtle light-curve features can be followed. Figure 1 shows a state-of-the-art photometric dataset obtained with a 1-m telescope. The target, HD 98851 shows a peak-to-peak variation of only 0.04 magnitude on a time-scale of about 80 minutes (Joshi et al., 2000).

Such high precision is indispensable if one wants to study stability of light curves and repeatability of subtle phenomena superimposed on the principal variation. For example, long period, Population II Cepheids (W Virginis type variables) are suspected in having light curves that vary from cycle-to-cycle but no sufficiently precise light curve covering several cycles has ever been obtained on any of them to study this behavior from observational point of view. Regularities can be found among seemingly irregular data and processes. The hydrodynamical models predict presence of low-dimensional chaos in the variations of W Virginis stars (Kolláth, 1998 and references therein).

The repetitive phenomena of the light curves, or more precisely, deviations from strict periodicity can be utilized to reveal stellar evolution on human time scale, as well as tiny variations in stellar properties that cannot be detected directly. Irregularities in the period (or cyclicity) can best be followed with the help of the O-C diagram. A detailed and up-to-date description of the method is given by Sterken (2000).

For singly periodic pulsating variables the monotonously changing period may refer to internal changes in the star, i.e. stellar evolution. In the case of classical Cepheids, the observed period changes correspond to values expected from evolutionary calculations. The interpretation of the observed period variations is, however, not always simple. Some examples based on Cepheids will be mentioned in the next section. Here I refer to the strangest case, which contradicts the general belief that secular change in the period is simply the result of restructuring in stellar interior, i.e. a consequence of stellar evolution. In the case of double-mode pulsators, structural changes in the star obviously have to cause simultaneous shortening or lengthening of the period for both excited modes. The RR Lyrae type variable star V53 in the globular cluster M15 pulsates both in the fundamental mode and the first overtone, and the period of the two excited modes vary in opposite sense (Paparó et al., 1998). This is a challenge for theoreticians and an appeal for observers: it is important to monitor other double-mode pulsators in order to study the general behavior of their period changes.

Double-mode pulsation is confined to very narrow regions in the HR-diagram, therefore the duration of the pulsation with two simultaneous periodicities is shorter than that of single-mode pulsation and the number of double-mode pulsators is small. The seventh double-mode RR Lyrae star in the Milky Way field, GSC 5002-0629, was just discovered (Garcia-Melendo et al., 2001), and this is the brightest RRd star in our galaxy to date which also shows how incomplete our census of variable stars is. It would be highly interesting to observe how a second period emerges or decays, or to reveal any sign of mode switching. This phenomenon would cause variability in the amplitude of the respective harmonics due to the mode energy change. It is, however, more difficult to point out any amplitude variability than to follow period changes.

There are many types of variable stars whose variability is induced by the sheer existence of a companion. All cataclysmic variables (novae, dwarf novae, symbiotic variables, etc.) are binary systems in which brightness variations are the results of interactions between the two components. Binaries are, however, found among the pulsating stars, too. A nearby companion may have an influence on the pulsation, especially on the modes that do not penetrate very deep in the star, i.e. on the p-mode oscillations. This kind of pulsation is confined to the subsurface regions of the star. The presence of a companion destroys the spherical symmetry of the oscillating star, and finding observational manifestation of such forced non-radial modes is a promising task that can be performed with small telescopes. There is one more fact why the observation of binary systems among pulsating stars is especially important: in several cases it is possible to determine the orbital elements of the system, and as a consequence, the most important physical properties, e.g. mass and radius of the components.

Depending on the orientation of the orbital plane and the separation between the components, even the orbital motion itself can be pointed out by simple photometry. The intrinsic periodicity of the variable star becomes modulated in phase because the star revolves around the common centre of mass of the binary. This light-time effect is caused by the finite speed of light and the O-C method is sufficiently sensitive to point out its occurrence if the brightness variations are covered well in time. The light-time effect can be discovered from modulation of any periodic phenomenon. Therefore, in addition to pulsating variables, it can be revealed in eclipsing binaries as well, if a distant third component belongs to the system and its orbital plane is oriented suitably.

Incidence of duplicity has not been studied for all types of pulsating variables. Cepheids are the most thoroughly investigated in this respect: the frequency of Cepheids with companions exceeds 50 per cent (Szabados, 1995). Implications of this frequent occurrence of binaries will be mentioned in the next section. One should keep in mind, however, that this value is not curiously large: the well-investigated region of the solar neighborhood shows that binarity is common among stars. In addition to Cepheids, all other types of pulsating variables have to be studied thoroughly from the point of view of binarity. There are only occasional studies on this important topic and the latest summary (Szatmáry, 1990) is obsolete. Though a companion can be readily pointed out by spectroscopy, the light-time effect based on pure photometric data is capable for pointing out faint companions whose presence is hardly noticeable in the spectrum of the variable star.

A remarkable example for the thoroughly studied light-time effect is the O-C diagram of SZ Lyncis (Paparó et al., 1988). As it is clearly seen in Figure 2, a wave-like pattern due to the light-time effect is superimposed on

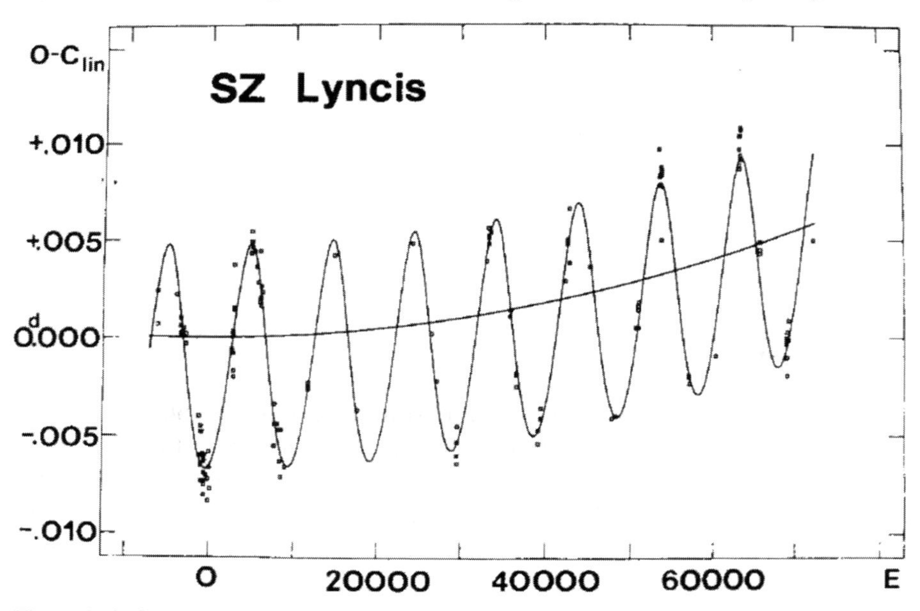

Figure 2. O-C diagram of SZ Lynxis showing clear-cut light-time effect (Paparó et al. 1988).

a parabola which testifies that the pulsation period of the bright, high amplitude δ Scuti type variable has been increasing. Whether this secular period change reflects stellar evolution or has to do with the presence of the companion is not clear yet. This question can be answered when such informative O-C diagrams of many pulsating stars belonging to binary systems are available.

The light-time effect interpretation of the O-C diagrams is justified when the cyclic pattern repeats many times. A satisfactory coverage needs regular photometric observations of the target stars, which means time consuming photometric projects for small telescopes.

As to the eclipsing binaries, if the light-time effect approximation of the O-C graph based on the moments of minimum brightness proves to be correct, there are a number of cases where a third star may be gravitationally bound to the eclipsing pair (Borkovits and Hegedüs, 1996). Moreover, the O-C diagram of eclipsing binaries can be used to detect apsidal motion of the system, either of relativistic origin or due to the non-spherical shape and envelope structure of the components. The apsidal motion also shows up as a long period modulation superimposed on the normal trend of O-C residuals. An additional feature characteristic of binary systems is that matter can move from one star to the other, depending on the size, evolutionary status and separation of the components. This mass transfer may alter the orbital period of the system according to the law of momentum conservation. This kind of period change can be considered as an observational proof of stellar evolution in binary systems. Again, simple photometric time series can lead to in-depth astrophysical studies.

An important topic of contemporary observational astrophysics is the study of stellar activity. Our Sun would be a photometric variable from a distance on various time-scales because of solar activity. As a matter of fact, more than one variability type can be assigned to the Sun: the nearest star is a pulsating, an eruptive and a rotating (spotted) variable simultaneously. Thorough studies of stars showing analogous behavior in any respect help understand the details that have not yet been elaborated concerning the mechanism what is going on in the Sun.

Activity cycles have been found in many solar type stars and it is crucial that reduced activity similar to Maunder minimum was also found. This topic is treated at length elsewhere in this volume (see the chapter written by Cristopher Johns-Krull). Here I only mention some less direct connections. Very strong flares or flashes were incidentally observed on seemingly normal stars (Schaefer, 1989), and superflares occur on ordinary solar-type stars (Schaefer et al., 2000). The Blazhko effect mentioned earlier is also explained as a consequence of an oblique magnetic rotator, and several-year-long cycles have been observed on the F giant RR Lyrae stars showing Blazhko effect. This means that analogues of solar cycle are not confined to solar type stars.

Presence of magnetic field is insufficient for developing a solar type cycle. The rapidly oscillating peculiar A stars (roAp stars) have a slightly higher effective temperature as compared with the RR Lyrae variables, and in spite of their magnetic field, no activity cycles seem to emerge. The roAp stars are also oblique pulsators, which are unique in the sense that the three axes which can be defined in these stars, viz. magnetic, rotation and pulsation axes are all inclined to each other (see Kurtz, 2001). The varying aspect angle offers more information on the non-radial modes excited in these stars. Observation of roAp stars requires high-speed photometry because the range of their oscillation periods is between 6 and 16 minutes. The photometric study of the roAp stars is a real challenge for the observers because the brightness variations are not only rapid, they are also tiny: peak-to-peak amplitudes of about 0.01 magnitudes or smaller are to be detected.

Although the roAp stars populate practically the same region of the classical instability strip as the δ Scuti stars, the driving mechanism of their pulsation is different. In roAp stars it is the partially ionized H I layer that maintains the oscillations, while in δ Scuti stars a deeper layer where singly ionized helium (He II) loses its second electron drives the pulsation.

In view of the fact that many types of pulsating variables were defined recently, δ Scuti stars are now considered as classical pulsators, similarly to Cepheids or RR Lyrae type variables. A particular type of variable stars usually corresponds to a definite phase of stellar evolution. Quite recently, however, the group of δ Scuti pulsators was supplemented with their younger counterparts, when the pulsation of pre-main sequence (PMS) stars in the instability strip was discovered. There are only nine known representatives of these youngest pulsators (see e.g. Kurtz and Müller, 2001 and references therein), therefore the coolest Herbig Ae/Be stars deserve a careful photometric check whether they show periodic brightness variations due to pulsation, superimposed on the erratic fluctuation caused by the variable transparency of the circumstellar dust. The theoretical calculations (Breger and Pamyatnykh, 1998) indicate a 10—100 times faster evolutionary period change for the PMS-pulsators than for their post-main sequence counterparts, so the youngest δ Scuti stars deserve regular observations to check this theoretical prediction within relatively short time.

According to the current observational strategy, people involved in the research of pre-main sequence variables usually obtain one or two photometric points in a single night. If gaps in such data sets are reasonably short, the long-term behavior can be ideally followed and periodicities longer than 1-2 days can be revealed. But short period (including δ Scuti type) brightness variations can only be discovered if data sampling is much more frequent. Small telescopes and patient observers are ideal for performing a continuous coverage of PMS-variables during several (not necessarily consecutive) nights.

The other extremum among the δ Scuti stars was also discovered quite recently (Koen et al., 2001). CP Octantis turned out to be an evolved Ap star and a δ Scuti type variable simultaneously. The study of periodicities in its oscillation and the mode identification are crucial for testing the relevant pulsation models.

The conventional strategy of observing pre-main sequence variables has been very productive, too. From several-year-long data sets it is possible to determine the rotation period of T Tauri stars. Owing to the unequal surface brightness of these stars, axial rotation appears as a periodicity with a cycle length of 1-20 days. It is, however, not easy to point out any periodicity in the brightness of these PMS stars because the low amplitude periodic brightness variations are superimposed on the persistent erratic variability inherent in such young stars.

A major photometric survey of T Tauri stars can lead to sound astrophysical results. Herbst (2001) demonstrates how the mass dependence of the stellar rotation rates can be determined from the extensive photometric series and even the stellar radii can be estimated.

Figure 3 is a light curve of the Be star 6 Cephei (V382 Cep) covering more than a decade. It shows that an uneventful, boring part in the history

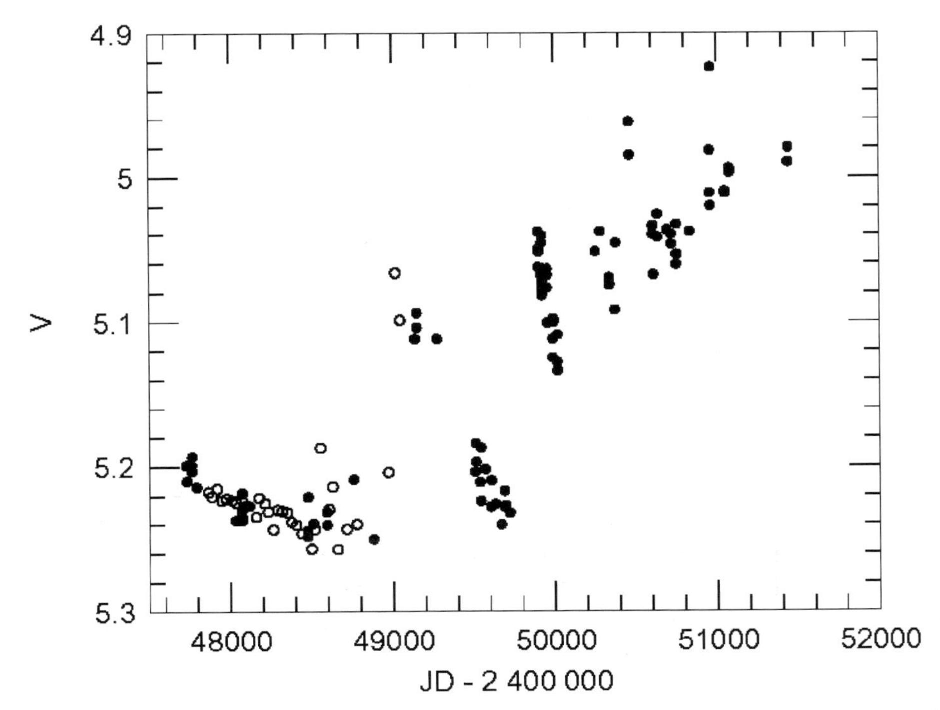

Figure 3. Light variation of the Be star 6 Cephei. The open circles are daily means of photometric data obtained during the Hipparcos project (ESA, 1997); filled circles are nightly means determined from unpublished data by Kun and Szabados.

of brightness variability of a star does not mean that observers should forget about that object.

To sum up this section, the realm of variable stars has been and will be explored by small telescopes. According to the common practice, it is in vain to apply for observation time on large telescopes for long term or continuous study of variable stars. For such time consuming tasks the main equipment is the small telescopes. They are easily accessible because small telescopes are usually available in each observatory.

3. NEW ISSUES CONCERNING VARIABILITY OF CEPHEIDS

In the previous section we gleaned the field of variable star research from the point of view of photometry, in order to exemplify that promising observational topics can be set for any type of stellar variability. That summary was inevitably superficial. In this section richness of the variable star photometry is demonstrated from another viewpoint: an arbitrarily chosen type—classical Cepheids—is investigated in depth. The conclusion will be the same: the research possibilities with a limited size telescope are practically unlimited.

Cepheids are important objects in both astrophysics and cosmology. Regularity of pulsation gives rise to various relations between the phenomenological and physical properties of these variables, and properly calibrated relationships are amply used in both disciplines. This means that fundamental stellar properties including mass and radius can be determined from simple photometric observations, and even the distance of the star can be deduced from the period-luminosity relationship.

Regularity is, however, not absolute: deviations from the perfect repetitiveness may occur, as well as subtle differences from star to star. Study of the individual behavior of each variable puts the observers in the possession of extra astrophysical information. Moreover, Cepheids are easy targets for research with small telescopes because they are supergiant stars and even if they are distant, they can be brighter than 11-12th magnitudes in optical bands (e.g. through Johnson B or V filters). Depending on the pulsation period of the Cepheid (the extreme values are 1.5 and 80 days in the Milky Way galaxy) a complete light curve can be obtained within a reasonably short time, in any case during one observing season. Below I briefly summarize how the light curves can be utilized for extracting the astronomical information.

Comparing the new light curve with the earlier ones, changes in the period and/or the light curve itself can be established. In the previous

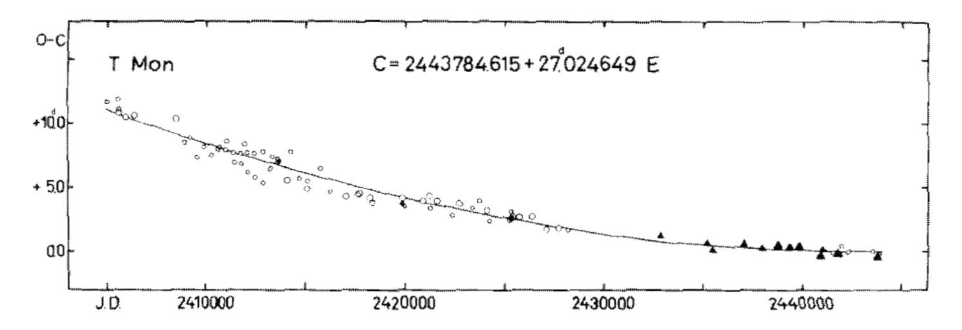

Figure 4. The monotonous period increase of T Monocerotis, due to stellar evolution. Subtracting the parabolic trend, the precise timings of brightness maxima reveal more subtle intrinsic period variations.

section I already mentioned that period changes of evolutionary origin have been pointed out for Cepheids (Fernie, 1984, Szabados, 1983). A typical parabolic O-C graph testifying continuous slow period variation is shown in Figure 4.

It is well known that larger mass stars evolve faster, and this fact has been confirmed by the period changes of Cepheids. According to the period-luminosity relationship, longer period Cepheids are more luminous and because evolutionary trajectories cross the instability strip nearly horizontally, higher luminosity corresponds to larger mass.

The parabolic pattern of the O-C graph is, however, not exclusive among the Cepheids. According to Fernie (1990), the pulsation period does not necessarily change in a linear fashion even if its only cause is stellar evolution. The O-C residuals of Y Ophiuchi can be best approximated by a cubic polynomial, yet this shape has been explained in terms of the evolution of the Cepheid.

Other features are also present in the O-C diagrams of Cepheids. In the case of the longest period Galactic Cepheids, cyclic variations are superimposed on the secular increase or decrease of the period (Berdnikov, 1994). Another interesting and still unexplained kind of period change appears at Cepheids. The manifestation of this phenomenon is that the Cepheid pulsates with two alternating periods, each value being valid for several years or decades. When the "second" period is in operation only for a very brief interval, the jump to the new value and the subsequent rejump to the original pulsation period looks as a phase jump in the O-C diagram. Figure 5 shows a typical example of the alternating periodicity. All Cepheids that show this phenomenon belong to binary systems. When studying incidence of this feature, no preferred value of the pulsation period was found. Clearly, regular photometric monitoring is necessary to study this subtle phenomenon in detail.

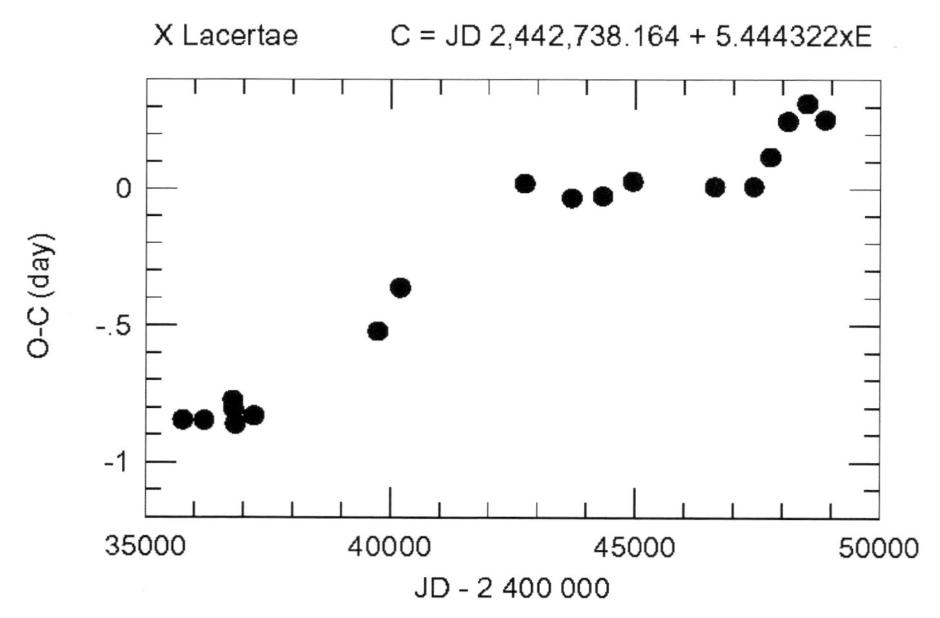

Figure 5. The O-C diagram of the Cepheid X Lacertae, showing two alternating values of the pulsation period.

Similarly, very precise and regular timing of brightness variation is necessary to reveal the light-time effect occurring at Cepheids belonging to binary systems. Since the Cepheids are supergiants, the shortest possible orbital period for a pair consisting of a Cepheid and a lower mass main sequence star is about one year. For such a short period system, however, the amplitude of the light-time effect would remain undetectable. The most clearly documented case of the light-time effect among Cepheids is that of AW Persei (Vinkó, 1993), the orbital period of this system being about 40 years.

In view of the high incidence of binaries among Cepheids, discovery of light-time effect is expected in many more cases. There are only two prerequisites for the success: the orbital inclination should not be too close to zero (an objective requirement) and regular photometric monitoring is necessary to show up the wave-like pattern of the O-C residuals. Even the mass of the components can be estimated with using the light-time effect, independently of the mass determinations based on the pulsational behavior of the Cepheid.

In addition to the (either intrinsic or apparent) period changes, the other kind of deviation from the perfect regularity is variability of the light curve. Double-mode pulsation obviously results in light curves of varying shape but there are intriguing examples for single-mode Cepheids showing secularly changing light curve, as well.

In the era of the millimagnitude precision photometry, it is not surprising that amplitude variation is found among Cepheids. It is extraordinary, however, that these stars seemingly do not have any sound reason for such behavior. A secular decrease in the amplitude of light variation is expected when the Cepheid is about leaving the instability strip. But none of the Cepheids subjected to slow amplitude decrease is near the border of the instability region.

Up to now three Cepheids have been found to pulsate with secularly decreasing amplitude: two in the Milky Way system (Polaris and Y Ophiuchi) and an extragalactic one (Hubble's Variable 19 in M33—Macri et al., 2001). The declining photometric and radial velocity amplitude of Polaris was revealed by Arellano Ferro (1983). According to the rate of amplitude diminution, complete cessation of pulsation was expected to occur in mid-nineties. The latest data, however, revealed that the decrease in amplitude had abruptly stopped and Polaris still pulsates with a peak-to-peak amplitude of 0.03 magnitude in V (Kamper and Fernie, 1998).

Even smaller amplitude Cepheids can exist. The first and only known such case, the Cepheid-like radial velocity variation of γ Cygni was discovered nearly ten years ago (Butler, 1992). The corresponding brightness variation was too small to be detected at that time but such discovery is possible with the state-of-the-art photometers. One cannot be much surprised if millimag-amplitude pulsation turns out to be a commonplace among Cepheids, too.

Returning to the normal amplitude Cepheids, the double-mode pulsators should be kept on the observing list. There are 20 known beat Cepheids in our Galaxy, all observable with small telescopes. It would be an interesting result to point out the energy change between the two excited modes, i.e. the secularly increasing pulsation amplitude of the surviving mode and the diminution of the other.

Another peculiar light-curve variation is the case of V473 Lyrae. The pulsation of this Cepheid is strongly modulated with a cycle length of 3.5 years (Burki et al., 1986). Since the pulsation period of V473 Lyrae is only 1.5 days (the shortest known value for a Galactic Cepheid), this modulation cannot be a beating of two radial modes. The ratio of the maximum and minimum amplitudes for both brightness and radial velocity variations is about ten, so this is not a Blazhko-type phenomenon, either. It is suspected that this strange behavior is caused by the resonant interaction between the second (or higher) radial overtone and a pulsationally stable non-radial mode (Van Hoolst and Waelkens, 1995). Photometric data can confirm this model, if the modulation turns out to be strictly periodic. Stability of the length of the modulation cycle has not been investigated yet.

Even if there is no noticeable variation in the pulsation amplitude, which is the case for the overwhelming majority of Cepheids, the light curve itself can be very informative. It is known that the shape of the light curve—quantified in terms of the Fourier parameters—depends on various properties of the stars; at least for RR Lyrae variables, a dependence on metallicity has been established (Jurcsik and Kovács, 1996). If such relationship also exists between the shape of the light curve and the [Fe/H] value for the classical Cepheids, pure photometric data can give information on the chemical evolution in different regions of the Milky Way. The absence of this relation is mainly due to the insufficient knowledge of the composition of Cepheid atmospheres. Seemingly, Cepheids do not belong to favorite targets of spectroscopic studies, although their spectra can be decisive in many respects.

A Cepheid can enter the instability strip several times at ever increasing luminosities. The first crossing occurs when the star leaves the main sequence and evolves into a red giant. During the subsequent evolutionary episodes, the internal changes in the red giant cause the star to perform blue loops that may enter or fully cross the instability strip. A blueward motion in the HR-diagram implies secular decrease in the pulsation period, while during the redward crossing the period is continuously increasing. Accordingly, the O-C diagram is suitable to differentiate between the direction of evolution, i.e. between the odd and even serial numbers of crossing. From the period changes alone, it is impossible to decide whether the odd crossing is the first, third, or fifth, just like whether the even crossing is the second or fourth one. During the red giant phase of evolution, however, heavy elements formed deep in the star dredge up. Therefore, in principle, it is possible to decide whether the actual chemical composition of the Cepheid atmosphere corresponds to the first, or second-third, or fourth-fifth crossings. Based on the long-term period change and the heavy element content, the crossing number can be unambiguously assigned to the Cepheid. Knowledge of the crossing number then sets the luminosity of the given Cepheid. The determination of the crossing number would be especially important for those Cepheids that serve as calibrating stars of the period-luminosity (PL) relationship. A part of the existing scatter in this relationship is explained by the fact that Cepheids of all crossings are intermingled in a unique PL-relation.

Another cause of the scatter in the PL-relation is that the photometric effect of the companion stars is not taken into account, partly because of negligence and insufficient information. The majority of Cepheids belong to binary systems (Szabados, 1995), and the contribution of a B or A type companion to the brightness of the system has to be taken into consideration when determining the apparent brightness of the Cepheid. Similarly, an unresolved blue companion alters the observable color indices, thus a false reddening correction can be obtained, if the effect of the companion is not removed (Szabados, 1997). When the goal is to calibrate the Cepheid PL-relationship as precisely as possible, all these subtle photometric effects have to be corrected for.

4. SPACE PROJECTS FOR VARIABLE STAR PHOTOMETRY

Significance of small space telescopes has never been doubted. It is remarkable that following the decade of great space observatories (the Hubble, Compton, and Chandra observatories, and the infrared space telescope to be launched), the near future will be the era of space projects involving both large and small telescopes. This is a further proof that small telescopes are also adequate to achieve top quality results, so it is worth to keep them in operation, especially on the ground where their maintenance is cheaper.

Without entering into the details and omitting the exact references, here I only remind of the main steps that demonstrate how photometry of variable stars became a space discipline.

The International Ultraviolet Explorer (1978-1996) was the first orbiting observatory whose fine error sensor (FES) was used not only for pointing and attitude control, but for obtaining photometric time series of individual stars, too. Since the FES was not constructed for such scientific purpose, the occasionally obtained sample light curves only demonstrated capability of the equipment, possibility of photometry from space and its advantages as compared with the ground-based photometry. Above the terrestrial atmosphere the length of the time series is not limited, and getting rid of seeing and transparency changes and time-dependent extinction, the quality of data is superior.

Another major progress in space photometry of variable stars was due to the Hipparcos astrometric satellite (1987-1991). This largest homogeneous sample of stellar brightness data is based on measurements carried out with a 29-cm mirror (by the way, the telescope of the IUE was of 45-cm diameter). More than 8000 stars turned out to be new variables among the targets of Hipparcos, and because of the magnitude limit of the pre-selected programme stars, all these new variables are bright.

A further big step for variable star research happened when the scientific program of the WIRE (Wide-Field Infrared Explorer) had to be modified soon after its launch in 1999. From an infrared observatory originally aimed at studying starburst galaxies, WIRE became an asteroseismological station because of the inevitable problems with the cryogenic cooling system. Since that time photometric observations of variable stars have been carried out in the visual region, and long time series of unprecedented accuracy have been obtained. This necessary change also testifies usefulness of a small (WIRE is equipped with a 30-cm mirror) orbiting telescope for photometric studies of variable stars.

As to the immediate future, there are three space projects for variable star photometry in the process of realization: the Canadian MOST (2003, 15-cm), the Danish MONS (2003, 34-cm), and the French COROT (2004, 27-cm). The year of the planned launch and the diameter of the telescope on board are given in brackets. The limited lifetime (2-3 years) of these space missions is compensated by the extreme photometric precision.

The number of the targets deserving regular or occasional photometric monitoring is so large that capacity of these telescopes would be exceeded by several orders of magnitude, if all interesting variable stars would be included in their program. Therefore these photometric projects do not jeopardize the future of ground-based photometry of variable stars. Instead, existence of these space projects also emphasizes that photometry with small telescopes is the cheapest means to achieve important results in astronomy.

5. ACKNOWLEDGMENTS

The author is indebted to Dr. Mária Kun for her comments on the manuscript. Financial support from the Hungarian OTKA grants T029013 and T034584 are gratefully acknowledged.

6. REFERENCES

Arellano Ferro A. 1983, ApJ 274, 755
Berdnikov L. N. 1994, Astr. L., 20, 232
Borkovits T. & Hegedüs T. 1996, Astron. Astrophys. Suppl., 120, 63
Breger M. & Pamyatnykh A. 1998, A&A 332, 958
Burki G., Schmidt E. G., Arellano Ferro A., Fernie J. D., Sasselov D., Simon N. R., Percy J. R., & Szabados L. 1986, A&A 168, 139
Butler R. P. 1992, ApJ 394, L25
Dorfi E. A. & Gautschy A. 2000, ApJ 545, 982
Duerbeck H. W., Liller W., Sterken C., Benetti S., van Genderen A. H., Arts J., Kurk J. D., Janson M., Voskes T., Brogt E., Arentoft T., van der Meer A., & Dijkstra R. 2000, AJ 119, 2360
ESA 1997, The Hipparcos and Tycho Catalogue, ESA SP-1200
Fernie J. D. 1984, in IAU Symp. 105, Observational Tests of Stellar Evolution Theory, ed. A. Maeder & A. Renzini (Dordrecht: Reidel), 441
Fernie J. D. 1990, PASP 102, 905
Garcia-Melendo E., Henden A. A., & Gomez-Forrellad J. M. 2001, Inf. Bull. Var. Stars, No. 5167
Herbst W. 2001, in ASP Conf. Ser. "Small-Telescope Astronomy on Global Scale", Proc IAU Coll. 183, ed. W.-P. Chen, C. Lemme, & B. Paczynski (San Francisco: ASP), 177
Joshi S., Girish V., Martinez P., Sagar R., Ashoka B. N., Gupta S. K., Seetha S., Kurtz D. W. & Chaubey U. S. 2000, Inf. Bull. Var. Stars, No. 4900
Jurcsik J. & Montesinos B. 1999, New Astr. Rev., 43, 415
Jurcsik J. & Kovács G. 1996, A&A 312, 111

Kamper K. W. & Fernie J. D. 1998, AJ 116, 936
Kaye A. B., Handler G., Krisciunas K., Poretti E., & Zerbi F. 1999, PASP 111, 840
Kazarovets E. V., Samus N. N., & Durlevich O. V. 2001, Inf. Bull. Var. Stars, No. 5135
Kearn K. S. & Herbst W. 1998, AJ 116, 261
Kilkenny D., Koen C., O'Donoghue D., & Stobie R. S. 1997, MNRAS 285, 640
Koen C., & Laney D. 2000, MNRAS 311, 636
Koen C., Kurtz D. W., Gray R. O., Kilkenny D., Handler G., Van Wyk F., Marang F., & Winkler H. 2001, MNRAS 326, 387
Kolláth Z. 1998, in Pulsating Stars - Recent Developments in Theory and Observation, ed. M. Takeuti & D. D. Sasselov (Tokyo: Universal Academy Press), 183
Kurtz D. W. 2001, in ASP Conf. Ser., "Small-Telescope Astronomy on Global Scale", Proc. IAU Coll. 183, ed. W.-P. Chen, C. Lemme, & B. Paczynski (San Francisco: ASP), 187
Kurtz D. W. & Müller, M. 2001, MNRAS 325, 1341
Macri L. M., Sasselov D. D., & Stanek K. Z. 2001, ApJ 550, L159
Paparó M., Szeidl B., & Mahdy H. A. 1988, ApSS 149, 73
Paparó M., Saad S. M., Szeidl B., Kolláth Z., Abu Elazm M. S., & Sharaf M. A. 1998, A&A 332, 102
Schaefer B. E. 1989, ApJ 337, 927
Schaefer B. E., King J. R., & Deliyannis C. P. 2000, ApJ 529, 1026
Shibahashi H. 2000, in ASP Conf. Ser. Vol. 203, The Impact of Large-Scale Surveys on Pulsating Star Research, ed. L. Szabados & D. W. Kurtz (San Francisco: ASP), 299
Sterken C. 2000, in Variable Stars as Essential Astrophysical Tools, ed. C. Ibanoglu, Kluwer, p. 529
Szabados L. 1983, ApSS 96, 185
Szabados L. 1995, in ASP Conf. Ser. Vol. 83, Astrophysical Application of Stellar Pulsation, Proc. IAU Coll. No. 155, ed. R. S. Stobie & P. A. Whitelock (San Francisco: ASP), 357
Szabados L. 1997, in Proc. Conf. Hipparcos Venice'97, ed. B. Battrick, ESA SP-402, p. 657
Szatmáry K. 1990, J. AAVSO, 19, No. 1, 52
Van Hoolst T. 2000, in ASP Conf. Ser. Vol. 203, The Impact of Large-Scale Surveys on Pulsating Star Research, ed. L. Szabados & D. W. Kurtz (San Francisco: ASP), 307
Van Hoolst T. & Waelkens C. 1995, A&A 295, 361
Vinkó J. 1993, MNRAS 260, 273

Chapter 14

Magnetic Activity and the Solar-Stellar Connection

Christopher M. Johns-Krull
Rice University
Houston, Texas USA

Jeff A. Valenti
Space Telescope Science Institute
Baltimore, Maryland USA

Abstract: It has long been challenging to reconcile the incredibly detailed (spatially and in wavelength coverage) data available from solar observations with the less detailed observations of stellar activity. Nevertheless, the Sun is a relatively inactive star and stellar data is necessary to explore phenomena spanning the full range of activity. Most of our current understanding and the recent advances in this field have been made with telescopes smaller than 4-m in aperture. This is also likely to be the case for the near future of this field. In this contribution, some of the recent advances in active star research will be reviewed and directions for future work will be described. Much of the focus will be on long-term monitoring programs and high-resolution spectroscopy, particularly in the infrared portion of the spectrum, a capability that is often omitted from large telescope projects.

Key words: stellar activity, magnetic fields, dynamo

1. INTRODUCTION

Magnetic fields almost certainly play a fundamental role in the physics of all late-type stellar atmospheres, and by analogy to the Sun, they are intimately related to the phenomena of stellar activity. Nearly all stars with outer convection zones show chromospheric activity (e.g., Linsky 1980). Magnetic buoyancy (Parker 1975) removes strong magnetic flux tubes from

T.D. Oswalt (ed.), The Future of Small Telescopes in the New Millennium, Vol. III, 225-243.

the convection zone on timescales short compared to the solar cycle period and global magnetic polarity reverses every 11 years, indicating that the observed solar magnetic field is regenerated by some sort of dynamo process. A similar dynamo is also believed to be responsible for magnetic activity in other late-type stars. Other contributions in this volume review the current and future state of the exquisitely detailed data on the Sun; however, the Sun is a relatively inactive star. While solar data are important for developing and testing theories of the dynamo generation of magnetic flux and its emergence, stellar data is required to test these theories as functions of stellar rotation, spectral type (convection zone properties), and possibly other parameters as well. Much of our understanding of stellar activity is the result of extrapolating the solar analogy, though Walter and Byrne (1998) argue that such extrapolation has led us badly astray in our picture of the most active stars. Thus, stellar studies remain critical for a complete understanding of magnetic activity.

In the context of the current volume, nearly all of the significant observational work on stellar activity to date has been done with telescope apertures less than 4-m. One reason this may continue to be true is the critical role of monitoring observations, which demand large amounts of telescope time that will likely be unavailable on larger telescopes in the foreseeable future. In this contribution we will review some of the highlights of work already accomplished, and then offer directions for future investigation.

2. THE MOUNT WILSON HK PROJECT

2.1 History

Dr. Olin Wilson initiated the HK Project on the Mount Wilson 100-inch telescope in the 1960s. Originally lasting just over a decade, Wilson monitored emission in a 1 Å bandpass centered on the chromospheric cores of the Ca II H and K lines of 91 main sequence stars. The original goal was to look for activity cycles akin to the 11 year solar cycle. Wilson (1978) presented the results of this study in a landmark paper, finding numerous stellar cycles with periods of about 7 years or longer.

After Dr. Wilson's retirement, Dr. Arthur Vaughan continued the HK Project, updating the instrumentation and moving the program to the Mount Wilson 60-inch telescope, where observing time was more plentiful. A new survey of several hundred solar neighborhood stars was started. During this time, it was noted that "noise" in the observations was greater than expected, hinting at another source of variability, likely on shorter timescales.

To explore this possibility, a 14 week observing run was conducted on the 60-inch telescope to closely monitor ~ 50 stars. Vaughan et al. (1981) reported rotation periods for 28 stars from the resulting analysis. From these data they also suggested that rotation, rather than initial conditions or age, is the primary factor determining chromospheric emission. The HK Project continues at Mount Wilson under the direction of Dr. Sallie Baliunas. The project primarily uses the 60-inch telescope, but the 100-inch telescope is also used occasionally to observe fainter targets.

2.2 Some Science Highlights of the HK Project

As mentioned above, observations made during the HK Project have demonstrated the rotation-activity connection that is now a staple of cool star research. Figure 1 shows a version of this relationship combining chromospheric (R'_{HK}) and coronal (L_X/L_{bol}) activity indices and plotting them as a function of reciprocal Rossby number (convective turnover time divided by stellar rotation period). Reciprocal Rossby number is related to the dynamo number of mean field dynamo theory (e.g. Durney & Latour 1978), and it does a better job than rotation alone of unifying the behavior of main sequence dwarfs and evolved stars (e.g. Basri 1987). Figure 1 shows that stellar activity increases as the rotation rate increases up to a "saturation" level, where the activity appears to no longer correlate with rotation (or convective turnover time). Saturation may represent a point where magnetic

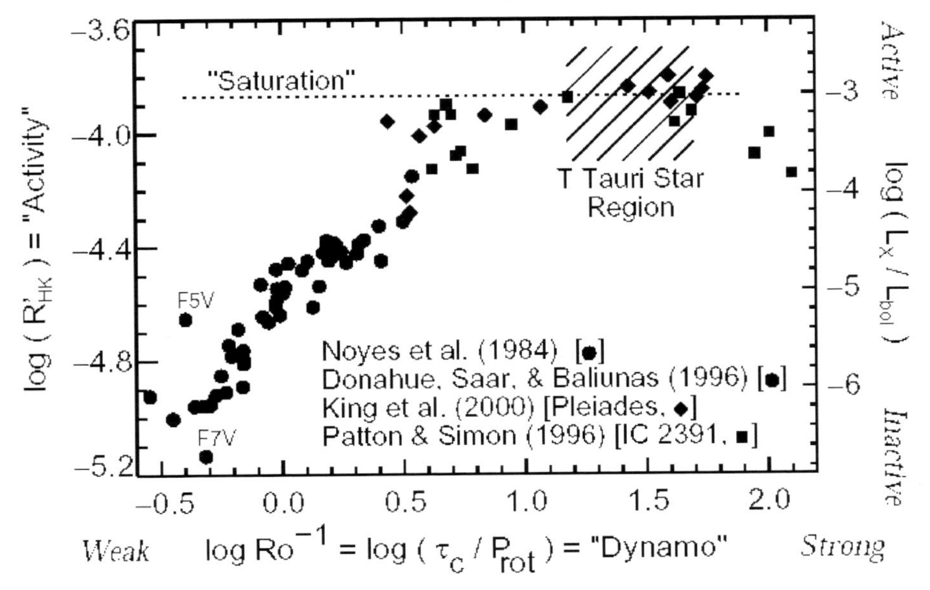

Figure 1. Rotation-Activity relationship with saturation for cool stars. R'_{HK} and L_X / L_{Bol} related using Sterzik & Schmitt (1997).

fields fill the entire stellar surface (though presumably the field strength can increase), or it may be the result of some dynamo limit, such as α-quenching (e.g. Leighton 1969). This rotation-activity relationship was first identified with data from the HK Project, but has since been shown in transition region data (e.g. Vilhu 1984) and coronal data (e.g. Stauffer et al. 1994) using satellite measurements.

The original goal of the HK Project was to search for activity cycles in other stars. Baliunas et al. (1995) report activity cycle periods for just over 50 stars out of 111 which have been monitored long term (~ 30 years). Examples of these activity cycles are shown in Figure 2. Baliunas et al. (1995) also

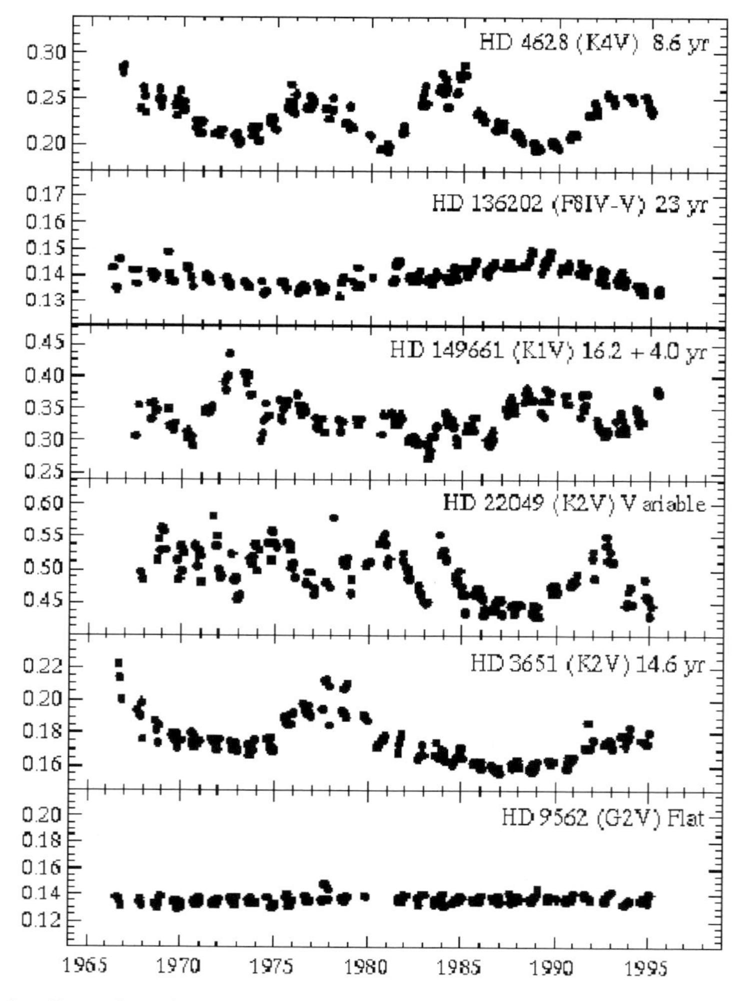

Figure 2. Examples of long term chromospheric behavior, including activity cycles, for several stars from the Mount Wilson HK Project. Monthly means of the Ca II index S (see Baliunas et al. 1995) are plotted versus time for approximately 30 years of observing.

report some general trends from these data: (1) young stars show relatively high activity levels, fast rotation, large stochastic variations, and do not exhibit Maunder minimum phases; (2) intermediate age stars (1-2 Gyr) show moderate activity and rotation rates with occasional smooth activity cycles; and (3) solar age and older stars are relatively inactive, rotate slowly, and typically have smooth cycles with occasional Maunder minimum phases.

Saar and Brandenburg (1999) have used data from the HK Project, supplemented with other sources, to look for relationships between activity cycle period, rotation period, and activity level. They find most stars obey a relationship between rotation and cycle period given by $P_{rot}/P_{cyc} \sim R_o^{-0.5}$, where R_o is the Rossby number. Stars can fall on either of two branches of this relationship, with the two branches separated by a factor of ~ 6 in P_{cyc}. These authors also find evidence that the α effect (part of the solar-like dynamo involving cyclonic motions which convert toroidal field into poloidal field) increases with mean magnetic field, contrary to the traditional α-quenching mechanism (e.g. Leighton 1969) for dynamo saturation. If borne out, these results will have strong implications for the nature of the cyclic dynamo in solar-like stars.

Another by-product of the HK Project is the detection of surface differential rotation on other solar-like stars (Baliunas et al. 1985, Donahue 1993). For some of the stars monitored at Mount Wilson, changes in the rotation period have been noted from one season to the next. As activity ramps up during the solar cycle, active regions emerge at progressively lower latitudes. If the sense of solar differential rotation applies to other low mass stars, we would expect apparent rotation period to decrease as stellar activity increases. This behavior has been observed in some stars, but the opposite behavior is equally likely. This suggests that for some stars, either the surface differential rotation pattern is opposite to that of the Sun, or that active regions first emerge near the equatorial regions and later appear at higher latitude as the cycle progresses.

In total, ~36 stars show evidence for differential rotation through changes in the Ca II monitored rotation periods, offering a substantial database against which to test dynamo models. To interpret these data correctly, however, we would like to know what the Sun would look like if monitored in the same way as stars in the HK Project. Donahue and Keil (1995) analyzed a time series of solar disk-integrated Ca II data (Keil & Worden 1984). The inferred solar differential rotation had the expected qualitative behavior, but the amplitude was too large. Active region growth and decay at a variety of latitudes is likely responsible, making the problem inherently two-dimensional. Hempelmann and Donahue (1997) have reinvestigated the problem using a wavelet analysis technique, recovering rotation parameters from the disk integrated data, which are in good agreement with spatially resolved measurements. Thus, there is promise of getting accurate surface differential rotation relationships for a large number of solar-like stars through long term monitoring such as the Mount Wilson HK Project.

3. STELLAR FLARE STUDIES

Approximately 10% of the stars in the Galaxy are flare stars (Haisch, Strong, & Rodonó 1991). Flare research is astrophysically interesting because flares probe the violent response of plasma to substantially nonequilibrium magnetic field configurations, in contrast to many other pseudo-equilibrium processes in astrophysics. Flares may also contribute significantly to heating of stellar chromospheres and coronae. While solar data are invaluable for testing flare models, the extremes reached in stellar flares offer important tests for a complete theory. Due to the low optical contrast of flares on G type and warmer stars, most stellar flare work has historically concentrated on dKe/dMe and RS CVn stars, but warmer stars and the Sun certainly flare as well. For example, the G0V star π^1 UMa has been observed to flare in X-rays by the *Exosat* satellite (Landini et al. 1986).

Early work investigated the temporal probability of stellar flares as a function of released energy, which is an important solar topic as well because micro-flares may contribute to heating of the solar corona (e.g., Priest 1999). Early studies also measured the temporal evolution of flare colors and modeled optical radiation due to recombination (Gershberg 1967; Moffett 1974; Cristaldi & Rodonó 1975; Kunkel 1975). These studies showed that optical white light (typically UBV photometry) emission from stellar flares is quite impulsive, similar in many respects to solar hard X-ray bursts. More recently, Hawley et al. (1995) have demonstrated a stellar "Neupert effect" (Neupert 1968, Hudson et al. 1992) in which coronal EUVE emission follows the integral over time of white light emission, formed presumably by the impact of accelerated particles on the stellar surface. In the case of the Sun, soft X-ray emission follows the integral over time of hard X-ray emission. Soft X-ray emission is interpreted as thermal emission from coronal loops filled by material that evaporates when energetic electrons strike the surface, generating hard X-rays. This stellar Neupert effect has also been reported by Güdel et al. (1996).

While much of the current focus in stellar flare research involves using satellite and radio data to study the resulting emission from the corona and transition region (Hawley et al. 1995; Gagné et al. 1998; Güdel et al. 1999; Franciosini, Pallavicini, Tagliaferri 2001; Hawley et al. 2001), optical observations (particularly high time resolution, multicolor photometry) still provide an important foundation for these studies. As described above, white light emission is currently the best indicator of flare onset and duration of the impulsive phase. White light emission is also the energetically dominant emission mechanism observed in stellar flares to date (e.g. Hawley & Pettersen 1991; Hawley et al. 1995). As a result, simultaneous optical photometry still bolsters the analysis of new satellite flare data and helps constrain new flare models.

Outstanding problems remain. In particular, the production mechanisms for the white light continuum emission during flares remain unknown despite decades of optical observations. Electron beam heating and X-ray heating are insufficient (e.g. Hawley & Fisher 1994). New models predict that this emission is driven by short bursts of heating during the impulsive phase (Abbett & Hawley 1999). Bursts are required so that heating can penetrate into the lower atmosphere before a shield of evaporated material enhances coronal densities and prematurely quenches heating of the photosphere and low chromosphere. Optical photometry at high time resolution is necessary to track the continuum response to the impulsive heating, which can be observed directly in high energy emission lines obtained with satellite data. In addition, long term heating of the chromosphere during the gradual phase of stellar flares cannot be caused directly by the impulsive heating mechanisms (electron or proton beams) generally assumed in flare models. Hawley and Fisher (1992) have predicted that the transition region and coronal emission lines will produce “back-warming” that powers this chromospheric heating. This prediction can only be tested with simultaneous satellite and ground based data.

Most of the ground based flare research described above has been done with moderate (1- to 2-m) sized telescopes, equipped with photoelectric photometers capable of high time resolution photometry. Similar data will continue to be an important aspect of future attempts to understand stellar flares. Photometers are disappearing faster than the smaller telescopes, and at present high speed CCD photometers have not yet filled the gap in capabilities. Telescopes up to 4-m in size are also valuable in flare studies for measuring chromospheric line broadening from high resolution spectra which can be used to trace dynamics during the flare, constrain densities in the flaring plasma, and provide detailed tests of current models (see Johns-Krull et al. 1997 for a solar example). As in the example of the Mount Wilson HK Project, stellar flare studies require large amounts of telescope time for target monitoring. The additional need to observe simultaneously with satellites also requires scheduling flexibility, which is generally easier to accomplish on moderate size telescopes. Telescopes with apertures of 4-m or less will continue to make valuable contributions to stellar flare research.

4. DOPPLER IMAGING

The Doppler imaging technique (see review by Rice 1996), inverting an observed time series of spectral line profiles to produce an "image" of the stellar surface, has been used by a variety of investigators to derive the distribution of photospheric spots on a number of active late-type stars (e.g. Vogt & Penrod 1983; Vogt & Hatzes 1991; Hatzes & Vogt 1992; Kürster et al. 1992). One of the principal discoveries of this work is the apparent prevalence of large polar spots on most of the rapidly rotating stars that have been imaged (see Figure 3). While the reliability of these polar spots has been seriously called into question, two hypotheses have been advanced to

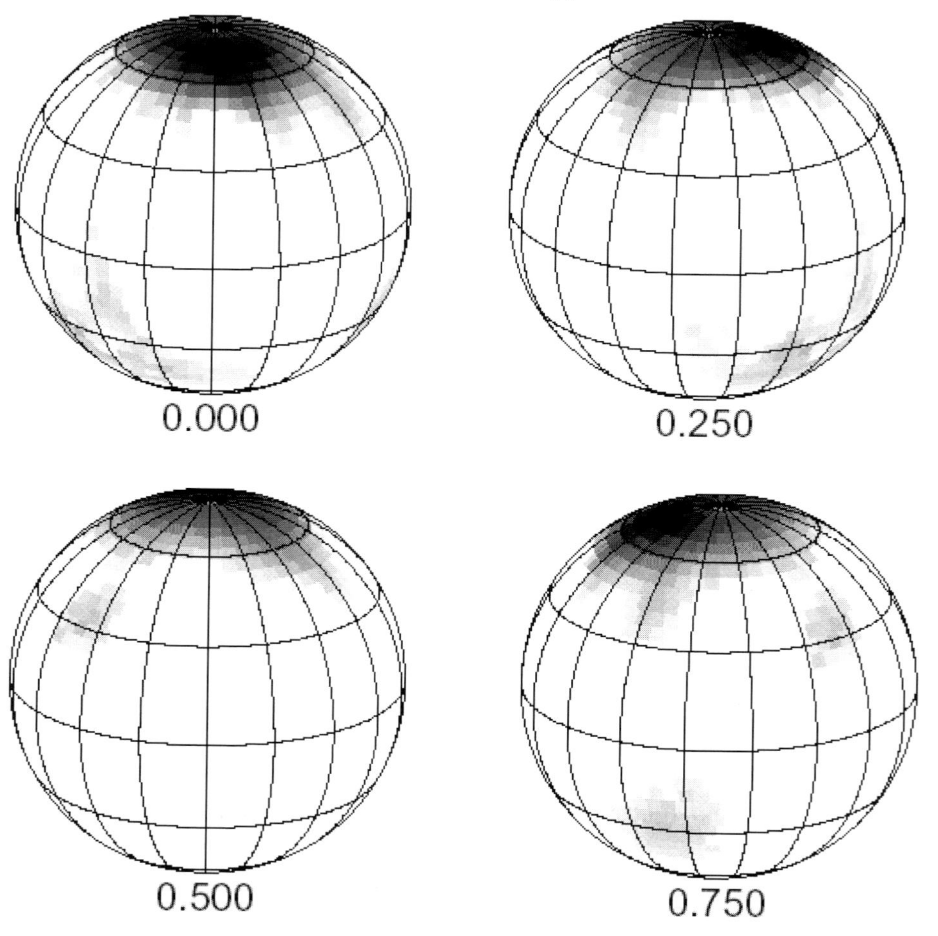

Figure 3. The Doppler image of the T Tauri star Sz 68 derived from fitting the line profile variations of Fe I 6190Å, Fe I 6400Å, Ca I 6439Å, and Li I 6708Å. The darkest regions (near the pole) are ~ 1000 K cooler than the photosphere and the lighter regions (more equatorial) are about ~ 300 K cooler than the photosphere.

explain their appearance. Vogt & Penrod (1983) have suggested that magnetic flux (traced by the spots) first emerges at equatorial latitudes and then migrates toward the rotation pole where it collects into large structures (polar spots). Schüssler & Solanki (1992—see also Schüssler et al. 1996) have argued that the Coriolis force in rapidly rotating stars with deep convection zones dominates the buoyancy force resulting in the emergence and maintenance of magnetic flux near the rotation pole. This phenomenon has also been studied in the solar context (e.g. Fan, Fisher, & McClymont 1994), and it can explain the range in latitude of active region emergence on the Sun as well as the sense and amount of tilt observed in active regions.

Most Doppler imaging studies of late-type stars have focused either on evolved stars (see references above), which derive their rapid rotation (a necessary ingredient for the technique) from a binary interaction, or on pre-main sequence T Tauri stars (Strassmeier, Welty, & Rice 1994; Joncour, Bertout, & Menard 1994; Joncour, Bertout, & Bouvier 1994; Hatzes 1995; Rice & Strassmeier 1996; Johns-Krull & Hatzes 1997). The rapid rotation of these young stars is a result of the star formation process and permits the study of surface features characteristic of the dynamo acting in a rapidly rotating, single star. As shown in Figure 3, Doppler images of T Tauri stars generally reveal large spots at or near the stellar pole. The Coriolis forces invoked by Schüssler et al. (1996) require the anchoring of magnetic flux tubes in a deep radiative zone, but most T Tauri stars are expected to be completely convective. Current theories place the cyclical solar dynamo (often referred to as an α- Ω dynamo) at the interface of the radiative core and the outer convection zone in the convective overshoot region (Gilman, Morrow, & DeLuca 1989). This is presumably the location of the dynamo giving rise to activity cycles seen in other cool stars observed in the HK Project. On the other hand, such a dynamo must function differently or not at all in a fully convective star, such as a T Tauri star or a late type dMe star, yet these stars are observed to be active, and as described below, direct measurements of magnetic fields are now available for many such stars.

Since rapid rotation is a necessary ingredient for Doppler imaging studies, relatively few main sequence stars have been imaged. One of the exciting examples of dwarfs that have been observed is the young star AB Dor (K0V). Donati and Collier-Cameron (1997) and Donati et al. (1999) have used the technique of Zeeman Doppler Imaging to map both temperature and magnetic structures on the stellar surface. This technique creates magnetic field maps from Stokes spectra obtained at many rotational phases. Rapid stellar rotation spectroscopically resolves surface magnetic structures, reducing polarization cancellation. Even with this advantage, many lines must be combined to detect a net circular polarization signal. This technique has allowed the detection of a solar-like differential rotation pattern on the surface of AB Dor. These observations were done using the 3.9-m Anglo-Australian Telescope and show that clever analysis techniques utilizing the wealth of data collected in high resolution echelle spectra can reveal substantial information on the surface structure of spatially unresolved stars.

5. MAGNETIC FIELD MEASUREMENTS

Magnetic activity can be broken up into the problems of (1) dynamo generation of magnetic flux and its emergence, and (2) the resultant magnetic heating of the outer stellar atmosphere. Actual observations of magnetic fields on other stars offer the chance to study these two phenomena separately, though the observational challenge is substantial. Ever since Robinson's (1980) seminal paper, measurements of the magnetic field strength, B, and surface filling factor, f, of magnetic regions on cool stars have been made by several investigators. As observational and analysis techniques have matured, reported values of the unsigned magnetic flux, Bf, have generally decreased. Valenti, Marcy, and Basri (1995) illustrate this trend for the active K dwarf ε Eri, and Rüedi et al. (1997) independently obtain a similar low value for the magnetic flux. While the early studies certainly detected the presence of magnetic fields on several late-type stars, more recent investigations are now providing reliable quantitative measurements of B and f on cool stars (Saar 1996a, b).

Virtually all measurements of stellar magnetic fields employ the Zeeman effect, manifested either as (1) differential Zeeman broadening of magnetically sensitive and insensitive lines observed in intensity spectra (see Figure 4), or (2) circular polarization of magnetically sensitive lines. For simple Zeeman triplets, a σ component splits to either side of the unshifted π component. The wavelength shift of a σ component is given by $\Delta\lambda = k\lambda^2 gB$; where k is a constant, g is the Landé g-factor of the specific transition, B is the strength of the magnetic field, and λ is the wavelength of the transition. One thing to note from this equation is the λ^2 dependence of the Zeeman splitting, compared with the λ dependence of Doppler line broadening mechanisms, meaning that observations in the infrared (IR) are generally more sensitive to the presence of magnetic fields than optical observations.

All existing measurements of stellar magnetic fields utilizing Zeeman broadening of photospheric absorption lines have been done on small to medium telescopes: the Kitt Peak 4-m (Saar & Linsky 1985; Valenti et al. 1995), the 3-m NASA Infrared Telescope Facility (Saar 1996a,b; Johns-Krull, Valenti, & Koresko 1999b; Johns-Krull & Valenti 2000), the 2.7-m Harlan Smith Telescope of McDonald Observatory (Johns-Krull & Valenti 1996, 2000), the 1.4-m Coudé Auxiliary Telescope of the European Southern Observatory (Rüedi et al. 1997), and the 4.2-m William Herschel Telescope

(Guenther et al. 1999) to mention a few. For some classes of stars where the field strengths are large enough (dMe and T Tauri stars), observations can separately determine B and f from high resolution (R ~ 120,000) optical (Johns-Krull & Valenti 1996, 2000) or moderate resolution (R ~ 30,000) IR (Saar 1996a, b; Johns-Krull et al. 1999b; Johns-Krull & Valenti 2000) data.

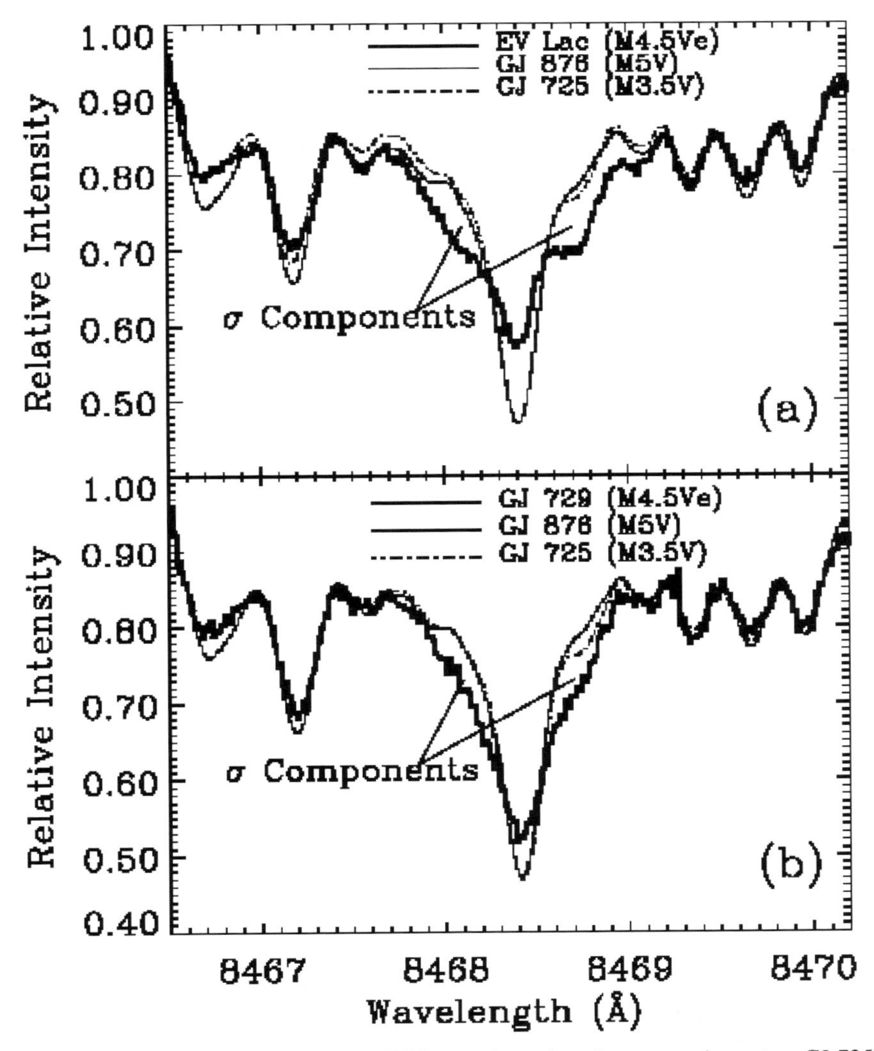

Figure 4. (a) Spectra of the flare star EV Lac and two inactive comparison stars GJ 725B and GJ 876 in the vicinity of the magnetically sensitive Fe I line at 8468.4Å. Note that the spectra of the inactive stars match closely the spectrum of EV Lac for all the nearby magnetically insensitive TiO lines at longer wavelength, but fail to match the σ components on either side of the Fe I line, demonstrating the presence of a magnetic field on the flare star. The mismatch at shorter wavelengths is partially due to the Ti I line at 8467.2Å, which is also Zeeman sensitive. (b) Same as in (a) but for the flare star GJ 729.

For G and K dwarfs, the field strengths are typically low enough that only the product *Bf* can be measured with optical data, but high quality IR data can still separate the two (Valenti et al. 1995).

Stellar magnetic field work is still somewhat in its infancy, particularly since analysis is hampered by a lack of good models for magnetic flux tubes on stars as a function of effective temperature, gravity, and flux tube size. Nevertheless, some interesting results deserve mention. A comparison of measured field strengths on the Sun and all but the youngest dwarfs implies that the dominant magnetic pressure in flux tubes scales with gas pressure in the confining quiet medium, a concept called "equipartition" (Saar 1996b). Clear evidence has been found, however, for distributions of magnetic field strengths on dMe and T Tauri stars (Saar 1992; Johns-Krull & Valenti 1996, 2000; Johns-Krull et al. 1999b). Optical spectra have inadequate Zeeman resolution to ascertain whether warmer stars have similar field distributions scaled to lower field strengths. Fully convective M dwarfs have been shown to possess strong fields (3 – 4 kG) covering roughly half the stellar surface (Saar 1996a, b; Johns-Krull & Valenti 1996, 2000), indicating that strong dynamo action can occur even in the absence of an internal radiative zone. The magnetic fields on T Tauri stars cover the entire surface and dominate gas pressure by up to an order of magnitude, inhibiting convection throughout the stellar photosphere and perhaps altering the atmospheric structure (Johns-Krull et al. 1999b; Johns-Krull & Valenti 2000).

Measuring circular polarization in magnetically sensitive lines is a more direct means of detecting stellar magnetic fields, but the method is also subject to several limitations. Emission parallel to a unipolar magnetic field generates no π component and two sets of circularly polarized σ components with opposite helicity and wavelength shift. The helicity of the σ components reverses as the polarity of the magnetic field reverses. Thus, on a star like the Sun that typically displays equal amounts of + and - polarity fields on its surface, the net polarization of the star as a whole is very small. If one magnetic polarity dominates the visible surface of a star, net circular polarization is present in Zeeman sensitive lines, resulting in a wavelength shift between line profiles observed through right- and left-circular polarizers. The magnitude of the shift scales with the surface averaged line of sight component of the magnetic field, which on the Sun is typically less than 2 G even though individual magnetic elements on the solar surface have field strengths of 1.5 kG in plage and ~ 3.0 kG in spots. Several polarimetric studies of cool stars have generally failed to detect circular polarization, placing limits on the disk averaged net magnetic field of 10-100 G (e.g. Vogt 1980; Brown & Landstreet 1981; Borra, Edwards, & Mayor 1984). One notable exception is the detection of circular polarization in segments of the line profile observed on rapidly rotating RS CVn stars, where Doppler broadening of the line "resolves" several independent strips on the stellar surface (e.g. Donati et al. 1997—see also the discussion of Zeeman Doppler Imaging above). More recently, Plachinda and Tarasova (1999, 2000) have pushed polarization limits on slowly rotating stars down to only a few Gauss, perhaps detecting fields on 3 solar-like stars, as well as a rotationally modulated field on ξ Boo A (G8V) with and amplitude of 30 G. If confirmed, these results suggest we may be able to directly observe magnetic cycles on stars other than the Sun.

6. FUTURE STUDIES WITH SMALL TO MODERATE SIZE TELESCOPES

In the previous sections, we provided examples of significant contributions by small and medium sized telescopes to our understanding of stellar activity, hinting that there is still much to be accomplished. Here, we look at a few specific examples of further advances that can be accomplished with telescope apertures up to 4-m.

6.1 Activity Monitoring

Saar and Brandenburg (1999) use their model to predict cycle periods for several stars of different types. Among the stars in the Mount Wilson HK Project, Saar and Brandenburg (1999) find evidence that less active stars can show cycles on both of the branches in their model.

The ~ 80 year Gleissberg (1971) cycle of the Sun appears to fit in with this scenario. Saar and Brandenburg (1999) predict secondary cycle periods ranging from 46 to 83 years for several stars in the HK Project which show long term trends in addition to their shorter (7 - 15 year) primary cycle periods. Continuation of the HK Project will test the existence of such long term cycles of magnetic activity. Saar and Brandenburg (1999) also suggest that activity cycles in young open cluster stars may be easier to detect using broad band photometric monitoring of spot cycles, rather than chromospheric diagnostics, since variations in surface fields have little effect in "saturated" diagnostics. The predicted cycle periods are a mere 10 - 40 days. Such a project would benefit from the wide field imaging characteristic of ~1-m class telescopes. Ideally, automated observations should be obtained each night for an entire observing season.

The long term behavior of solar like stars is potentially important for fully understanding the impact of solar activity on the Earth's climate. During the Maunder minimum of solar activity in the mid-17th century (Maunder 1890; Eddy 1976), sunspot activity essentially vanished and Europe experienced the "Little Ice Age." More detailed correlations have been found between the long term magnetic behavior of the Sun and the Earth's climate (e.g. Friis-Christensen & Lassen 1991; Baliunas & Jastrow 1993), making studies of solar and stellar activity quite relevant to humanity. In this context, clusters provide a very useful tool for understanding how activity evolves with age for solar-like stars. The solar age cluster M67 is a key astrophysical laboratory that has been the subject of a few snapshot studies (Baliunas et al. 1995a; Giampapa et al. 2000) to estimate the range of activity in solar age stars. While these studies show that the current range of activity of solar-like (mass and age) stars is larger than the range currently displayed by the Sun, long term monitoring is necessary to study how frequently and dramatically stars change their activity level. Previous studies have demonstrated that such a project can be performed efficiently on a 4-m class telescope with a multi object high-resolution spectrometer, such as the 3.5-m WIYN telescope on Kitt Peak with the Hydra multi-fiber spectrometer. While M67 is particularly important in relation to Sun-Earth connections, other clusters should also be monitored to track the evolution of activity with age.

6.2 Turbulent Dynamos

Most attempts to interpret stellar activity using dynamo theory have focused on cyclical dynamos of the sort observed in the Sun and other solar-like stars. As described above, cyclical α - Ω dynamos are believed to operate in the convective overshoot region immediately below the outer convection zone, where helioseismology has detected substantial radial sheer (e.g. Gilman et al. 1989). However, we have already seen that fully convective stars (T Tauri and later type dMe stars) are observed to have strong, pervasive magnetic fields on their surface. Even in stars that are not fully convective, the need for a non-standard dynamo has become apparent in recent years. For main sequence F, G, and K stars, chromospheric emission in diagnostics such as Mg II never drops below a minimum "basal" level. (e.g. Rutten et al. 1991, Mathioudakis and Doyle 1992). Basal heating, which is present even in stars with no detectable rotation or periodic activity, may be due to acoustic heating (see review by Schrijver 1995; and objections by Judge & Carpenter 1998) or magnetic fields produced by a noncyclical dynamo. Even on the Sun there is evidence that the total magnetic flux outside of active regions hardly varies during the solar cycle (Rabin et al. 1991), suggesting the operation of a noncyclical dynamo.

Work by Durney, DeYoung, and Roxburgh (1993) suggests that turbulence alone can produce magnetic fields throughout stellar convection zones without the need for radiatively stable anchor points. If true, perhaps an α - Ω dynamo drives cyclical magnetic activity that is concentrated into active regions on F, G, and K dwarfs, while a turbulent dynamo acts throughout convection zones generating small scale, possibly weaker magnetic activity. High angular resolution magnetograms of solar "quiet" regions reveal small-scale magnetic fields (the so called "magnetic carpet") that evolve on very short timescales (Title & Schrijver 1998), lending credence to the notion that a separate magnetic component might exist in all convective stars. More recently, Cattaneo (1999) convincingly demonstrated with very high resolution simulations of convection, in a non-rotating reference frame, that turbulent motions can generate small scale magnetic fields with a surprisingly large magnetic energy density. He found the kinetic and magnetic energy densities eventually reach a statistical equilibrium, with the magnetic energy roughly 25% that of the kinetic energy.

What are the expected and observed properties of turbulent dynamos? Durney et al. (1993) argue that rotation enhances the production and scale of magnetic flux in a turbulent dynamo, but is not essential. Inactive M dwarfs do rotate slowly, but Hα emission strength in active M dwarfs apparently does not depend on $v \sin i$ (Hawley, Gizis, & Reid 1996). Explaining this dichotomy is challenging. What about the presence of activity cycles? No dwarfs later than M2 were included in the original Mount Wilson survey, so very little is known about long term activity variations on fully convective stars. Mavridis and Avgoloupis (1986) report fluctuations in the quiescent brightness and flare frequency of EV Lac (M3.5V) with a period of ~ 5 years. Additionally, Weiss (1994) reports possible periodic signals in the photometric behavior of Gl 213 (M4V - 2.7 years) and Gl 876 (M4V - 2.9 years).

We propose here a project similar to the Mount Wilson HK project, but targeting M dwarfs, to observationally constrain turbulent dynamos. Moderate resolution spectroscopy ($R \sim 10{,}000$) with a 4-m telescope would be adequate to record the necessary activity indices. While Hα is an important diagnostic, interpretation in terms of activity is complicated because formation of high excitation lines in cool stars is itself complicated (e.g. Byrne 1993). Using a 4-m telescope would allow one to monitor a large sample of solar neighborhood M stars, recording emission in the cores of Ca II H and K, Hα, and the Ca II infrared triplet lines. Accurate M dwarf rotation periods are necessary to quantitatively study turbulent dynamos. By monitoring activity, rather than broadband photometric variations, greater sensitivity to rotational modulation can be achieved, as in the original HK project. In addition, as progress is made in Doppler imaging techniques, the more rapidly rotating dMe stars can be imaged to study their spot distributions. Following the work of Schüssler et al. (1996), this may provide clues on the location of the dynamo in these fully convective stars.

6.3 Magnetic Field Measurements

Reliable magnetic field measurements are possible when Zeeman broadening is at least comparable to the combination of all other line broadening mechanisms. When field strengths are high and rotation is modest, as in active M dwarfs, optical measurements suffice. When rotation is more significant (as in T Tauri stars) or field strengths are lower (as in F, G, and K dwarfs), the greater sensitivity of infrared diagnostics are required. In all cases, spectrograph resolution must be high enough to resolve Zeeman broadening, which in practice means resolutions of at least 60,000 in the optical and 30,000 in the infrared. Spectroscopy at higher resolution for at least some stars is required to measure and understand the distribution of surface magnetic field strengths for various types of stars. Analyses of multiple spectral lines spanning a range of Zeeman sensitivity helps mitigate systematic errors. A high resolution IR echelle with broad wavelength coverage would be ideal for obtaining additional magnetic field measurements. Given that even very bright stars have yet to be studied, such an instrument fed by a 4-m telescope could, for example, measure the magnetic field properties of all stars in the Mount Wilson HK project.

As a homogeneous sample of magnetic parameters for a large sample of F, G, and K stars with well determined rotation periods and chromospheric fluxes is built, it will be possible to directly and quantitatively test a number of predictions. Examples include: (1) magnetic flux tube models that predict B vs. spectral type (Spruit & Zweibel 1979; Galloway & Weiss 1981); (2) chromospheric and coronal heating mechanisms that depend on Bf (Ulmschneider & Stein 1982; Muzielak & Rosner 1988); (3) empirical relations that predict Bf from UV and X-ray fluxes (Vilhu 1984; Montesinos & Jordan 1993); (4) dynamo theories that predict B and f given rotation and convection parameters (Durney & Robinson 1982; Durney et al. 1993); and (5) angular momentum models that give spin down rates as a function of mass loss and Bf (MacGregor & Brenner 1991).

The recent claims of Plachinda and Tarasova (1999; 2000) suggest that it may be possible to study the large scale component of magnetic fields even on stars as inactive as the Sun (~ 2 G mean line-of-sight field modulations through the activity cycle). Using a 2.6-m telescope, these authors are striving to achieve a spectrapolarimetric precision of ~ 10 G. For ξ Boo A, they claim to have detected a rotationally modulated net magnetic field of ~ 30 G, coherent for hundreds of rotation periods. Data in this initial study are too sparse to draw definite conclusions, but the prospects are certainly interesting to consider. For stars observed at inclinations far from 90°, observed polarization may be produced primarily in polar regions dominated by the dipole component of the field. Polarity might then change sign over the course of the activity cycle, as on the Sun. It would also be very intriguing to see how this large scale component of the field evolves with stellar age, especially since open field lines associated with the large scale component play a key role in stellar angular momentum evolution. Currently, there are few reported attempts to achieve sensitivity at the few Gauss level, but the main impediments are systematic effects related to instrument calibration and stability and analysis techniques, as opposed to signal-to-noise limits posed by telescope aperture. Telescopes of 4-m size or less can make important contributions in this field over the next several years.

7. CLOSING

Modest (by today's standards) aperture telescopes have made many significant contributions to the study of stellar activity. Much is still left to be done; however, significant investments of telescope time to continue and initiate new monitoring programs will be required to make progress on many fronts. Nevertheless, ground based optical and IR telescopes with apertures of 4-m or less will continue to make many important discoveries in the field of stellar activity and magnetism.

8. REFERENCES

Abbett, W. P. & Hawley, S. L. 1999, ApJ, 521, 906

Baliunas, S. L. et al. 1985, ApJ, 294, 310

Baliunas, S. L. et al. 1995a, ApJ, 438, 269

Baliunas, S. L., Donahue, R. A., Soon, W., Gilliland, R., & Soderblom, D. R. 1995b, AAS, 186, 2109

Baliunas, S. L. & Jastrow, R. 1993, Energy, 18, 1285

Basri, G. 1987, ApJ, 316, 377

Borra, E. F., Edwards, G., & Mayor, M. 1984, ApJ, 284, 211

Brown, D. N. & Landstreet, J. D. 1981, ApJ, 246, 899

Byrne, P. B. 1993, A&A, 272, 495

Cattaneo, F. 1999, ApJ, 515, L39

Cristaldi, S. & Rodonó, M. 1975, IAU Symp. 71, Variable Stars and Stellar Evolution, 75

Donati, J.-F. & Collier Cameron, A. 1997, MNRAS, 291, 1

Donati, J.-F., Collier Cameron, A., Hussain, G. A. J., & Semel, M. 1999, MNRAS, 302, 437

Donati, J.-F., Semel, M., Carter, B. D., Rees, D. E., & Cameron, A. C. 1997, MNRAS, 291, 658

Donahue, R. A. 1993, PhD Thesis, New Mexico State University

Donahue, R. A. & Keil, S. L. 1995, Solar Phys., 159, 52

Donahue, R. A., Saar, S. H., & Baliunas, S. L. 1996, ApJ, 466, 384

Durney, B. R. & Latour, J. 1978, Geophys. Astrophys. Fluid Dynamics, 9, 241

Durney, B. R., DeYoung, D. S., & Roxburgh, I. W. 1993, Sol. Phys., 145, 207

Durney, B. R. & Robinson, R. D. 1982, ApJ, 253, 290

Franciosini, E., Pallavicini, R., & Tagliaferri, G. 2001, A&A, 375, 196

Friis-Christensen, E. & Lassen, K. 1991, Science, 254, 698

Galloway, D. J. & Weiss, N. O. 1981, ApJ, 243, 945
Gagné, M., Valenti, J., Johns-Krull, C., Linsky, J., Brown, A., & Güdel, M. 1998, ASP Conf. Ser. 154: Cool Stars, Stellar Systems, and the Sun, 10, 1484
Gershberg, R. E. 1967, Astrofizika, 3, 127
Giampapa, M. S., Radick, R. R., Hall, J. C., & Baliunas, S. L. 2000, AAS/Solar Physics Division Meeting, 32, 002120
Gilman, P. A., Morrow, C. A., & DeLuca, E. E. 1989, ApJ, 338, 528
Gleissberg, W. 1971, Sol. Phys., 21, 240
Güdel, Benz, A. O., Schmitt, J. H. M. M., & Skinner, S. L. 1996, A&A, 471, 1002
Güdel, M., Linsky, J. L., Brown, A., & Nagase, F. 1999, ApJ, 511, 405
Haisch, B., Strong, K. T., & Rodonó, M. 1991, ARAA, 29, 275
Hatzes, A.P. & Vogt, S.S. 1992, MNRAS, 258, 387
Hatzes, A. P. 1995, ApJ, 451, 784
Hawley, S. L. et al. 1995, ApJ, 453, 464
Hawley, S. L. & Pettersen 1991, ApJ, 378, 725
Hawley, S. L. & Fisher, G. H. 1992, ApJS, 81, 885
Hawley, S. L. & Fisher, G. H. 1994, ApJ, 426, 387
Hawley, S. L., Gizis, J. E., & Reid, I. N. 1996, AJ, 112, 2799
Hawley, S. L., Johns-Krull, C. M., Fisher, G. H., Abbett, W. P., Seiradakis, J. H., & Avgoloupis, S. I. 2001, ASP Conf. Ser.: Cool Stars, Stellar Systems, and the Sun, 12, in press
Hempelmann, A. & Donahue, R. A. 1997, A&A, 322, 835
Hudson, H. S., Acton, L. W., Hirayama, T., & Uchida, Y. 1992, PASJ, 44, L77
Joncour, I., Bertout, C., & Bouvier, J. 1994, A&A, 291, L19
Joncour, I., Bertout, C., & Menard, F. 1994, A&A, 285, L25
Johns-Krull, C. M. & Hatzes, A. P. 1997, ApJ, 487, 896
Johns-Krull, C. M., Hawley, S. L., Basri, G., & Valenti, J. A. 1997, ApJS,112, 221
Johns-Krull, C. M. & Valenti, J. A. 1996, ApJ, 459, L95
Johns-Krull, C. M. & Valenti, J. A. 2000, ASP Conf. Ser. 198: Stellar Clusters and Associations: Convection, Rotation, and Dynamos, 371
Johns-Krull, C. M., Valenti, J. A., Hatzes, A. P., & Kanaan, A. 1999a, ApJ, 510, L41
Johns-Krull, C. M., Valenti, J. A., & Koresko, C. 1999b, ApJ, 516, 900
Judge, P. G. & Carpenter, K.G. 1998, ApJ, 494, 828
Keil, S. L. & Worden, S. P. 1984, ApJ, 276, 766
King, J. R., Soderblom, D. R., Fischer, D., & Jones, B. F. 2000, ApJ, 533, 944
Kunkel, W. E. 1975, in Variable Stars and Stellar Evolution, ed. V. E. Sherwood & L. Plant, (Dordrecht:Reidel), 15
Kürster, M., Hatzes, A.H., Pallavicini, R., & Randich, S. 1992, ASP Conf. Series, 26, Cambridge Workshop on Cool Stars, Stellar Systems, and the Sun, 7, 24
Landini, M., Monsignori-Fossi, B. C., Pallavicini, R., & Piro, L. 1986, A&A, 157, 217
Leighton, R. B. 1969, ApJ, 156, 1
MacGregor, K. B. & Brenner, M. 1991, ApJ, 376, 204
Mathioudakis, M. & Doyle, J. G. 1992, A&A, 262, 523
Mavridis, L. N. & Avgoloupis, S. 1986, A&A, 154, 171
Montesinos, B. & Jordan, C. 1993, MNRAS, 264, 900
Moffett, T. J. 1974, ApJS, 29, 1
Muzielak, Z. E., & Rosner, R. 1988, ApJ, 329, 376
Neupert, W. M. 1968, ApJ, 153, L59
Noyes, R. W., Hartmann, L. W., Baliunas, S. L., Duncan, D. K., & Vaughan, A. H. 1984, ApJ, 279, 763

Patton, B. M. & Simon, T. 1996, ApJS, 106, 489

Plachinda, S. I. & Tarasova, T. N. 1999, ApJ, 514, 402

Plachinda, S. I. & Tarasova, T. N. 2000, ApJ, 533, 1016

Priest, E. R. 1999, Ap&SS, 264, 77

Rabin, D. R., Devore, C. R., Sheeley, N. R., Jr., Harvey, K. L., & Hoeksema, J. T. 1991, in Solar Interior and Atmosphere, (Tucson: Univ. of Arizona), 781

Rice, J.B. 1996, in Stellar Surface Structure, IAU Symp. 176, K.G. Strassmeier & J.L. Linsky (eds.), (Dordrecht: Kluwer), p. 19

Rice, J.B. & Strassmeier, K.G. 1996, A&A, 316, 164

Robinson, R. D. 1980, ApJ, 239, 961

Rüedi, I., Solanki, S. K., Mathys, G., & Saar, S. H. 1997, A&A, 318, 429

Rutten, G. G. M., Schrijver, C. J., Lemmens, A. F. P., & Zwaan, C. 1991, A&A, 252, 203

Saar, S. H. 1992, ASP Conf. Ser. 26: Cool Stars, Stellar Systems, and the Sun, 7, 252

Saar, S. H. 1996a, Stellar Surface Structure, K. G. Strassmeier & J. L. Linsky (eds.), IAU Symp. 176, 237

Saar, S. H. 1996b, Magnetodynamic Phenomena in the Solar Atmosphere - Prototypes of Stellar Magnetic Activity, Y. Uchida et al. (eds.), (Dordrecht: Kluwer), 367

Saar, S. H. & Brandenburg, A. 1999, ApJ, 524, 295

Saar, S. H. & Linsky, J. L. 1985, ApJ, 299, L47

Schrijver, C. J. 1995, A&ARev, 6, 181

Schüssler, M. & Solanki, S.K. 1992, A&A, 264, L13

Schüssler, M., Caligari, P., Ferriz-Mas, A., Solanki, S.K., & Stix, M. 1996, A&A, 314, 503

Spruit, H. C. & Zweibel, E. G. 1979, Sol. Phys., 62, 15

Stauffer, J. R., Caillault, J.-P., Gagne, M., Prosser, C. F., & Hartmann, L. W. 1994, ApJS, 91, 625

Sterzik, M. F. & Schmitt, J. H. M. M. 1997, AJ, 114, 1673

Strassmeier, K. G., Welty, A. D., & Rice, J. B. 1994, A&A, 285, L17

Title, A. M. & Schrijver, C. J. 1998, ASP Conf. Ser. 154: Cool Stars, Stellar Systems, and the Sun, 10, 345

Ulmschneider, P. & Stein, R. F. 1982, A&A, 106, 9

Valenti, J. A., Marcy, G. W., & Basri, G. 1995, ApJ, 439, 939

Vaughan, A. H., Preston, G. W., Baliunas, S. L., Hartmann, L. W., Noyes, R. W., Middelkoop, F., & Mihalas, D. 1981, ApJ, 250, 276

Vilhu, O. 1984, A&A, 133, 117

Vogt, S. S. 1980, ApJ, 240, 567

Vogt, S. S. & Penrod, G. D. 1983, PASP, 95, 565

Vogt, S. S. & Hatzes, A. H. 1991, in The Sun and Cool Stars: Activity, Magnetism, and Dynamos, IAU Col. 130, I. Tuominen, D. Moss, & G. Rüdiger (eds.), (Berlin: Springer), p. 297

Walter, F. M. & Byrne, P. B. B. 1998, ASP Conf. Ser. 154: Cool Stars, Stellar Systems, and the Sun, 10, 1458

Weis, E. W. 1994, AJ, 107, 1135

Wilson, O. C. 1978, ApJ, 226, 379

Chapter 15

The Whole Earth Telescope: International Adventures in Asteroseisomology

Steven D. Kawaler
Iowa State University
Ames, Iowa USA

Abstract: Today, we are beginning to probe the interiors of stars through the new science of stellar seismology. Certain stars, ranging from our own Sun to white dwarfs, undergo natural vibrations that can be detected with sensitive time-series photometry and/or spectroscopy. Since the signal we seek is an unbroken time-series to allow determination of the vibration frequencies, data from a single-site is usually incapable of uniquely identifying the pulsation modes, no matter how large the telescope being used. In many cases, the observational goals can be achieved using smallish telescopes in well-coordinated global networks. Here, I briefly describe the work of one such international network of observatories and scientists known as the Whole Earth Telescope (WET). With the WET, we have sounded out the interiors of a large number of nonradially pulsating stars. Over the past 14 years, WET has observed dozens of stars in 21 separate observing campaigns. Our team has wide span of interests, and has observed several other classes of objects such as δ Scuti stars, cataclysmic variables, pulsating sdB stars, and rapidly oscillating Ap stars.

Key words: WET, white dwarfs, photometry, network

1. INTRODUCTION - A LITTLE WET HISTORY

In 1986, astronomers from the University of Texas established a worldwide network of cooperating astronomical observatories to obtain uninterrupted time-series measurements of variable stars. The technological goal was to resolve the multi-periodic oscillations observed in these objects into their individual components; the scientific goal was to construct accurate theoretical models of the target objects, constrained by their observed

T.D. Oswalt (ed.), The Future of Small Telescopes in the New Millennium, Vol. III, 245-256.

behavior, from which their fundamental astrophysical parameters could be derived (Nather et al. 1990). This approach has been extremely successful, and has placed the fledgling science of stellar seismology at the forefront of stellar astrophysics.

This network, now known as the Whole Earth Telescope (WET), is run as a single astronomical instrument with many operators. The collaboration includes scientists from around the globe in data acquisition, reduction, analysis, and theoretical interpretation. For the first decade of its existence, the WET was headquartered at the University of Texas in Austin. After 1994, with Dr. J. Christopher Clemens as a Hubble Fellow at Iowa State University, and with support from the International Institute of Theoretical and Applied Physics (IITAP), headquarters for WET runs began to be held at Iowa State. Tables 1 and 2 list all WET observing runs through 2000.

WET co-founder Dr. Edward Nather retired as director in 1997. The directorship is now shared by Darragh O'Donoghue (at SAAO) and the author; responsibility for coordinating WET runs moved to Iowa State University. Fortunately, the founders, Drs. Ed Nather and Don Winget, continue active involvement in WET science.

Through May 2001, the WET has managed 21 observing runs (called "Xcovs"). The principal targets have been pulsating white dwarfs, ranging from the coolest (ZZ Ceti stars) through the pulsating central stars of planetary nebulae. WET has also observed a pulsating sdB star (PG~1336). In November 2000, the WET observed a rapidly oscillating Ap star.

WET observations have played a central role in over 13 Ph.D. dissertations, generated over 35 refereed publications, and been highlighted in 5 dedicated international workshops. Proceedings from the past four have been published by the journal *Baltic Astronomy*. Active WET members number approximately 50, with home institutions in 18 countries.

2. WET SCIENCE GOALS AND TECHNICAL CHALLENGES

The science goals of WET revolve around fully resolving the pulsation spectra of multiperiodic nonradially pulsating stars. Once fully resolved, we attempt to match the observed pulsation periods with those of stellar models. Success in doing so provides a determination of global properties of the pulsating stars (such as mass, rotation rate, luminosity, and distance) and, more importantly, gives us a window into the stellar interior structure.

Table 1.
WET runs 1988 – 1994

Run/Date	Target	Type	PI	Status
Xcov 1	PG 1346	CV	Winget, Provencal	published
Mar 88	V803 Cen	O'Donoghue		published
Xcov2	G29-38	ZZ Ceti	Winget	published
Nov 88	V471 Tau	CV	Clemens	published
Xcov 3	PG 1159	GW Vir	Winget	published
Mar 89				
Xcov 4	AM CVn	CV	Solheim, Provencal	published
Mar 90	G117-B15A	ZZ Ceti	Kepler	published
Xcov 5	GD 358	DB	Winget	published
May 90	GD 165	ZZ Ceti	Bergeron	published
Xcov 6	PG 1707	GW Vir	Clemens	in analysis
May 91	GD 154	ZZ Ceti	Vauclair	published
Xcov 7	PG 1115	DB	Barstow	in analysis
Feb 92	G226-29	ZZ Ceti	Kepler	published
	WET-0856	δ Scuti	Handler	published
Xcov 8	PG 2131	GW Vir	Kawaler, Nather	published
Sep 92	G185-32	ZZ Ceti	Moskalik	in analysis
Xcov 9	PG 1159	GW Vir	Winget	published
Mar 93	FG Vir	δ Scuti	Breger	published
Xcov 10	GD 358	DB	Nather, Bradley	published
May 94				
Xcov 11	RX J2117	PNN	Vauclair, Moskalik	published
Aug 94				

Table 2.
WET runs 1995 – 2001

Run/Date	Target	Type	PI	Status
Xcov 12	PG 1351	DB	Hansen	in analysis
Apr 95	L19-2	ZZ Ceti	Sullivan	in prep
Xcov13	RE 0571+14	CV	Marar, Seetha	in prep
Feb 96	CD-24 7599	δ Scuti	Handler	published
Xcov 14	PG 0122	GW Vir	O'Brien	published
Sep 96	WZ Sge	CV	Nather	in clouds
Xcov 15	DQ Her	CV	Nather	published
Jul 97	EC 20058	DB	Sullivan	in analysis
Xcov 16	BPM 37093	ZZ Ceti	Kanann, Nitta	in analysis
May 98				
Xcov 17	PG 1336 (N)	sdB	Kilkenny, Reed	in press
Apr 99	BPM 37093 (S)	ZZ Ceti	Nitta	in analysis
Xcov 18	HL Tau 76	ZZ Ceti	Dolez	in analysis
Nov 99	PG 0122	GW Vir	O'Brien	in analysis
Xcov 19	GD 358	DB	Kepler	in analysis
Apr 00	PG 1159	GW Vir	Kepler	in analysis
Xcov 20	HR 1217	roAp	Kurtz	published
Nov 00				
Xcov 21	PG 1336	sdB	Reed	in analysis
Apr 01				
Xcov 22	PG 1456	DB	Handler	in analysis
May 02	PG 1159	GW Vir	O'Brien	in analysis

For us to succeed in meeting these goals for the stars we are interested in, several technical requirements need to be met. To avoid the unresolvable confusion caused by 1-cycle/day aliases in data from a single terrestrial site, we must obtain uninterrupted time-series photometry of these rapidly variable stars. Our "instrument" must be sensitive in the temporal frequency range between 700 and 50,000 μHz (i.e. to periods between 20 and 1400 seconds). To obtain a frequency resolution of ~1 μHz (though we sometimes wish for even better) a run must last at least one week. To see the pulsations, which are small amplitude, we require an amplitude sensitivity of less than one millimagnitude for $V < 17$—that is, we need telescopes, at good sites, of 1-m aperture or more. Finally, we try to keep the maximum 1-cycle/day alias amplitudes below 20%.

We meet these technical challenges by attempting to obtain global coverage through coordinated observations at up to 15 observatories. If the weather cooperates, we can obtain 24-hour/day coverage and squash the 1-cycle/day alias. The weather rarely cooperates to this degree. However, the weather is a random variable—and with persistence, we can obtain data from all sites around the globe in a typical 2-week WET run. This means that while the coverage may not be continuous, we still can get a high duty cycle (frequently better than 70%) and, most importantly, cover all phases of the 24-hour rotation period of our Earth. Nather et al. (1990) illustrate the significant reduction of these 1-cycle/day alias structures that we can achieve with WET.

Success also requires using detectors of uniform wavelength sensitivity, and care in combining data from different telescopes. These chores are minimized with a "standard" photometer design (Kleinman et al. 1996) and a standard set of software tools. By working with astronomers who are interested in the science of the WET and who are also those at the telescopes during the observations, we are able to meet the technical challenges. Because of the sensitive nature of the signal and instrumentation needed to measure it, we have found that automating the data collection is not practical or wise. We really need to have observers—observers who care about the data—at the telescopes.

3. WET OPERATIONS

The WET is an organization that allows astronomers from around the world work together—first in planning an observing run, next in obtaining the data during the actual observing campaign, and finally in analyzing the data and publishing the results. With over 50 astronomers from more than a dozen countries, this mode of operation would be nearly impossible without the use of the Internet to link observers and scientists. By using computers linked through the Internet, such "virtual collaborations" are viable. In the case of the WET, we have shown that not only can we have a virtual collaboration, we can actually do the science in a networked fashion—thus the term "virtual collaboratory."

3.1 Target Selection and Observing Time

The WET can only obtain high-quality data on one (or maybe two) targets at a time. The resources (in both time and money) needed for a WET run are considerable. Given these facts, we can only observe a limited number of targets, and selection of which stars to observe is an important and difficult assignment.

Approximately 8 WET members serve on a "Council of the Wise" or simply COW. Once or twice a year, proposals for WET targets are solicited from the entire community by the COW. The COW then decides, based on the science case as well as the case for the need of the peculiar strengths of the WET, which targets are to be observed. The successful proposer becomes the PI for that target, and prepares a generic scientific justification for use in applying for telescope time. The PI works with the Associate Director for WET Operations (or ADWO) in coordinating proposals for the various WET sites.

The WET has no dedicated telescopes. When the target is approved, all members of the WET collaboration prepare and submit proposals to their local observing facility for time to participate in the run. These proposals are coordinated so that the separate time allocation committees are aware of the collaborative nature of each proposal. Observing sites range from smaller instruments that are local to the home research institutions, through national facilities such as KPNO, CTIO, OHP, and SAAO.

3.2 The WET Control Center

With up to 15 observers around the world observing the same target, there is ample room for confusion and waste of resources. To maximize the effectiveness of our collaborative observations, during a WET run (which typically lasts 2-3 weeks) we maintain a continually staffed control center. At the control center (or HQ) are WET members with significant experience in observing, data reduction and analysis. HQ personnel usually include the PI for the target of the run.

The HQ staff communicates with all observers daily to update them on the progress of the run, and to help solve any problems that may arise at a site.

They also provide an independent test of the observatory and data acquisition clocks. Throughout the run, as an observer completes his/her night of data collection, s/he sends the data via e-mail or ftp to the HQ. HQ staff then reduces the data and examines it for quality, timing errors, etc.

With a fully staffed HQ, we can maintain an up-to-date light curve and compute current pulsation spectra for analysis. The reductions and analysis are made available to all members of the WET team, including those not directly involved in the run, via the WWW. The WWW site for the run-in-progress, linked to the main WET website (currently the URL is http://wet.iitap.iastate.edu), shows the log of each individual run, the combined lightcurve, and amplitude spectrum. To help plan the immediate future, and assess the run, the website also has weather updates and other status variables from each site. An open discussion page allows collaborators to discuss the results so far, speculate on new discoveries, etc.

Another important role of the HQ staff is allocating observatories to secondary targets. When overlapping sites are both active, the HQ staff can move one of the observers to a secondary target, allowing the WET to obtain data on more than one target at a time. Occasionally, we have been able to get nearly 48 hours of data in one 24-hour period!

At the end of a run, the HQ computers have accumulated all of the raw data. In addition to the raw data, we archive the first-look reductions by the HQ staff, plots of light curves, spectra, and other reductions, e-mail messages from all sites pertaining to the run, and the current version of all reduction and analysis software. A copy of this archive, which fits easily on a standard CD, is taken home by the PI for a more leisurely and thorough analysis.

3.3 Data and Publication

Despite the apparent simplicity of time-series photometry (i.e. low data volume compared to imaging or spectroscopy) correct reduction and analysis of WET data is very difficult. The PI has the responsibility to handle the data and draft a paper (or papers) for publication, but this can take years. To try to stimulate our members, the WET has adopted an 18-month proprietary period, wherein the PI has all rights to publishing the data. Typically the first paper based on a WET run will have as authors all observers and HQ staff that participated in the run. Author order questions are solved by the nature of the collaboration—first-author honors go to the PI and the person or people who did the analysis after the run was over. Authorship continues starting at the geographic location of the HQ, and working west. Subsequent papers that use the data have an author list that includes only those who participated in the additional analysis.

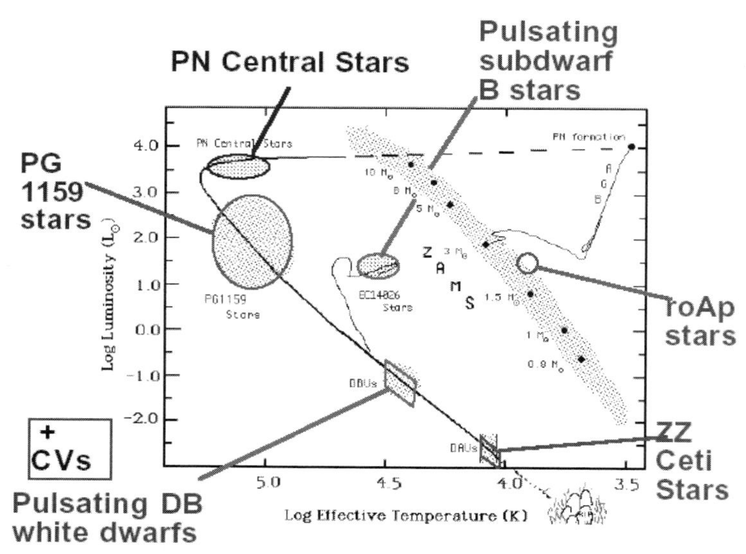

Figure 1. A schematic Hertzsprung-Russell diagram including representative main sequence and evolutionary tracks. Indicated on this diagram are the classes of variable stars that have been examined as targets of WET observations (after Kawaler 1986).

After the proprietary period is over, or the first paper is published, the rest of the collaboration is free to publish independent analysis of the data in whatever form they choose. Of course, all WET collaborators have free access to all the data from the time of the run forward. The raw data and all reduction software are publicly available after the proprietary period is over.

4. WET SCIENCE

In this brief review, there is no chance to fully describe the many exciting science results of the WET, or to adequately describe the wide range of impact of WET (and WET-style) data. This section simply gives a few "headlines" about these sample targets, with references to papers in the literature. Reviews of many WET results can be found in the proceedings of the last three WET workshops (Meistas & Vauclair 2000, Meistas & Moskalik 1998, Meistas & Solheim 1995). See also Kawaler & Dahlstrom (2000) for a broad overview of white dwarf stars and the impact of WET.

Figure 1 illustrates the range of variable stars that the WET has investigated in terms of the HR diagram. Clearly, we have covered a large portion of this map, with the inclusion of white dwarf stars over six orders of magnitude in luminosity, roAp and δ Scuti stars, and the pulsating sdB variables.

Sample amplitude spectra of WET observations are shown in Figure 2. This figure shows the remarkable uniformity in the pulsation spectra of a wide variety of stars. It is this similarity in photometric behavior that gives WET such a wide range of targets. It is an instrument designed with white dwarfs in mind, but applicable to a much larger range of targets.

4.1 PN central stars and pre-white dwarfs

WET has observed one central star in detail: RX J2117. This star, first identified as an evolved star by its identification as a ROSAT X-ray source, was found to be variable in 1993. This star is a complex pulsator—a paper with analysis of the WET data is in preparation (Moskalik et al. 2001).

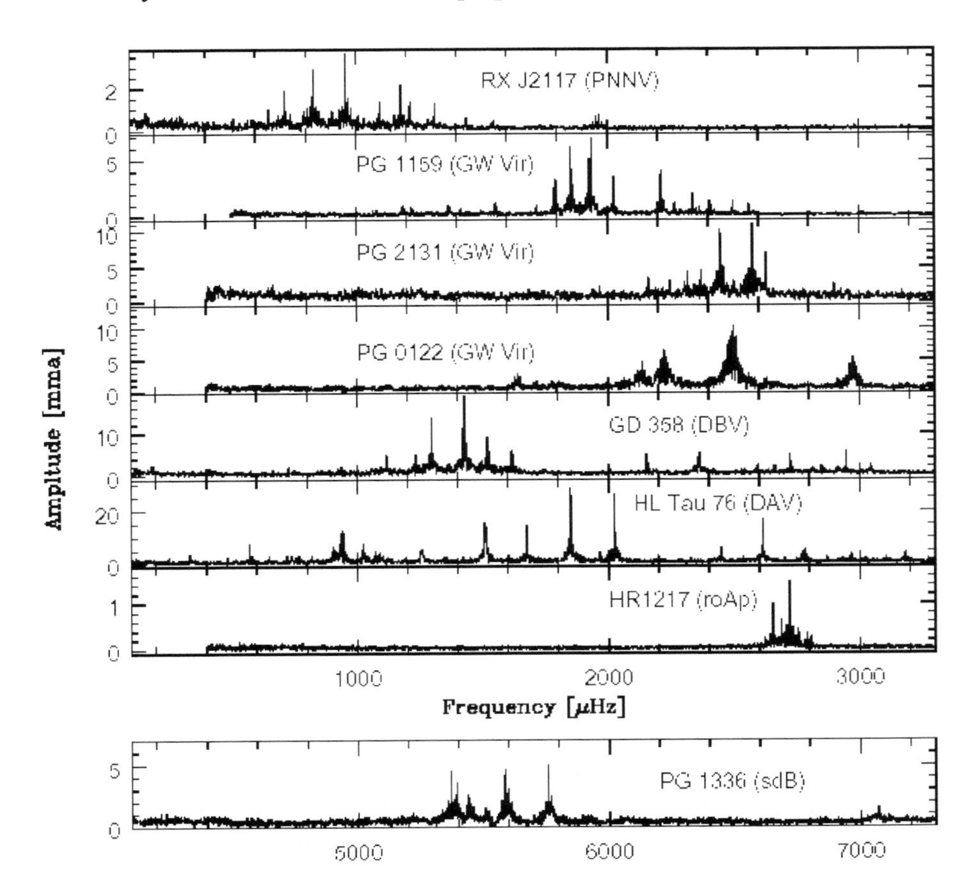

Figure 2. Pulsation spectra from WET. Panels show a PN central star (PNNV), pre-white dwarfs (GW Vir stars), pulsating DB white dwarf (DBV), and ZZ Ceti star (DAV). Bottom two panels show spectra for a rapidly oscillating Ap star and subdwarf B star.

Arguably the most productive WET target has been PG 1159, a pulsating hydrogen-deficient pre-white dwarf. Observations in this star from 1991 have yielded accurate determinations of its mass, distance, and subsurface composition profile (Winget et al. 1991, Kawaler & Bradley 1994). These data have been combined to reveal a secular period change that challenges theoretical models of these stars (Costa et al. 1999). We continue monitoring PG 1159—and it continues to bear fruit.

Another pre-white dwarf, PG 2131, has provided an important test of asteroseismological analysis method (Reed et al. 2000). Perhaps the acme of asteroseismological inference is the distance to the pulsating star. All of the many analysis steps must be correct to derive an accurate distance. PG 2131 has a close companion that is a low-mass main sequence star. The companion has been resolved (with a separation of 0.3 arcseconds) in a pair of images obtained by the Hubble Space Telescope. Reed et al. (2000) analyze these images and obtain a spectroscopic parallax distance to the system that they compare with the seismological distance determined through a reanalysis of the WET data of Kawaler et al. (1995). Reed et al. (2000) find that the two distance agree within the 1σ level, with the seismological distance having a small formal error of about 10%.

Stellar probes of other areas of physics now include the coolest of the pre-white dwarfs, PG 0122. The evolution of this star should be dominated by energy loss through neutrino emission—the neutrino luminosity exceeds its photon luminosity by a factor of about three. O'Brien & Kawaler (2000) show how to use WET observations of this star, coupled with single-site monitoring, to measure the influence of neutrinos on the change of the pulsation periods.

4.2 Cooler White Dwarfs: The DB and DA (ZZ Ceti) Pulsators

For the pulsating DB white dwarfs, the principal object has been GD 358, first observed with the WET in 1990 (Winget et al. 1994). Results for GD 358 show it to be the most prolific of the pulsators, with over 185 separate modes visible. This high density of modes provides a good probe of the outer layers (Bradley & Winget 1994, Metcalfe et al. 2000), and constraints on models of white dwarf chemical evolution (Dehner & Kawaler 1995) Its pulsation patterns change dramatically from year to year, providing grist for the nonlinear pulsation theorist's mill (Vuille et al. 2000).

ZZ Ceti stars have been observed throughout the history of the WET. Results for ZZ Ceti stars at the hot end of the instability strip are collected in Clemens (1994), while Kleinman (1995) did a comprehensive study of those at the cool end of the strip. One of the most exciting individual targets has been BPM 37095, a ZZ Ceti star with a mass high enough for it to have a crystalline C/O core (Montgomery & Winget 1999); observations by WET are included in the Ph.D. dissertation of A. Nitta (2000). Finally, WET observations have played a key role in the determination of the cooling rate of the ZZ Ceti star G117-B15A by Kepler et al. (2000b).

4.3 Cataclysmic variables, subdwarf B stars, and main sequence stars

One of the early classes of targets was CVs that showed short-period phenomena. See, for example, the results for PG 1346 (Provencal et al. 1997), and AM CVn (Solheim et al. 1998).

WET has been used to look at short-period δ Scuti stars—because we do relative photometry, our pulsation spectra are unreliable for periods longer than 20 minutes or so. Table 1 lists those stars that have had successful observations, including one ("WET-0856") that was discovered to be a variable during a WET run because it was used as a comparison star (Handler et al. 1996). In 1999, we observed a pulsating sdB star within a short-period eclipsing binary. The most recent WET run had as its target the rapidly oscillating Ap star HR 1217.

5. CONCLUSIONS

The WET, a network of "small" telescopes for global photometry of variable stars, has enabled a broad range of inquiry in stellar astronomy. The first 14 years of the WET have been extremely fruitful. This "new" way of observing produces front-line data using modest instrumentation employed in creative ways. WET results have shaped our view of white dwarf evolution ranging from the formative stages down through the ZZ Ceti instability strip.

At the recent European Workshop on White Dwarfs, for example, most of a day contains papers directly related to WET observations and analysis, as well as results from WET that are directly incorporated into atmospheric and evolution studies. Many current white dwarf researchers earned their Ph.D.s working on WET or using WET data and results—and have gone on to expand their careers well beyond pulsating white dwarfs.

Sites on the WET network are testing CCD photometry systems that can provide rapid photometry (cycle times of 10 seconds or less) of much fainter stars for a given aperture. Additional "expansion" of the WET network has already included simultaneous observations with time-resolve Hubble Space

Telescope spectroscopy (Kepler et al. 2000a). With new classes of objects, of which the WET has only made some tentative observations, there is great promise for it to continue its legacy of providing entirely new views of the interiors and environments of interesting stars.

6. ACKNOWLEDGEMENTS

The author expresses thanks to the National Science Foundation for support of Whole Earth Telescope science under Grant No. AST-9876655 to Iowa State University, and support from Iowa State University, through the International Institute for Theoretical and Applied Physics.

7. REFERENCES

Bradley, P.A. & Winget, D.E. 1994, ApJ, 430, 850
Clemens, J.C. 1994, Ph.D. dissertation, University of Texas
Costa, J.E S., Kepler, S.O., & Winget, D. E. 1999, ApJ, 522, 973
Dehner, B.T. & Kawaler, S.D. 1995, ApJL, 445, L141
Dreizler, S. & Heber, U. 1998, \aa, 334, 618
Handler, G. et al. (the WET collaboration) 1996, \aa, 307, 329
Kawaler, S.D. 1986, Ph. D. dissertation, University of Texas
Kawaler, S.D. & Bradley, P.A. 1994, ApJ, 427, 415
Kawaler, S.D. et al. (The WET Collaboration) 1995, ApJ, 450, 350
Kawaler, S.D. & Dahlstrom, M. 2000, American Scientist, 88, 498
Kleinman, S.J., Nather, & T. Phillips, 1996, PASP, 108, 356
Kleinman, S.J. 1995, Ph.D. dissertation, University of Texas at Austin
Kepler, S.O., Robinson, E.L., Koester, D., Clemens, J.C.,
Nather, R.E., & Jiang, X.J. 2000a, ApJ, 539, 379
Kepler, S.O., Mukadam, A., Winget, D.E., Nather, R.E.,
Metcalfe, T.S., Reed, M.D., Kawaler, S.D., & Bradley, P.A. 2000b, ApJ, 543, 185
Meistas, E. & Solheim, J.E. 1995, *The Third WET Workshop Proceedings*, Baltic Astron., 4, 104
Meistas, E. & Moskalik, P. 1998, *The Fourth WET Workshop Proceedings*, Baltic Astron., 7, 1
Meistas, E. & Vauclair, G. 2000, *The Fifth WET Workshop Proceedings*, Baltic Astron., 9, 1
Metcalfe, T.S., Nather, R.E., & Winget, D.E. 2000, ApJ, 545, 974
Montgomery, M.H. Winget, D.E. 1999, ApJ, 526, 976
Moskalik, P. et al. (the WET Collaboration) 2001, in press
Nather, R.E., Winget, D.E., Clemens, J.C., Hansen, C.J., & Hine, B.P. 1990, ApJ, 361, 309
Nitta, A. 2000, Ph.D. dissertation, University of Texas at Austin
O'Brien, M.S. & Kawaler, S.D. 2000, ApJ, 539, 372
Provencal, J. et al. (the WET Collaboration) 1997, ApJ, 480, 383
Reed, M.D., Kawaler, S.D., & O'Brien, M.S. 2000, ApJ, 545, 429
Solheim, J.E. et al. (the WET Collaboration) 1998, \aa, 332, 939
Vuille, F. et al. (the WET Collaboration) 2000, MNRAS, 314, 689
Winget, D.E. et al. (the WET Collaboration) 1991, ApJ, 378, 326
Winget, D.E. et al. (the WET Collaboration) 1994, ApJ, 430, 839

Chapter 16

WIYN Open Cluster Study: The Next Generation

Ata Sarajedini
University of Florida
Gainesville, Florida USA

Robert D. Mathieu
University of Wisconsin
Madison, Wisconsin USA

Imants Platais
Universities Space Research Association
Washington, D.C. USA

Abstract: We summarize the present status and discuss future directions of the WIYN Open Cluster Study (WOCS). The goals of the WOCS collaboration are 1) to establish a multidimensional database of properties for stars in selected open clusters, including photometric (colors, magnitudes, magnitude variations), astrometric (positions, proper motions), and spectroscopic (chemical abundances, radial velocities) characteristics, and 2) to investigate a broad set of astrophysical questions through the study of open clusters. In the area of astrometry, WOCS has underway an ambitious program to combine existing 4-m plate collections with newly acquired images from large-format CCD mosaics. This effort will continue into the next decade as we complete proper-motion studies of northern and southern open clusters while at the same time expanding the temporal baseline over which astrometric studies of these clusters are performed. The WOCS photometric effort is currently focused in the optical (UBVRI) regime; we plan to extend the photometry to the infrared (JHK) passbands. Optical and infrared photometry of comparable depth and precision will facilitate a number of scientific investigations that are currently

T.D. Oswalt (ed.), The Future of Small Telescopes in the New Millennium, Vol. III, 257-272.

not possible. The WOCS radial-velocity program has obtained several thousand radial-velocity measurements of samples containing hundreds of stars in each of four open clusters. The number of clusters being surveyed will grow, and with an increase of a factor of a few in precision the project will expand into the detection of brown dwarf and large planet companions. Finally, WOCS is committed to improving the accuracy of metallicity determinations in fundamental clusters by a factor of a few through careful control of systematic errors. These abundance analyses also include lithium, which will prove an ideal probe of diffusion, mixing, mass loss, and other processes in stellar evolution. The WOCS observations have primarily been made with the WIYN 3.5-m and 0.9-m telescopes, and we anticipate that these facilities will continue to play the central role for northern WOCS studies.

Key words: cluster, WOCS, metallicity

1. INTRODUCTION

Studies of star clusters define the entire range of fundamental stellar properties, which themselves serve as the observational foundation for much of astrophysics. Perhaps most importantly, the common origin of cluster stars ensures the same age and chemical composition for a wide range of stellar masses. This exceptional feature fully justifies the designation of star clusters as 'astrophysical laboratories'.

The subjects of this discussion – open star clusters – provide a unique environment where the properties of stellar evolution, binary stars, stellar rotation and activity, etc. can be observed and understood to an extent that is perhaps impossible anywhere else. In addition, open clusters play a crucial role in defining the Initial Mass Function and tracing the chemical enrichment history throughout the Galaxy. Understanding these issues has broad implications for the formation and evolution of galaxies, including those at the highest redshifts which are being observed with the Hubble Space Telescope and the new generation of large telescopes.

To fully exploit the potential offered by open clusters, it is essential to create a multidimensional database of cluster star properties. An invaluable archival approach to attaining this goal has been realized in the WEBDA[1] — an extensive database for stars in open clusters developed by J.-C. Mermilliod. However, the user of this database will quickly find that, except for a few well-studied clusters, the available data are not adequate to address the issues mentioned above to the fullest extent possible.

Recent advances in instrumentation and telescope design provide an opportunity to improve dramatically the scope and quality of observational

[1] WEBDA is available on the WWW at the URL http:/obswww.unige.ch/webda/

data. One such facility is the WIYN 3.5-m telescope located on Kitt Peak near Tucson, Arizona. Its two primary instruments, a CCD mosaic imager and a multiobject spectrograph, are optimal for systematic and comprehensive studies of open clusters (Mathieu 2000). A no less important aspect is the guaranteed access to this telescope by the members of the WIYN consortium (University of Wisconsin - Madison, Indiana University, Yale University and National Optical Astronomy Observatory). To optimally exploit the data collecting ability of this facility and other supporting telescopes, a collaboration was established which combines a diverse set of skills and interests into a pointed and coordinated effort known as the WIYN Open Cluster Study (WOCS).

The goals of WOCS are twofold:

- To create a multidimensional database of properties for stars in selected open clusters, including photometric (colors, magnitudes, magnitude variations), astrometric (positions, proper motions), and spectroscopic (chemical abundances, radial velocities) characteristics. All of these various types of data are being obtained with the best instrumentation of its kind and are reduced and analyzed using uniform techniques. As a result, this database provides coverage to much fainter magnitudes and, consequently, to lower masses, in comparison with the existing studies. The hallmarks of the WOCS database are comprehensiveness, completeness and precision.
- To investigate a broad set of astrophysical questions through the study of open clusters. Subjects that are actively being pursued include: detailed testing of core convective overshoot and the implications for stellar lifetimes; photometric monitoring of periods to probe angular momentum evolution; white dwarf sequences as a method of age-dating clusters; Fe, CNO, and Li abundances for studies of internal stellar processes (e.g. mixing, diffusion), Galactic chemical evolution, and primordial abundances; binary populations, stellar evolution in close binary systems, and mass functions (initial and present-day).

This paper describes the present status of the WOCS program and the expected future advances in astrometry, photometry and spectroscopy of open clusters. In each case, we make an effort to highlight the role of small telescopes in enabling the various aspects of WOCS. Furthermore, we provide a look into the next decade and outline the directions in which WOCS is likely to go.

Table 1.
Open Cluster Targets

Cluster	l	b	$(m\text{-}M)_0$	E(B-V)	[Fe/H]	Age (Gyr)
NGC 188	122.8	+22.5	11.4	0.09	0.0	7.0
NGC 1039	143.6	-15.6	8.6	0.1	-0.1	0.2
NGC 1817	186.1	-13.1	12.2	0.27	-0.3	0.8
NGC 2158	186.6	+1.8	14.5	0.42	-0.2	2.0
NGC 2168	186.6	+2.2	10.3	0.20	-0.2	0.1
NGC 2264	203.0	+2.2	9.4	0.07	0.0:	0.003
NGC 2420	198.1	+19.7	12.1	0.05	-0.3	2.5
NGC 2451	252.0	-7.0	6.4	0.00	-0.5:	0.04
NGC 2506	230.6	+9.9	12.6	0.05	-0.4	2
NGC 2682	215.6	+31.7	9.8	0.04	0.0	4.5
NGC 6791	70.0	+11.0	13.4	0.15	+0.3	8
NGC 6819	74.0	+8.5	12.3	0.16	+0.1	2.4
NGC 7092	92.5	-2.3	7.3	0.03	-0.1	0.3
NGC 7789	115.5	-5.4	12.5	0.29	-0.1	1.2

2. WOCS CLUSTER SELECTION

To ensure the WOCS consortium achieves its goals, we applied the following criteria in selecting the primary program clusters: 1) a range of ages from a few Myr to ~ 10 Gyr; 2) a range of metallicities from [Fe/H]= -0.5 to +0.3; 3) as rich as possible to provide large samples of stars at each point in the metallicity-age domain; 4) distances ~1 kpc to ensure complete spatial coverage within the 1 deg-square area typical of modern imagers and multi-object spectrographs; 5) a range of right ascension so as to distribute observations throughout the year. Other open clusters of special scientific interest are also included in the WOCS observational program.

Our selection of program clusters for this study is given in Table 1 along with the galactic longitude and latitude (l,b), distance modulus $(m - M)_0$, reddening $E(B{-}V)$, metallicity [Fe/H], and an estimated age in Gyr. All of these data are taken from the most recent publications.

3. ASTROMETRY

3.1 Proper Motions and Cluster Membership

An average open cluster typically contains less than a thousand cluster members, although some of the WOCS clusters are very rich including several thousand stars. Usually located in the Galactic disk, open cluster stars are blended in projection with large numbers of field stars, which in many respects have properties indistinguishable from those of the cluster members. This implies that any selection of cluster members could be contaminated by field stars. This contamination is usually negligible at the bright end of the main sequence but inevitably increases towards the fainter magnitudes mainly because of a much larger spatial volume encompassing field stars at these magnitudes. Color-magnitude diagrams presented in Figure 1 (see Section 4) exemplify this effect. The use of proper motions is one way to substantially lower the fraction of field stars among the selection of cluster members. However, inspecting color-magnitude diagrams like those presented in Figure 1, it may appear that astrometric membership is not that critical. In the case of rich open clusters *and* their central areas, this may be partially true; however, the CMD alone is considerably less useful in poorly populated clusters, in the outer parts of clusters and in the sub-solar mass range. In fact, a complete Present Day Mass Function can only be determined using a reliable astrometric membership study, ideally in combination with photometric and radial-velocity information. In addition, the absolute proper motion of the whole cluster provides two velocity components necessary in calculating the galactic orbit of the cluster.

Any proper motion study requires stellar positions at a minimum of two different epochs. The precision of ground-based astrometry is limited by instrumental factors and atmospheric turbulence; hence, to obtain high-precision proper motions, a large epoch difference ranging between 20 to >100 years is a necessity. The larger the epoch difference, the more precise the proper motions. Because of this, the old photographic plates taken as early as the 1890s with the 13-inch so-called normal astrographs are still a valuable source of first epoch observations. The astrometric cluster membership studies based upon such plate material from a variety of astrographs and refractors are summarized by van Leeuwen (1985) and van Altena et al. (1993). However, these studies are always constrained by the limiting magnitude of the old plates, usually no deeper than 13th to 16th photographic magnitude. The beginning of the era of large reflectors and Schmidt cameras around the 1950s pushed this limiting magnitude down to V=20 and beyond. While the large-scale Schmidt plates are very useful in studies of nearby clusters, e.g. the Pleiades (Schilbach et al. 1995), the plate material from various large reflectors is largely untapped for astrometry. A probable explanation for this might include the relatively recent epochs of such plate collections, the presence of large geometric distortions in the focal plane of reflectors, and a sweeping change from photographic plates to CCD detectors in the late 1980s. The last two reasons combined actually complicate considerably the process of astrometric reductions. The pioneering efforts of WOCS are aimed at applying state-of-art reduction procedures to solve these complex issues.

The astrometric part of the WOCS project is one of the most extensive attempts to make use of the KPNO and CTIO 4-m telescope plate collections for open cluster studies. However, these plates comprise only one epoch. To get the second epoch, the only choice for direct imaging in the last several years has been large-format CCD mosaics (Luppino, Tonry, & Stubbs 1998; Groom 2000). We should note that the astrometric performance of a CCD mosaic as an integrated imaging device (as opposed to individual CCD chips) is not known very well. It is clear though that knowledge of the mosaic's metrology and its stability with time are two critical parts of precision astrometry (Platais et al. 2001). We attempted to acquire this information using specifically constructed astrometric standard fields. The Lick (Mt. Hamilton, California USA) and Yale-San Juan (El Leoncito, Argentina) 51-cm double astrographs served to create such standard fields in both hemispheres, thus providing a fruitful link between these very different telescopes. Once a good distortion-free CCD 'plate' is generated, the remaining reductions follow the standard procedure for deriving the proper motions. On average, the resulting proper motions are good to ±0.5 mas yr^{-1} (standard error) and go as faint as $V \approx 22$. The final products are cluster membership probabilities to a limit at least six magnitudes fainter (Platais et al. 2000) than nearly all existing proper motion studies. These deep cluster membership estimates provide an excellent separation between the cluster and field stars.

3.2 Cluster Astrometry in the Next Decade

In terms of ground-based astrometry we do not expect a radical improvement in the quality of proper motions. What certainly will increase is the number of clusters having excellent determinations of proper motions. To date, only a hundred out of >1,200 catalogued open clusters have at least one determination of their relative proper motion. Within this hundred, only a couple of dozen can be considered as having comprehensive sets of high-quality proper motions. WOCS is in a good position to considerably increase the number of clusters with high quality astrometry. In the near future, the WOCS goal is to complete five northern hemisphere clusters (NGC 188, 2264, 2682, 6791, 6819) and then continue to work on a similar-size-selection of the southern hemisphere open clusters. We note that those observatories having appropriate telescopes and CCD mosaic devices should make a concerted effort to calibrate the metric of CCD mosaics, archive the CCD frames and make them publicly available. A decade later these frames will serve as first epochs much like the photographic plates do today. In terms of limiting magnitude no planned space astrometry mission can compete with the CCD imaging obtained even at a 1-m class ground-based telescope. This rather exceptional situation may last well over a decade.

4. PHOTOMETRY

4.1 The Present

In its current incarnation, WOCS photometric study of a given cluster is comprised of three elements: deep (V~26) UBVRI photometry of the cluster core with the WIYN 3.5-m telescope, intermediate-depth (V~21) UBVRI photometry of the entire cluster, and photometric monitoring to provide a comprehensive survey for variable stars. Examples of the first two elements for the open cluster NGC 188 can be found in von Hippel & Sarajedini (1998) and Sarajedini et al. (1999). The upper panels of Figure 1 show the NGC 188 color-magnitude diagrams (CMDs) from these two studies while the lower panels of the same figure display analogous diagrams for the open cluster M35 (NGC 2168; Deliyannis et al. 2002; von Hippel et al. 2002).

Reductions are done via iterative multiple point-spread-function fitting using the DAOPHOT II software suite (Stetson 1994). Multiple images yield between 5 and 10 measures of the magnitude of each star in each filter. These are then combined using a weighted average giving us a magnitude and an error. This allows us to attain the 1% relative precision that is the standard of WOCS photometry.

Equally importantly is the photometric calibration of both the 3.5m deep core photometry and the 0.9-m wide-field photometry. As noted above, we utilize the well-measured calibration standards from the list of Landolt (1992) typically observing between 50 and 100 such stars per night. A given night of observations is deemed to provide all-sky photometric performance if the color-corrected instrumental magnitudes of the standard stars exhibit a variation of less than 1.5% (3% in the U band) relative to their standard values. However, based on our experience and anecdotal evidence gathered by others, systematic errors of order ~2% to ~3% (on top of the random errors inherent in the transformation equations) can occur in the determination of the extinction coefficients, color terms, and zeropoints so that typically two to three photometric nights are needed to attain the 2% level of accuracy that we have set as the goal for our absolute photometric scale.

With photometry of this kind (areal coverage, depth and quality) in the UBVRI Johnson-Kron-Cousins system, we have been able to study a number of important aspects of open clusters. For example, careful analysis of our photometry has yielded cluster reddenings and distances

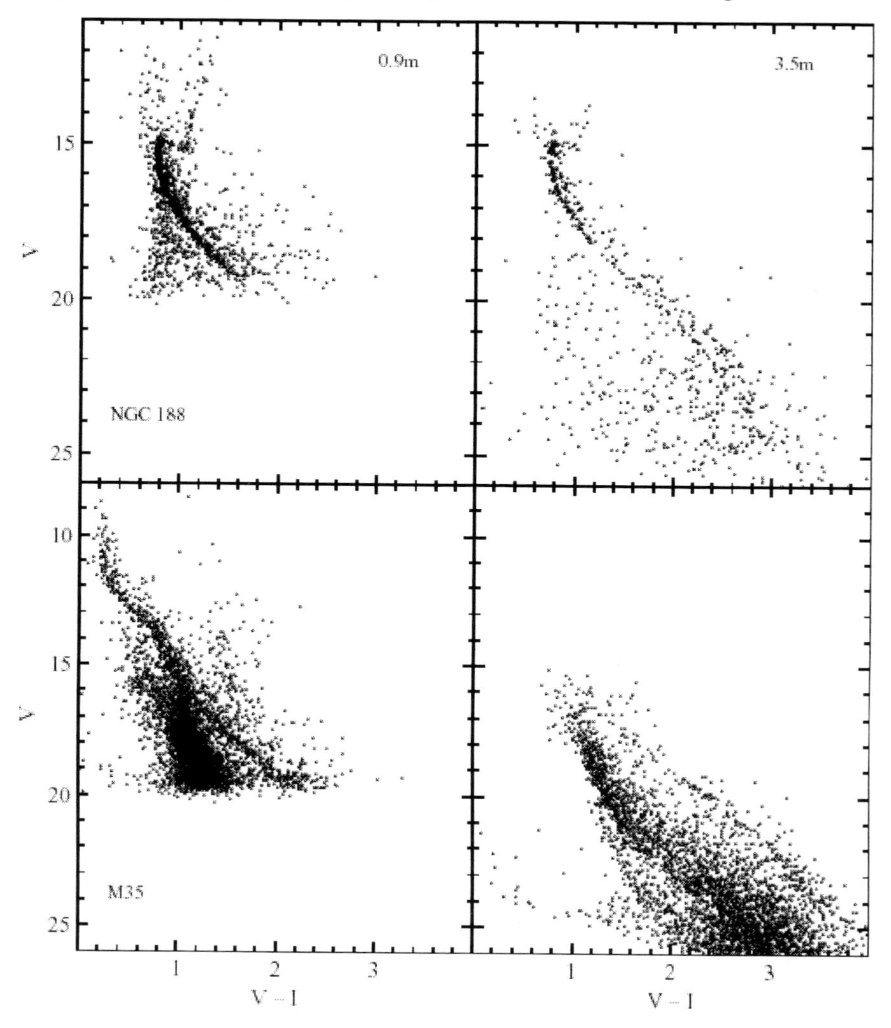

Figure 1. Color-magnitude diagrams in the VI passbands of NGC 188 (top) and M35 (NGC 2168, bottom) based on 0.9-m (left) and 3.5-m (right) telescope WOCS observations.

with unprecedented accuracy. Our precise stellar colors and the cluster reddenings are used to set the effective temperature in spectroscopic abundance studies; the results of these investigations yield high quality metallicities; these are in turn used along with the distances and reddenings to determine cluster ages via comparison with theoretical isochrones (Sarajedini et al. 1999).

4.2 The Future

The existing photometric effort of WOCS has been focused in the optical realm (UBVRI). However, infrared photometry of open clusters also holds great promise. Thus far, most of this effort has been geared toward the study of embedded young star clusters in which the primary goal has been to investigate the star (and planet) forming properties of these systems. The infrared is a natural choice in this regard because of its ability to minimize dust obscuration ($A_K \sim 0.1A_V$), which is often a problem in these objects as they are still enshrouded in their molecular clouds.

An often overlooked use of infrared photometry is in the study of 'normal' open clusters - that is to say, clusters that do not necessarily suffer from extreme extinction (e.g. most of the clusters in Table 1). There is very limited published work on this topic even though infrared CMDs of these clusters would be useful. Houdashelt et al. (1992) present JHK photometry for ~100 red giants and clump stars in eight open clusters – an average of ~15 stars per cluster. They couple these IR data with optical measurements in order to investigate the temperatures and luminosities of the stars as well as the metal abundances of the clusters (see also Tiede et al. 1997). More recently, there has been a flurry of publications from the Padova group presenting IR photometry for several open clusters (Carraro et al. 2001, and references therein). These efforts represent the current state-of-the-art in open cluster IR photometry. However, in the next 5 to 10 years, with larger format and higher throughput IR arrays coming online and available on small telescopes (i.e. <4-m), it will be possible for infrared photometry to approach the optical data in terms of areal coverage, photometric depth, and precision.

Thus, in the near future, many of the clusters in Table 1 will possess UBVRIJHK photometry extending to V=K=21 with photometric precisions of ≤1% over most of their spatial extent. As a taste of what to expect, we have extracted JHK photometry of the open cluster M35 from the Second Incremental Release of the 2MASS Point Source Catalog[2] and matched it to our existing WOCS wide-field optical data to produce the CMDs shown in Figure 2. The left panel displays the (V,U–I) CMD of M35 based on the optical photometry of Deliyannis et al. (2002). The right panel shows our combined WOCS+2MASS (V,U–K) CMD plotted on the same scale. In

[2] http://irsa.ipac.caltech.edu

interpreting these diagrams, we must keep in mind that the photometric precision of the optical data is significantly higher than that of the IR data, which will become steadily more precise as time goes on. We first note that the color scale is roughly 5 magnitudes in (U–I), but is extended to 7 magnitudes in (U–K), which, for the first time, rivals the scale of the vertical (magnitude) axis. In addition, the (U–K) diagram provides a better definition of the principal cluster sequences (especially the binaries located brightward of the main sequence) as well as a better separation of the cluster stars from the field stars. With higher quality IR photometry, we can expect these advantages to be even more evident in the future. There are a number of scientific questions that one can investigate with data such as these.

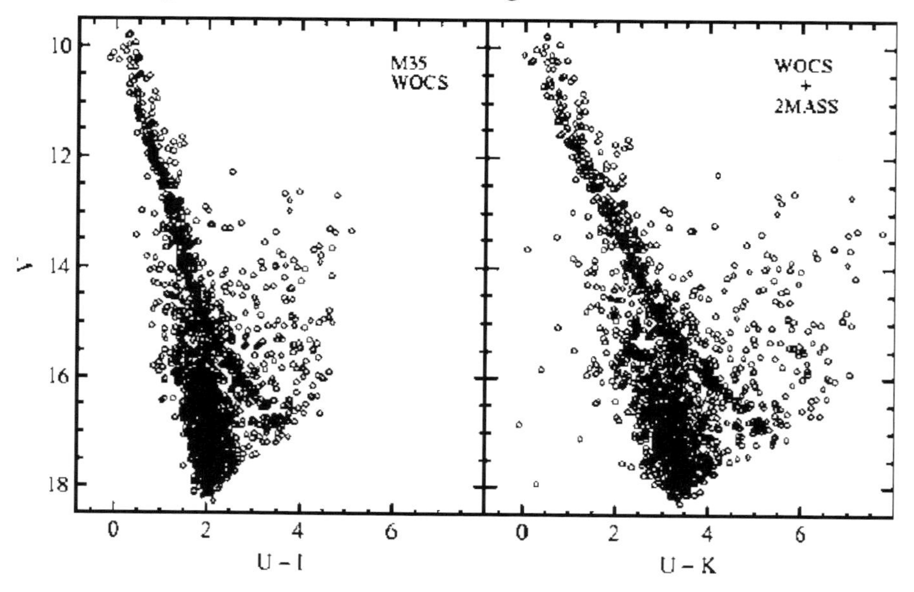

Figure 2. Color-magnitude diagrams for M35 based on our WOCS optical photometry in the (V, U-I) plane (left) and the combined WOCS and 2MASS photometry in the (V, U-K) plane (right).

First, these data would allow us to more accurately test theoretical stellar evolutionary models - in particular, the color-T_{eff} transformation especially via
the temperature sensitive V–K color index. This is facilitated by the fact that we will have optical/IR photometry for clusters with a significant range of metallicities (+0.3 ≥ [Fe/H] ≥ –0.5) and ages (0.003 Gyr ≤ Age ≤ 8 Gyr) containing stars that span a range of temperatures and evolutionary stages. In addition, we will already possess well-determined values of distances, reddenings, and metal abundances for the clusters from WOCS optical photometry and high dispersion spectroscopy.

Second, IR photometry of 'nearby' stellar populations with a range of ages and chemical compositions (open clusters along with globulars) is *crucial* to interpreting the properties of distant (high-redshift) galaxies observed with ground-based (e.g. Gemini, VLT) and space-based (e.g. SIRTF, NGST) facilities. These observations will consist primarily of integrated photometry and spectroscopy in the *infrared* portion of the electromagnetic spectrum. Because of the range of redshifts, to interpret these observations, we will need to understand the properties of stellar populations with a range of ages and metallicities in both the optical and infrared regimes. Having an existing database of such data for open (and globular) clusters is an important asset in studies such as these.

Another direction that would be fruitful for WOCS to take would be a search for old open clusters (t > 1 Gyr) in the infrared wavelength regime. This would be similar to the enormously successful effort undertaken by Phelps et al. (1994) except in the JHK passbands. This survey could be started by simply analyzing the 2MASS point source catalog. One would look for stellar clustering on small scales and construct CMDs for any candidate open clusters. The candidates would then be observed with larger telescopes to obtain deeper photometry. This has already been done to a limited extent by Hurt et al. (2000) who discovered a spiral galaxy and two globular clusters near the Galactic plane using the 2MASS survey products. In the future, one would like to conduct an all-sky IR imaging survey with a deeper limiting magnitude than that of 2MASS (K~14). From such a deeper survey, building upon the work performed on the 2MASS catalog, there are likely to be a significant number of old open clusters discovered.

5. SPECTROSCOPY

The spectroscopic studies of WOCS comprise three components: stellar radial-velocity measurements, abundance measurements, and stellar activity studies. We discuss each of these in turn.

5.1 The Present – Radial Velocities

Since 1996 we have obtained precise radial velocities with the WIYN 3.5-m telescope and the Hydra Multi-Object Spectrograph (MOS) for stars in the initial set of primary WOCS clusters – NGC 188, NGC 2168, NGC 2264 and NGC 6819 – with an emphasis on late-type stars ($(B - V) > 0.4$) brighter than V ≈ 16.5. The configuration of the spectrograph with an echelle grating provides a resolution of 20000 at a central wavelength of 513 nm and over a wavelength range of 25 nm which includes a rich array of narrow absorption lines and the Mg B triplet. The typical precision of a measurement is 0.4 km/s; more details on the system performance can be found in Meibom et al. (2000). The status of the survey effort is summarized in Table 2.

Table 2.
The radial-velocity survey status as of August 2001

Cluster	Stars Observed	Measurements	Binary candidates
NGC 188	511	5529	85
NGC 2264	354	2021	26
NGC 2168	1068	3932	103
NGC 6819	915	3550	59

These data are presently being used for studying a) cluster membership, b) binary populations, c) stellar evolution in close binaries, d) initial and present-day mass functions, and e) stellar dynamics in clusters with rich binary populations. For example, in NGC 188 we have orbital solutions for more than 70 binaries and have defined the tidal circularization cut-off period at 14.9 days, establishing that main-sequence tidal circularization does operate (cf. Mathieu et al. 1992); demonstrated the presence of mass segregation in the binary population; identified four spectroscopic binaries among the blue stragglers, two with unusual orbital properties; and identified a candidate sub-subgiant as also found in M67 (Mathieu et al. 2002).

We note that for clusters out of the Galactic plane, multi-object spectroscopy now makes possible high-precision radial-velocity surveys of *all* stars in a 1-deg field to $V \approx 16.5$. Even for clusters in the Galactic plane with substantial field contamination, the selection of all candidate cluster members from a spatially complete CMD results in a sample that can be comprehensively surveyed via radial velocities. As such radial-velocity surveys are poised to join proper-motion studies in comprehensive kinematic cluster membership studies. This is particularly significant for those open clusters which do not have deep early-epoch plates. Of course, the combination of proper motions and radial velocities is most powerful (e.g. Platais et al. 2001).

5.2 The Future – Radial Velocities

Given a primary goal of obtaining binary orbit solutions, the timescales for the WOCS radial-velocity studies are longer than most other WOCS investigations. As such, there is virtue in the stability of the program. Comprehensive radial-velocity surveys of a dozen WOCS clusters will require a decade of observation. Nonetheless, there are several directions of evolution which are commensurate with these ongoing surveys.

In collaboration with S. Barden (NOAO) and C. Pilachowski (Indiana) we have begun to explore significant improvement of radial-velocity measurement precision. The heart of this capability rests on the scrambling properties of optical fibers and the self-calibration that can be achieved with multiple simultaneous measurements. Recent tests of both system performance and data acquisition techniques make us confident that our limiting precision can be reduced by a factor of 2-3 without substantial modification of the spectrograph or other hardware. This level of improvement will allow us to begin higher precision surveys that will be targeted towards defining the populations of brown-dwarf and higher-mass planetary companions to stars in open cluster environments.

Studies of stellar rotation via surveys of periodic stellar variability due to spot modulation recently have become a major component of WOCS. The combination of these surveys with radial-velocity surveys provides a powerful new means to comprehensively study angular momentum exchange between stellar rotation and binary orbital motion.

More broadly speaking, radial-velocity programs of the future can be greatly expanded in the context of 4-m class telescopes. For cluster work multi-object spectrographs will remain the key instruments. Two major challenges are setup time and focal-plane scale. Major gains in the former are already happening. The MMT Hectochelle Spectrograph, presently under construction, is capable of positioning several hundred fibers in a few minutes, as compared to 94 fibers in 25 minutes with the existing Hydra.

Thus instruments like Hectoechelle will substantially increase the rate of radial-velocity measurements even independent of the larger aperture of the upgraded MMT.

The issue of scale is a bit more complex. It is the nature of dynamically relaxed gravitational systems to be centrally concentrated. Because of crowding, only a limited number of spectrograph apertures can be located in the domain of most of the stars in a cluster. This inefficiency can be reduced by targeting nearby clusters, but of course at the loss of comprehensive spatial coverage and a large sample of clusters. Small telescopes will always have the advantage of wide fields of view, but both excellent seeing and technological designs that allow densely packed spectrograph apertures will need to be developed to efficiently study cluster cores. Ultimately integrated field unit (IFU) technology or Fabry-Perot techniques may be the solution.

Finally, as with most time-domain science, small telescopes can play a crucial role in the studies of binary populations. The scientific gains in going from binary identification to binary orbital solution are large and broad. A "smart" robotic radial-velocity machine could play a defining role in deriving orbital solutions for spectroscopic binaries identified by multi-object surveys. In this case, "smart" would incorporate independent determinations of optimal phasing and prioritization of observations.

5.3 Stellar Abundances

The immediate goals of WOCS abundance studies are two-fold: a) to determine open cluster metallicities, and b) to use lithium as a diagnostic of important yet poorly understood physical processes occurring in the stellar interior.

Although knowledge of cluster metallicities are of vital importance to improving our knowledge of stellar structure and stellar evolution, Galactic structure and evolution, the cosmic distance scale, and a whole host of other key astrophysical problems, there exist only a handful of cluster abundance papers based on detailed analysis of weak lines using modern instrumentation (e.g. Cayrel et al. 1985; Boesgaard 1989; Boesgaard & Friel 1990). In the next few years we plan to expand the set of clusters with well-determined metallicities by a factor of several (e.g. Barrado et al 2001; Margheim et al. 2001). Several steps are being followed in an attempt to minimize both systematic and random errors. First, samples are restricted to late-F and G stars, to make as direct a comparison to the Sun as possible. At the same time, multi-object spectroscopy allows many stars to be used in each cluster so as to improve the statistics. Second, only slow rotators with sharp lines are used. Third, the full power of multiple WOCS diagnostics for binarity is employed to isolate probable single stars. Fourth, only weak lines are chosen, which are minimally affected by uncertainties in the stellar parameters. Fifth, precision photometry from multiple colors is employed to determine precise effective temperatures. Assuming sufficiently accurate photometry and knowledge of foreground reddening, it is hoped that cluster metallicities can be determined to within 0.03-0.05 dex. In the future, we might experiment with additional wavelength regions that provide a larger mixture of neutral and ionized lines of different excitation potentials.

It is becoming increasingly clear that physical mechanisms not included in the standard theory, such as rotation, diffusion, mass loss, and perhaps magnetic fields, have very interesting and important effects in the evolution of solar-type stars (Deliyannis et al. 1990; Bahcall & Pinsonneault 1995; Deliyannis 2000; Jeffries 2000). To study these mechanisms, and using as many stars as possible (with Hydra) in a given cluster, WOCS is determining how Li surface abundance depends on stellar mass within a given cluster. Furthermore, knowledge of the evolution of this Li-mass relation and its dependence on metallicity can elucidate the relative importance of each mechanism, and the details of how each acts. Interesting questions being addressed include, is the Sun normal? Does mixing cause the Li gap and if so how? What causes the decline of the cluster Li plateau, and what are the implications for Big Bang cosmology? Although using Li alone can teach us a lot, it has been demonstrated that progress can be made far more efficiently when additional probes are employed such as beryllium (Deliyannis et al. 1998). With improvements to Hydra, it is hoped that WOCS can add Be to its arsenal in the future.

6. ACKNOWLEDGEMENTS

The authors are grateful to the National Science Foundation for supporting this work through grants AST-9731302 (Wisconsin), AST-9819768 (Florida), AST-9819777 (USRA), and AST-9812735 (Indiana). We would like to thank Constantine Deliyannis for his important contribution to Section 5.3 of this paper. A.S. is grateful to Aaron Grocholski for helping to produce the combined optical/IR data of M35.

7. REFERENCES

Bahcall, J. N., & Pinsonneault, M. H. 1995, Phys Rev 67, 781
Barrado y Navascues, D., Deliyannis, C. P., & Stauffer, J. R. 2001, ApJ 549, 452
Boesgaard, A. M. 1989, ApJ 338, 875
Boesgaard, A. M., & Friel, E. D. 1990, ApJ 351, 467
Carraro, G., Hassan, S. M., Ortolani, S., & Vallenari, A. 2001, A&A, 372, 879
Cayrel, R., Cayrel de Strobel, G., & Campbell, B. 1985, A&A, 146, 249
Deliyannis, C. P. 2000, ASPCS 198, 235
Deliyannis, C. P., Demarque, P., & Kawaler, S. K. 1990, ApJS, 73, 21
Deliyannis, C. P., Boesgaard, A. M., Stephens, A., King, J. R., Vogt, S. S., & Keane, M. J. 1998, ApJ, 498, L147
Deliyannis, C. P. et al. 2002, in preparation
Groom, D.E. 2000, in Optical and IR Telescope Instrumentation and
Detectors, eds. M. Iye, & A.F.M. Moorwood, Proc. of SPIE, 4008, 634
Houdashelt, M. L., Frogel, J. A., & Cohen, J. G. 1992, AJ, 103, 163
Hurt, R. L., Jarrett, T. H., Kirkpatrick, J. D., Cutri, R. M., Schneider, S. E., Skrutskie, M., & Driel, W. 2000, AJ, 120, 1876
Jeffries, R. D. 2000, ASPCS 198, 245
Luppino, G.A., Tonry, J.L., & Stubbs, C. 1998, in Optical Astronomical Instrumentation, ed. S. D'Odorico, Proc. of SPIE, 3355, 469
Margheim, S. J., King, J. R., Deliyannis, C. P., \& Platais, I. 2000, BAAS 37, 742
Mathieu, R. D. 2000, in Stellar Clusters and Associations: Convection, Rotation, and Dynamos, ASP Conf. Ser., Vol. 198. Edited by R. Pallavicini, G. Micela, & S. Sciortino (ASP: San Francisco), 517
Mathieu, R. D., Duquennoy, A., Latham, D. W., Mayor, M., Mazeh, T., & Mermilliod, J. -C. 1992, in Binaries as Tracers of Stellar Formation. Edited by A. Duquennoy & M. Mayor (Cambridge University Press: Cambridge), 278
Meibom, S., Barnes, S. A., Dolan, C., and Mathieu, R. D. 2001, ASP conf. ser. (eds. Andre, P. and Montmerle, T.), vol. 243, in press
Phelps, R., L., Janes, K. A., & Montgomery, K. A. 1994, AJ, 107, 1079
Platais, I., Kozhurina-Platais, V., Demarque, P., Girard, T.M., van Altena, W.F., & Klemola, A.R. 2000, BAAS, 32, 741
Platais, I. et al. 2001, AJ 124. 601
Sarajedini, A., von Hippel, T., Kozhurina-Platais, V., & Demarque, P. 1999, AJ, 118, 2894
Schilbach, E., Robichon, N., Souchay, J., & Guibert, J. 1995, A&A, 299, 696
Tiede, G., P., Martini, P., & Frogel, J. A. 1997, AJ, 114, 694-698
van Altena, W.F., Platais, I., Cudworth, K.M., Lasker, B.M., Russell, J.L., & Johnston, K. 1993, in Workshop on Databases for Galactic Structure, eds. A.G. Davis Philip, B. Hauck, & A.R. Upgren (Davis Press: Schenectady), 215

van Leeuwen, F. 1985, in Dynamics of Star Clusters, IAU Symp. 113, eds. J. Goodman, & P. Hut, (Reidel:Dordrecht), 579
von Hippel, T. A. et al. 2002, AJ 124, 1555

Chapter 17

Automated Spectral Surveys of Clusters and Field Star

Ted von Hippel
University of Texas
Austin, Texas USA

Abstract: I propose a survey that would collect ~10^5 high signal-to-noise, high resolution spectra of F- and G-type stars at a 4-m class telescope, along with supporting intermediate band photometry collected at a 1-m class telescope. The goals of the survey are to greatly refine our understanding of Galactic chemical evolution, the ages and dynamical histories of Galactic subpopulations, star formation, and lithium in stellar surface layers. This survey would be made possible by development of a 4-channel spectrograph behind a multi-fiber system at a 4-m class telescope.

Keywords: clusters, spectral surveys, instrumentation

1. INTRODUCTION

I will describe a particular stellar spectral survey and supporting photometric survey designed to address a few important Galactic and stellar structure problems. The techniques and instrumentation are not specialized to this problem, however, and can be generally applied to a range of problems, including building the next generation of stellar spectral libraries, radial velocity programs searching for Galactic subpopulations, and Galactic structure via spectroscopic parallaxes and metallicities. Many such programs, including the one described below, provide an excellent preparation for the upcoming high precision trigonometric satellites, such as the Space Interferometry Mission (SIM) and the Global Astrometric Interferometer for Astrophysics (GAIA). The distances and proper motions which these satellites should deliver in ten to fifteen years would greatly enhance any of these programs and conversely these data sets would help to quicken the realization of many of the science goals set for the parallax satellites.

T.D. Oswalt (ed.), The Future of Small Telescopes in the New Millennium, Vol. III, 273-280.

2. SCIENCE GOALS

The science goals of the program described here are all based on acquiring high-resolution spectroscopy for > 10^5 Galactic F- and G-type stars. The science goals are to advance the studies of:

- Galactic chemical evolution,
- The ages and dynamical histories of Galactic subpopulations,
- Star formation processes, and
- Lithium in stellar surface layers.

Our understanding of Galactic chemical evolution currently is based upon a few samples of hundreds up to approximately 10^3 stars (e.g., Ryan & Norris 1991; Carney et al. 1994; Norris 1986; Beers & Sommer-Larsen 1995; Beers et al. 2000) with high resolution spectroscopic abundances and kinematics. While halo, bulge, thick disk, and old and young disk stars have been studied in this way, the number of objects available are still too small to tease out some of the subtler features of the Galaxy's chemical evolution, such as the time scale of enrichment, the details of the enrichment sources, the range of abundances within some of the Galactic populations, and whether there are abundance pattern subpopulations within the major Galactic populations (e.g., see Burkert, Truran, & Hensler 1992; Cowan, Thielemann, & Truran 1991; Smecker-Hane & Wyse 1992; Sneden et al. 1994). In addition, other Galactic populations may exist and may only be apparent in larger studies. Likewise, the study of the ages and dynamical histories of Galactic subpopulations would also be greatly advanced by an expanded high-precision database. Ages and kinematics for > 10^5 stars should reveal whether the thick disk and halo are single age populations or whether they were formed over an extended period (e.g., Beers & Chiba 2001). This is particularly important for an understanding of how the Galaxy formed, and is connected to both the merger histories of the Galaxy and the relationship between angular momentum loss and star formation.

A long-standing star formation question is whether stars in open clusters are identical to those in the Galactic field with the same age and metallicity. We know that the metallicity distribution is broader for the Galactic halo field stars than for the globular clusters. We expect denser globular clusters and those on more favorable orbits to survive longer. But how does this survival bias translate into the different field and globular cluster metallicity distributions? Such a comparison is harder for the Galactic disk. Too few open clusters have known metallicities and the age range and uncertainties within the disk population makes such a comparison more complicated. Nonetheless, open cluster survival should also depend on mass, density, and Galactic orbits, possibly creating differences between the cluster and field population. Any differences between the cluster and field population allude to the nature of the clusters in which field stars were born and at least a subset of these clusters are likely to be particularly rare. Matching cluster types to the field stars would elucidate the distribution of properties of star formation events.

Finally, the program mentioned here would target lithium abundances in a large number of open clusters and field stars in a wide range of metallicities. Lithium was created in the early Universe (e.g., Thomas et al. 1994; Suzuki, Yoshii, & Beers 2000) but is destroyed in stellar interiors. Stars cooler than F-type convect their surface material downward to layers hot enough to destroy Li, and thereby a study of surface Li abundances as a function of stellar temperature is an excellent way to understand stellar structure, in particular surface convection and rotationally-induced mixing. A large sample of stars with a range of metallicities and ages would map out convection and other near-surface physical properties. Additionally, the advancements in understanding stellar evolution would lead to a refined primordial Li abundance (see, e.g., Barrado y Navascués, Deliyannis, & Stauffer 2001).

So, how are all of these science goals achieved? The key is high-resolution ($R \geq 15000$), high signal-to-noise ($S/N \geq 200$) spectroscopy in the Li, C, O, and Mg regions of a large sample of F- and G-type stars. The high resolution and high S/N are required for precision abundances for these species as well as iron, and will simultaneously yield radial velocities with precision < 1 km s^{-1}. Since many of the above studies have been attempted with a few dozen up to 10^3 stars, real advances require substantially more stars, and I set 10^5 as a realistic and useful goal for these projects. Modern 4-m class telescopes equipped with good instrumentation can work at this resolution and signal-to-noise down to $V = 14$ in reasonable exposure times, i.e., ≤ 4 hours.

A spectroscopic survey of $> 10^5$ to $V = 14$ without any pre-selection other than color to constrain the stars to F- and G-type stars, and a sample of a few x 10^3 open cluster stars would yield > few $\times$ 10^3 thick disk stars and $> 10^3$ halo stars. Such samples would be far larger than anything that exists today with high S/N and high resolution.

3. PRIMARY INSTRUMENTATION

How can one obtain $R \geq 15000$, $S/N \geq 200$ spectroscopy in four spectral regions for $\geq 10^5$ stars in a reasonable amount of time? In the following, I describe a 4-camera echelle multi-fiber spectrograph, which would be capable of providing these spectroscopic data. Such a spectrograph could be envisioned as an upgrade to Hydra, the 100 fiber-fed bench spectrograph at

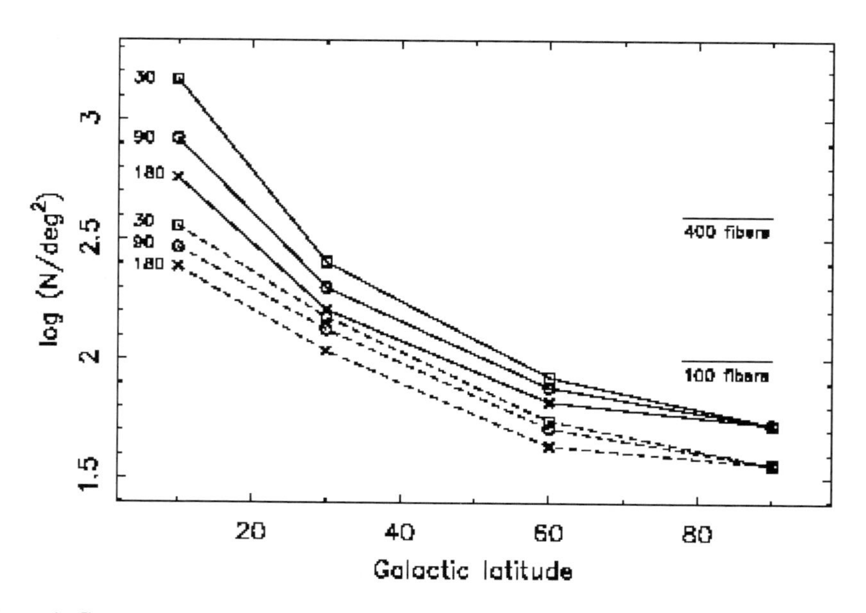

Figure 1. Cumulative star counts down to V = 14 as a function of Galactic latitude, based on the Reid & Majewski (1993) model Galaxy. Three different Galactic longitudes are represented, and are indicated by the numbers just to the left of the distributions. The solid lines indicate the total stellar count as a function of Galactic location, whereas the dashed lines indicate the number of F- and G-type dwarfs expected for these same Galactic locations. All fields are assumed to have no line-of-sight absorption, which is essentially true for most fields with |b| > 10. Fields with substantial reddening would have fewer stars than plotted here. The numbers of stars observable in a single field observation, assuming no fiber collisions, for a spectrograph with 400 (upper line) and 100 (lower line) fibers are also indicated.

the WIYN 3.5-m telescope on Kitt Peak or the Blanco 4-m telescope on Cerro Tololo. The WIYN telescope, for example, has a 1-degree diameter field of view where the input end of the 100 Hydra fibers is placed. The fibers deliver light to a bench spectrograph where a number of gratings are available, including an echelle. While the echelle gives the high resolution desired, it requires the obvious trade-off of limited spectral coverage given limited detector (and optical element) size. While CCD detectors and mosaics of detectors are steadily becoming larger, detectors will not be large enough to cover the range of wavelengths described here for some time. I propose instead that a four-arm version of the current Hydra bench spectrograph be developed with multiple beam splitters, echelle gratings, and cameras to simultaneously observe the Li, C, O, and Mg regions near 6708, 7200, 7770, and 5180 Å, respectively. An alternative approach applicable at lower resolution would be to use a stacked set of virtual phase (holographic) gratings to disperse and separate multiple beams which would then be directed to multiple cameras. The multiplexing of such an advanced instrument would increase the efficiency of the proposed survey by a factor of four, and it would similarly increase the efficiency of a wide range of non-survey programs. In addition, the current generation of multi-fiber spectrographs with 100 fibers should be upgraded to 400 or more fibers. Since the survey would target stars at nearly all Galactic latitudes, the survey could efficiently begin at high Galactic latitude (see Figure 1) and phase observations at lower latitudes with the fiber number upgrade(s). Even without an upgrade above 100 fibers, however, a spectrograph such as Hydra is well matched to $V \geq 14$ stars (see Figure 1) and $\sim 2 \times 10^5$ stars would be observed in a survey which obtained two fields per night and 100 clear weather nights on a 4-m class telescope over the next decade.

4. SUPPORT INSTRUMENTATION

The spectroscopic survey would be supported by photometry at a 1-m class telescope to provide initial object selection in the denser stellar fields, as well as stellar effective temperatures and surface gravity, and line-of-sight extinction. I suggest that the intermediate-band Strömgren photometric system be used for such a study because of its clear advantage in precision determinations of these quantities. Large mosaic cameras on 1-m telescopes (e.g., the KPNO 0.9-m telescope) already obtain one square degree per exposure, a field size well matched to the one-degree diameter spectroscopic fields of 4-m telescopes such as WIYN.

Strömgren photometry requires high precision in order to determine quality stellar parameters, but even S/N = 100 photometry can be easily obtained in these filters down to $V = 14$ on a 1-m telescope. Expected exposure times are of order 10 seconds in *vby* and of order 45 seconds in *u*. Additional use of the Strömgren Hβ filter set for reddening would require of order 120 seconds per field. The equivalent of 3 to 5 exposures per filter per field would be used to increase precision. With the current configuration of most mosaic cameras such a program, based on simple point-and-shoot imaging, would be dominated by read out time. A straightforward procedure to avoid high read out overheads is to park the telescope and read out the CCD at the sidereal rate, or otherwise scan the telescope plus CCD system (Zaritsky, Schectman, & Bredthauer 1996; Sabbey, Coppi, & Oemler 1998). Assuming such a procedure reduced the observing overhead down to ~25%, the exposure time required per field would be of order 600 to 1000 × 1.25 = 750 to 1250 seconds, or approximately 20 minutes. This would probably be broken up into 7 to 10 individual scans across a strip of 4-m fields. In this way, perhaps 15 fields plus many standards could be obtained on a photometric night. The 2000 4-m fields could be covered in approximately 150 photometric nights. While non-photometric nights could be used, given the repeated scans on the same field, such an approach would require extensive internal cross-calibration and boot-strapping as a function of time and field position, and would likely be costly in reduction time.

5. AUTOMATING REDUCTION AND ANALYSIS

How does one reduce and analyze photometry and high-resolution spectroscopy for > 10^5 stars? Clearly automation is required. The large telescope projects and space-based missions have pushed pipeline reductions forward past the point required for the survey proposed here. The Sloan Digital Sky Survey (SDSS, Gunn et al. 1998; York et al. 2000), for instance, requires a much more elaborate reduction pipeline since their imaging and spectroscopic data set is larger and their astrometric demands higher. Large telescopes such as Gemini and the VLT are incorporating pipeline reductions for instruments with many more observing modes than would be used in the survey proposed here. In fact, this survey requires just one spectrographic and one imaging mode, ideal for developing and operating reduction pipelines. Once the reduction pipelines produce photometry, radial velocities, and wavelength- and flux-calibrated spectra, these data would be passed to automated analysis algorithms such as artificial neural networks. A number of studies (e.g., von Hippel et al. 1994; Vieira & Ponz 1995; Weaver & Torres-Dodgen 1995, 1997; Bailer-Jones et al. 1997, 1998) have demonstrated that artificial neural networks can efficiently and accurately classify stellar spectroscopic data in multi-dimensional space based on a subset of standards. These classifications can be made in the astrophysically important domains of effective temperature, surface gravity, and abundance (e.g., Bailer-Jones 2000; Snider et al. 2001) and can even include error estimation and a flag for potential problems (see Bailer-Jones et al. 1998, and references therein). Field overlap within the imaging survey and ~10% re-observation within the spectroscopic program would ensure good internal calibration.

6. ADDITIONAL SCIENCE FROM SPECTROSCOPIC AND SUPPORT DATA

While the survey mentioned here was designed for specific stellar studies, it is clear that a number of other good science programs could be based at least in part on the same observational data. The extensive Strömgren photometry would be well suited to studying population scaleheight and normalization as a function of metallicity (e.g. von Hippel & Bothun 1993). Strömgren photometry is also an efficient white dwarf discovery tool as these stars lie along different loci from main sequence stars in color-color diagrams (e.g. von Hippel 1992). In fields first obtained and studied photometrically, the survey could be broadened somewhat by applying a subset of fibers to objects with unusual colors, increasing the probability for discovering unusual objects, much as has been done with the SDSS (see Krisciunas, Margon, & Szkody 1998; Strauss et al. 1999; Richards et al. 2001) and the Two Micron All Sky Survey (2MASS, Kleinmann et al. 1994; Burgasser et al. 1999; Finlator et al. 2000). Cross-correlating objects in the Strömgren photometric database with those from the SDSS and 2MASS would likely also yield variable stars and possible high proper motion objects.

7. REFERENCES

Bailer-Jones, C. A. L. 2000, A&A 357, 197
Bailer-Jones, C.A.L., Irwin, M., Gilmore, G., & von Hippel, T. 1997, MNRAS, 292, 157
Bailer-Jones, C.A.L., Irwin, M., & von Hippel, T. 1998, MNRAS 298, 361
Barrado y Navascués, D., Deliyannis, C. P., & Stauffer, J. R. 2001, ApJ 549, 452
Beers, T. C., & Chiba, M. 2001, in ASP Conf. Ser. 245, Astrophysical Ages and Time Scales, ed. T. von Hippel, C. Simpson, & N. Manset, (San Francisco: ASP), 149
Beers, T. C., Chiba, M., Yoshii, Y., Platais, I., Hanson, R. B., Fuchs, B., & Rossi, S. 2000, AJ 119, 2866
Beers, T. C., & Sommer-Larsen, J. 1995, ApJS 96, 175
Burgasser, A. J., et al. 1999, ApJ 522, L65
Burkert, A., Truran, J. W., & Hensler, G. 1992, ApJ 391, 651
Carney, B. W., Latham, D. W., Laird, J. B., & Aguilar, L. A. 1994, AJ 102, 2240
Cowan, J. J., Thielemann, F.-K., & Truran, J. W. 1991, ARA&A 29, 447
Finlator, K., et al. 2000, AJ 120, 2615
Gunn, J. E., et al. 1998, AJ 116, 3040
Kleinmann, S. G., et al. 1994, ExA 3, 65
Krisciunas, K., Margon, B., & Szkody, P. 1998, PASP 110, 1342
Norris, J. E. 1986, ApJS 61, 667
Richards, G., et al. 2001, AJ 121, 2308
Ryan, S. G., & Norris, J. E. 1991, AJ 101, 1835
Sabbey, C. N., Coppi, P., & Oemler, A. 1998, PASP 110, 1067
Smecker-Hane, T. A., & Wyse, R. F. G. 1992, AJ 103, 1621
Sneden, C., Preston, G. W., McWilliam, A., & Searle, L. 1994, ApJ 431, L27
Snider, S., Allende Prieto, C., von Hippel, T., Beers, T. C., Sneden, C., Qu, Y., & Rossi, S. 2001, ApJ 562, 528
Strauss, M., et al. 1999, ApJ 522, L61
Suzuki, T. K., Yoshii, Y., & Beers, T. C. 2000, ApJ 540, 99
Thomas, D., Schramm, D. N., Olive, K. A., Mathews, G. J., Meyer, B. S., & Fields, B. D. 1994, ApJ 430, 291
Vieira, E. F., & Ponz, J. D. 1995, A&ApS 111, 393
von Hippel, T. 1992, AJ 104, 1765
von Hippel, T., & Bothun, G. D. 1993, ApJ 407, 11
von Hippel, T., Storrie-Lombardi, L.J., Storrie-Lombardi, M., & Irwin, M.J. 1994, MNRAS 269, 97
Weaver, W.B. & Torres-Dodgen, A.V. 1995, ApJ 446, 300

Weaver, W.B. & Torres-Dodgen, A.V. 1997, ApJ 487, 847
York, D. G., et al. 2000, AJ 120, 1579
Zaritsky, D., Schectman, S. A., & Bredthauer, G. 1996, PASP 108, 104

Chapter 18

Monitoring Extremely Young Clusters: Past, Present and Future

William Herbst
Wesleyan University
Middletown, Connecticut USA

Abstract: CCD cameras attached to small telescopes have been powerful and cost effective tools for studying the variability of pre-main sequence stars. By monitoring extremely young stellar aggregates, such as the Trapezium Cluster in Orion, it has been possible to learn a great deal about solar-like and lower-mass stars during a time when planet formation is active. Here we briefly review the results of such studies, concentrating on the rotation of young stars. The short time scales (hours to days) and complex nature of the phenomena involved dictate the need for intensive monitoring campaigns, which benefit from coordinated observations obtained at a variety of longitudes spanning the globe. Hence, international cooperation is essential. The instrumentation and manpower needs of future programs in this area are actually less demanding than the organizational needs. However, wider field CCD cameras, higher degrees of automation, improved communication across great distances and, especially, the development of CCD astronomy in Asia, will contribute to increased scientific productivity in this exciting arena. The Large-aperture Synoptic Survey Telescope, proposed by the Astronomy and Astrophysics Survey Committee of the U. S. National Research Council, will, if constructed, open a new window on the "time frontier" which will create increased opportunities and demand for follow-up work by small telescopes.

Key words: CCD astronomy, T Tauri stars, clusters, time, Asian astronomy

1. INTRODUCTION

It is widely recognized and often noted that science advances when new technology gives us the ability to explore, for the first time, some previously remote frontier. For example, large telescopes allow us to probe the faint, distant (in time and space) Universe, as well as the low luminosity objects

T.D. Oswalt (ed.), The Future of Small Telescopes in the New Millennium, Vol. III, 281-294.

closer to us. Space telescopes have opened frequency "windows" and, along with ground-based active optics and interferometers, given us ultra-high resolution images of the sky. For some reason, it is less often noted and probably not sufficiently widely recognized that there is another important domain in observational astronomy besides brightness, frequency and resolution – namely, time. Parts of the Universe are variable on human time scales from as short as milliseconds, in the case of pulsars, to as long as our collective memory, in the case of many variable stars and active galactic nuclei. To explore variable phenomena and exploit them to learn more about the Universe requires — *time*.

There must be, in the first place, enough time on appropriate telescopes and instruments to permit monitoring a single portion of the sky with a duty cycle high enough to resolve the variable phenomena. Since no large telescope or space facility will meet this criterion in the foreseeable future, it will obviously require small or medium-sized, ground-based telescopes. Except in the special case of polar instruments, it also requires coordinated efforts from several widely spaced longitudes if the time scales involved are of the order of one day. Here we discuss the advantages of exploring the time domain in one field – star formation – illustrating what has been learned and what important questions remain. The future needs and, in particular, the impact of the proposed Large Aperture Synoptic Survey Telescope (LSST) as well as policy proposals in the U.S. Decadal Study, chaired by McKee & Taylor (2001) are also discussed.

Star formation is generally considered one of the outstanding problems in astrophysics today. Obviously, stars form and planets, at least sometimes, form with them. It is also well known that low-mass stars form more frequently than high-mass stars and that most of the baryonic matter in typical galaxies appears to be in the form of low-mass stars. Beyond that, however, we know surprisingly little. The details of how an individual solar-like star and planetary system come into existence are sketchy, at best. The processes that determine exactly when a star will form within an interstellar cloud and what the resulting distribution of stellar masses will be are uncertain. The central question of angular momentum evolution – how a star manages to rid itself of its excessive angular momentum – and the role that disks play in this are debatable. Of course, as in most of astrophysics, the role that magnetic fields play in all of this is very uncertain. The nature of the star, when it first becomes luminous and its interaction with a disk, possibly containing proto-planets is poorly known. As a result of this lack of detailed understanding of star formation, we have considerable uncertainty about how our own planet, star and solar system came into existence and how they evolved to their present state. We, likewise, have trouble understanding differences between the early Universe, when star formation was much more common but of different character, and the Universe of today. The important effects that environment can have on the products of star formation are far from understood.

A useful approach to resolving some of these issues is to study one product of star formation – stars – at the earliest possible phases in which they can be seen. For optical astronomers this is the so-called T Tauri phase for solar-like stars, and the Herbig Ae/Be (HAEBE) star phase for more massive stars. Largely as a result of the pioneering work of George Herbig (1960,1962), these objects have come to be recognized as stars with ages between 0.1 and 10 My that are mostly in their pre-main sequence phase of evolution. It is at this critical juncture in their lives that they are dealing with a variety of key phenomena which will dictate their futures, including magnetic field evolution within the central star and disk, angular momentum evolution, accretion, formation of proto-planets and other aspects of disk evolution. It also turns out that time scales for some of these phenomena are very short – typically hours or days, but occasionally as short as minutes – and the amplitude of brightness changes is often 10% or more. Finally, some of the processes (e.g. accretion and stellar magnetic field evolution) are nearly stochastic, while others (e.g. rotation and orbital phenomena) are quite regular. We have been studying T Tauri stars for about 50 years now and, while much has been learned, they are still enigmatic in fundamental ways. The complexity of the phenomena exhibited demands a large data base if its underlying causes are to be discovered and this, in turn, requires a good deal of monitoring, which only small- and moderate-sized telescopes can provide in the foreseeable future. We explore this important area for astrophysics in the next section, focusing on extremely young clusters, since the returns for monitoring any particular solid angle are highest in those cases.

2. TIME-DEPENDENT PHENOMENA IN CLUSTER T TAURI STARS

The basic components of a T Tauri star, relevant to its optical variability, have been clearly described by many authors including Herbst et al. (1994), Hartmann (1998), Mahdavi & Kenyon (1998), Shu et al. (2001) and are briefly summarized here. The star itself is normally between 0.1 and 1.5 solar masses and is on the Hayashi track, descending more or less vertically on a theoretical HR diagram. It is fully convective and powered mostly by gravitational potential energy as it contracts quasi-statically towards the main sequence. In some cases there may be contributions to the energy production from deuterium burning and/or accretion. A typical size for the star is 1-3 solar radii, and this is not highly dependent on mass, for stars of the same age. The work to date has concentrated primarily on a few nearby associations or clusters, particularly the Taurus-Auriga T association, IC 348, NGC 1333, the Orion Nebula Cluster, other subgroups of Ori OB I, and NGC 2264. These are characterized by ages of 0.5-3 My for the bulk of the stars. The age spread within a cluster is still debated by, for example, Palla & Stahler (1999) and Hartmann (2001) and may be as small as 1 My (or even less) or as large as 10 My. Images of stars in Taurus and in the Orion Nebula cluster clearly reveal that a significant fraction of these stars are embedded in disks. In Orion, these so-called "proplyds" (for "proto-planetary disks") are being rapidly evaporated by the central OB stars which comprise the Trapezium, but the process is not over yet for most cluster members, as O'Dell (2001) discusses.

Evidence from photometric monitoring programs, one of which is described in Section III, indicates that most, if not all, T Tauri stars have large cool spots on their surfaces, which can remain stable in size and shape for weeks to months (Herbst et al. 1994). If the spots are centered on the rotation axis or are distributed widely and regularly, or if the line of sight is parallel to the rotation axis, then no variations may occur. Otherwise, a relatively stable spot pattern will produce a periodic signal in the measured brightness of the star which can be used to infer its rotation period. Rotation periods of T Tauri stars span a range of about 0.5 – 20 days, and have a curious dependence on mass and age which is discussed further by Herbst et al. (2000). Examples of spotted star rotation curves, from the work of Herbst et al. (2001) are shown in Figure 1.

The role of the disk, and its interaction with the star through accretion and magnetic fields, is also central to the study of T Tauri stars. The magnetic star-disk interplay is thought to be at the root of the energetic and highly collimated winds and jets that are observed to emanate from the younger T Tauri stars. It plays a role in the accretion process as well, funneling matter onto hot spots or zones on the photosphere. Light curves of 8 spotted stars in the Orion Nebula Cluster from Herbst et al. (2001). The star's identification and rotation period (in days) are given.

The intrinsic brightness variation of these zones, as well as their rotation into and out of the line of sight is another major source of variability for T Tauri stars. Finally, it is possible that parts of the disk are optically thick and can cause variable obscuration of the photosphere. All of these processes separately, or in combination, can act to produce variability in the observed brightness of T Tauri stars on time scales of minutes to weeks.

What have we learned about T Tauri stars by monitoring their variations and what do we still need to learn? To date, most of the progress has been in the area of rotation. In part, this is because it is the simplest physical effect involved and can be characterized by a single number, the stellar angular velocity, or rotation period. Fourier techniques can be applied to sufficiently lengthy and accurate data sets to allow periods to be found. The Lomb-

Scargle periodogram technique (Scargle, 1982) is widely used in this area because the data points are always irregularly spaced in time to some degree. Caution must be exercised in interpreting these data since it is not easy to properly assess the probability of "false alarms", i.e. apparent periodicities which are actually the result of chance. Techniques for doing this have been discussed by Horne & Baliunas (1986), Herbst & Wittenmyer (1996) and, most recently, Rebull (2001).

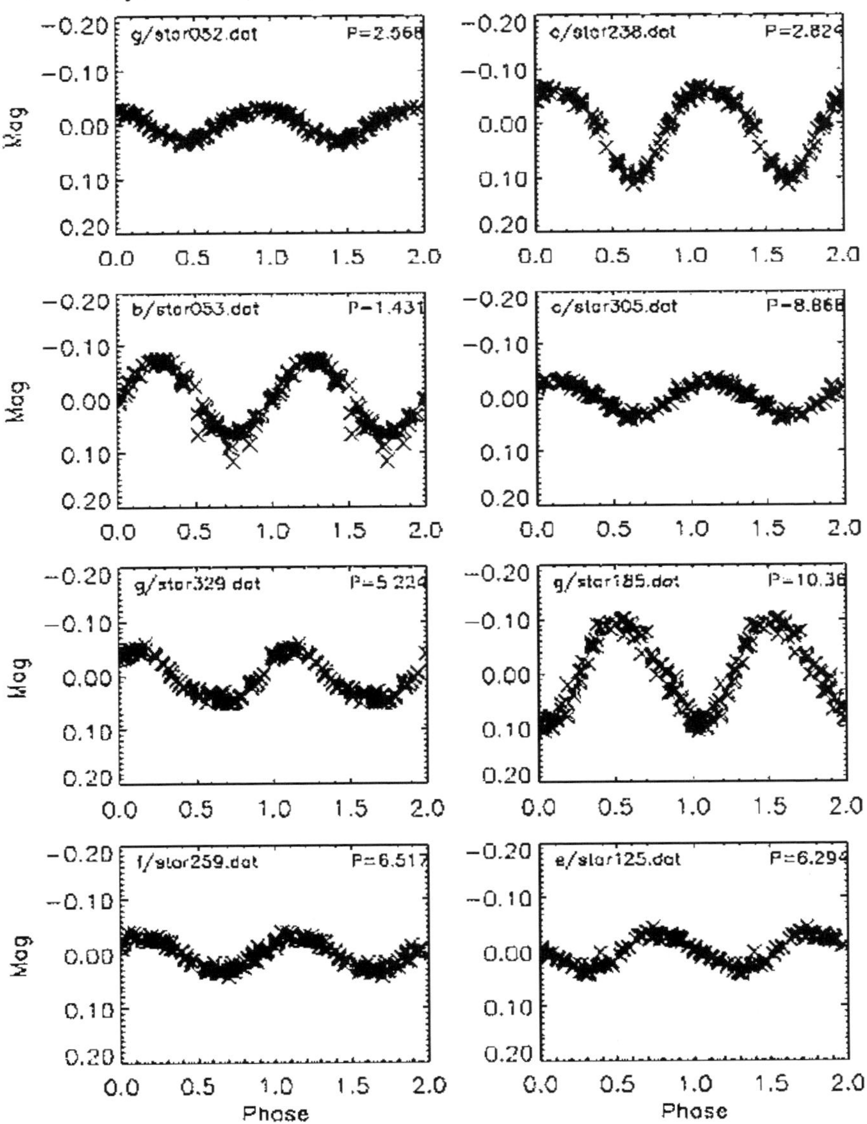

Figure 1. Light curves of eight spotted stars in the Orion Nebula Cluster, from Herbst et al. (2001). The star's identification and rotation period (in days) are given.

Even when there is no doubt that a periodic component to a star's light curve exists, it is not certain that the period is caused by rotation alone. If data are obtained at a single longitude, then there is a "natural frequency" in the problem as well – namely, the rotation of the Earth. A beautiful example of how observing T Tauri stars once per night from a single longitude, can lead to "beating" between the star's rotation period and the Earth's rotation period is provided in a recent near infrared study of the Orion cloud by Carpenter, Hillenbrand & Skrutskie (2001). They used the 2Mass instrument to observe Orion once per night for a month, always at about the same time. While they found excellent agreement with previously determined periods which exceeded 2-3 days, for the shorter periods they generally discovered a beat period, not the true rotation period. This result emphasizes the importance of obtaining data at a variety of times of night and, preferably, longitudes, when variability on time scales of a day or so is involved.

Another complication of interpreting the data comes from the fact that T Tauri stars probably rarely have only a single large spot. More commonly their spot distributions are complex. It might even be the norm for stars to have large spot clusters near the same latitude but in opposite hemispheres. This can give rise to complex light curves and even double-peaked light curves. Periodogram techniques, which are equivalent to fitting sine curves to the data, will never reveal fundamental periods of double-peaked light curves. This can only be ferreted out by looking carefully at the data, plotting light curves not only with the periods found from the Lomb-Scargle technique but also with harmonics of those periods. An example of a double-peaked light curve is provided by V410 Tau, as shown by Vrba, Herbst & Booth (1988) and Herbst (1989). The complicating phenomena of beat periods and harmonics can only be corrected for by the acquisition of a lot of data. It is generally not enough to monitor a particular star for only a few rotation periods. To be sure of what is happening requires good data spanning many cycles and, preferably, at least a couple of years. Since spot patterns change on time scales of weeks or months, it is also true that some stars which did not show a periodic component one year (owing to a complex or rapidly evolving spot pattern or to interference from accretion events) may well appear periodic the next.

3. PRODUCTIVITY OF THE T TAURI MONITORING PROGRAM AT VAN VLECK OBSERVATORY

A program to follow the optical variability of T Tauri stars was initiated by the author at Van Vleck Observatory in 1980. At first, this was intended purely as a pedagogical exercise for students. However, we soon learned that by exploiting the one advantage we had at VVO – time – we could make an important contribution to the understanding of T Tauri stars. Initially, we used a photomultiplier tube attached to the 0.6-m Perkin telescope on Wesleyan's campus. In 1990, this was replaced with a CCD camera and we made the transition from working on brighter field T Tauri stars (i.e. mostly in the Tau-Aur association) to cluster studies. The obvious reason for this was to increase efficiency. We could literally study dozens of T Tauri stars at once by focusing on clusters. A pilot program in 1990 targeted the Trapezium cluster, at the center of the Orion Nebula and resulted in the first detection of periodic variables there, reported by Mandel & Herbst (1991). This work is described more fully by Herbst (2001). Here I would like to focus on the cost and productivity of this effort, as an example of what can be done with small telescopes.

The total cost of operating this telescope and its CCD cameras has been computed for the one decade period of 1990 to 2000. It can be broken down as follows: operation, maintenance and equipment upgrades. It does not include the capital costs of the telescope and dome, which were donated to Wesleyan in 1972 by the Perkin-Elmer Corporation. Operating expenses for the telescope are, as one can imagine, quite small. What expense there is, is covered entirely by the University budget, and probably amounts to no more than $500 per year, mainly for electricity and small amounts of supplies. Maintenance items include regular aluminization of the mirror (on a five-year schedule), lubrication of the telescope and dome, minor instrument and telescope problems, etc. all of which amount to no more than $1,000 per year. The fact that the University has a skilled staff of machinists and employs an electronics expert helps us keep these costs low, since most work can be done in-house. The largest expense in maintenance is the services of a part time computer support and facilities manager, which costs us $4,000 per year. Most of the costs of maintaining the instrument and facility have been covered by external grants to the author, principally through NASA's Origins of the Solar System program. Finally, there is the issue of renewing instrumentation (CCD cameras) on about a 5-7 year time scale. A CCD camera and the associated hardware, including filter wheels, computer, software, etc. has cost us about $50,000 each time we construct one, or about $10,000 per year over the past decade.

The total cost of operating the 0.6-m telescope at VVO is, therefore, estimated to be about $16K per year, or less than $50 per night. For a fair comparison with telescopes at remote sites, we should add that only about one night in three is photometric in Connecticut, so the cost of the telescope could also be stated as $150 per clear night. To compare with the costs of operating a small telescope at a remote site, we take the example of the WIYN 0.9-m telescope, recently purchased from NOAO. Wesleyan is a partner in this operation and our share of expenses has amounted to $32K over a three-year period, during which we are entitled to 16 nights per year. This is $666 per night or, assuming that 2/3 of the nights are clear, $1000 per clear night. Evidently, the 0.6-m telescope at VVO has been a bargain, but has it been productive?

To address the productivity issue, we have counted the number of papers published in refereed journals which have been based on data obtained with the Perkin telescope from 1982 until present. A total of 31 such papers was found, leading to an average publication rate of about 1.5 papers/year. This, in turn, corresponds to an average cost per paper of approximately $10K. Not included in this estimate are senior theses done by Wesleyan students (at a rate slightly above 1/yr) and abstracts of presentations to meetings (also at a rate of above 1/yr). If these were included, it would reduce the cost per paper to approximately $3K. For comparison, KPNO has reported in a newsletter that, in 1999, there were 272 scientific papers published in refereed journals which were based on observations obtained with KPNO telescopes. Given its annual operating budget of about $7 million, this corresponds to about $25K per paper, or 2.5 times greater than VVO. While it is probably pointless to push this comparison too far, the data do make it quite clear that small, locally operated telescopes can be a real bargain for science, even before their educational mission is taken into account. The question of relevance to this monograph, of course, is – will they continue to be a bargain, and have a role to play, in the future. Before addressing that head-on, I would like to comment on the importance to the world of developing astronomy in Asia.

4. INTERNATIONAL COOPERATION AND THE IMPORTANCE OF ASIAN ASTRONOMY

The International Astronomical Union recently sponsored Colloquium 183, entitled Small Telescope Astronomy on Global Scales and the Proceedings of that meeting, edited by Paczynski, Chen & Lemme (2001) will be of interest to all readers of this book. It is, perhaps, obvious, but nonetheless worth stating that, to study objects which are time variable on scales of 0.3 - 3 days it is not only advantageous, but necessary, to obtain data at a variety of longitudes. Examples of scientific results from this approach include the discovery of a rotation period for T Tau itself by Herbst et al. (1986) and the discovery of a number of short period rotating stars in the Orion Nebula Cluster by Stassun et al. (1999). As the GONG network has so nicely illustrated, there are also real advantages to having continuous or nearly continuous coverage of astronomical events that occur on even shorter timescales. For ground-based study this is only possible from polar sites or from a coordinated string of telescopes around the world. The Whole Earth Telescope (Kawaler, 2001) is another fine example of how this coordination can work to the benefit of all. As may be expected, the longitudes where one is in most need of observations are those covering central Asia, for objects only visible from the northern hemisphere. IAU Colloquium 183 was an important first step in building partnerships between Asian astronomers and their western counterparts. Hopefully, it will spur the development of a number of small to medium-sized telescopes in that portion of the world, which will be of great value to the studies described here and that will arise from the survey which the LSST will carry out (See next section). Here I would like to briefly profile one extraordinary site which is already available, in Uzbekistan, and could benefit from the support of the world astronomical community.

The site is Maidanak Observatory in southern Uzbekistan. It was selected during the 1960's by scientists from the Soviet Union as one of the premiere sites to develop, and several telescopes were installed during the last two decades. The astro-climate at Mt. Maidanak is superb, as detailed by Ehgamberdiev et al. (2000) and there are several operating telescopes, including a 1.5-m equipped with a CCD camera. The head of variable star research at the home institution of Mt. Maidanak, the Tashkent Astronomical Institute, was Valery Shevchenko. An expert on Herbig Ae/Be stars, Shevchenko directed the successful ROTOR program, which has provided more photometric data on HAEBE stars than all other programs in the world combined. A recent compilation of their data and some analysis of it was presented by Herbst & Shevchenko (1999). This work is being carried on by the current head of variable stars, Konstantin Grankin, and his co-workers. Also active in the study of variable phenomena is Mansur Ibraghimov, pictured in Figure 2.

Given its location and excellent astro-climate, as well as the dedicated and experienced staff and existing instrumentation, Mt. Maidanak would be an excellent site for future development, as the interest in time variable phenomena grows. The important role that a new project - the LSST - will play in that growth is profiled in the next section.

5. THE IMPACT OF THE LSST, NVO AND OTHER RECOMMENDATIONS OF THE DECADAL STUDY ON SMALL TELESCOPE ASTRONOMY

As it does each decade, the U.S. astronomical community has recently completed a study, led by the Astronomy and Astrophysics Survey Committee (AASC) for the National Research Council. Their report was published in book form by the National Academy Press of Washington, D.C. (McKee & Taylor 2001; also see the Chapter by those authors in volume 1 of this series) and contains, among many items, recommendations of priorities for U.S.

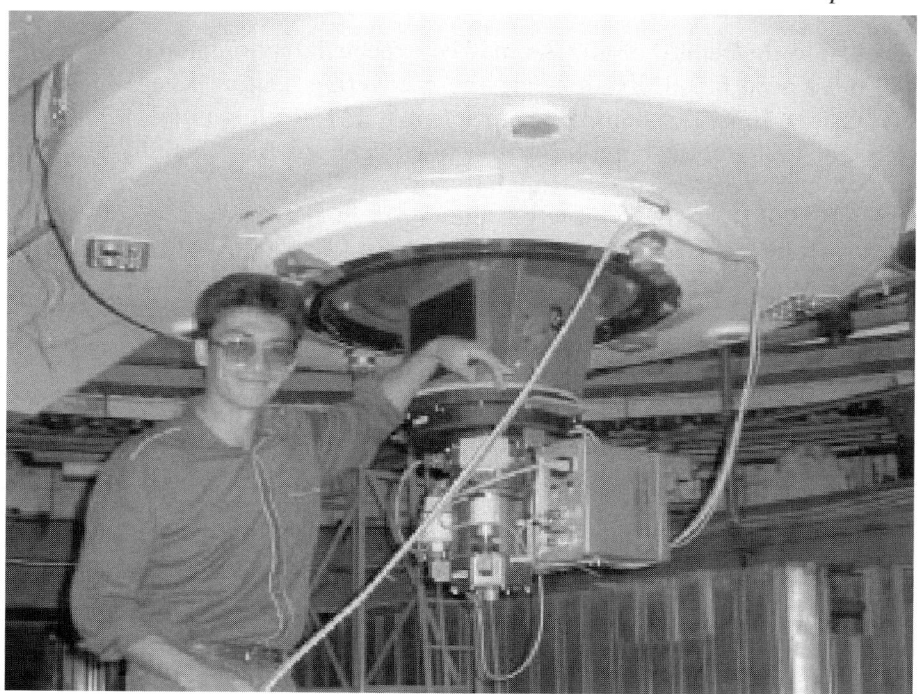

Figure 2. Mansur Ibraghimov, an astronomer at the Tashkent Astronomical Institute in Uzbekistan stands next to the 1.5-m telescope on Mt. Maidanak, which is equipped with a CCD camera.

ground-based astronomy initiatives in the first decade of the new millennium. This "Decadal Study" has great influence on U.S. policy and will undoubtedly affect the way astronomy can and will be done in the near future. A number of the proposals are directly relevant to the science discussed in this chapter and throughout this book. Here I would like to comment on three particular initiatives which are most relevant, in my mind, to the future of small telescope astronomy. These are: the Large-aperture Synoptic Survey Telescope (LSST), the National Virtual Observatory (NVO) and the proposed policy on access to small and moderate-sized telescopes within the U.S. "system" of telescopes.

The LSST is proposed as a single telescope or suite of telescopes which will survey the entire night sky visible from its site to about 24th magnitude on a time scale of about three days to one week. Then, it will repeat the process. Obviously, such an instrument will accumulate an enormous amount of data, and the plan is that these will be made available to the entire (world) astronomical community *via* electronic means essentially as they are taken. This is a revolutionary concept which, according to a number of experts, is not far beyond our technical capabilities at present. In some scenarios, first light for the LSST could occur as early as 2010. If this comes to pass, it will certainly revolutionize the way some astronomy is done. Its principal scientific drivers are the search for supernovae and the detection and cataloging of solar system objects, including near-Earth asteroids, but it will also discover thousands of T Tauri stars in clusters and associations and many other time variable objects. There is simply no doubt that this will be an enormously valuable project and its inclusion in the AASC report is clearly warranted and will, hopefully, lead to its eventually becoming a reality.

How will the LSST impact small telescope astronomy? Some have argued that it will put an end to the need for small telescopes. I argue exactly the opposite – that the need for small telescopes will, in fact, be greater once the LSST goes into operation. While it may be able to do its own follow-up on some objects (e.g. supernovae) which evolve on relatively longer time scales, the LSST will never be able to pursue most of its discoveries, if it is to maintain its survey mission, and it is fundamentally excluded from follow-up of some variable objects by the time scales involved. I believe the LSST will create, therefore, an increased need for telescopes which can exploit the time domain if we are to take full advantage of its survey. It should be obvious, from the discussion above, that a telescope at a single longitude will never be able to achieve time coverage in a manner completely appropriate to the requirements of objects which vary on a time scale of 0.3 - 3 days. Furthermore, it is not possible, even in the most optimistic scenarios, for the LSST to have the time resolution necessary to investigate periodic or aperiodic phenomena occurring on time scales less than about a week; the sampling theorem places a fundamental limit on that. Also, of course, the LSST will have a set range of wavelengths and bandpasses that it can employ. If one wishes, for example, to study photometric behavior in other bandpasses, or to study spectral changes directly, it will be necessary to use other small or intermediate-sized telescopes.

Far from being an end to small telescope science, the LSST should create an enhanced opportunity and, indeed, demand for such instruments. In an earlier era, Schmidt telescopes provided the raw material for follow-up by 4- to 5-m telescopes and now the Sloan Digital Sky Survey provides the raw material for follow-up with 8- to 10-m telescopes. In the same way, the LSST will provide raw material (discoveries of interesting objects) in the time domain for small to medium-sized telescopes (whose advantage is the time available on them) to follow-up with intensive study. Not to do so or even be able to do so, because of the lack of sufficient time on small telescopes, will result in a partial waste of the scientific potential of the LSST. Since the telescopes required for follow-up studies would represent only a small fraction of the cost of the LSST one might hope that, as the plan evolves, it will be seen as prudent to invest in some small telescopes, new or existing, at scattered longitudes around the world to do the follow-up work. (I note, in passing, that Maidanak Observatory might be an excellent choice for one or more such telescopes.)

The NVO is sometimes also championed as a creation which will eliminate the need for small and medium-sized telescopes. Once again, it is arguably more likely that the opposite will be true. The NVO will have limited ability to display the time behavior of the Universe. How, exactly, it will handle moving objects, such as solar system members is unknown. With time variable stars or AGN's the situation may be equally complex, especially if the character of the variations change with wavelength, as in the case of T Tauri stars. The most useful functions of the NVO may be for calibration of fields and for identification of interesting targets. The NVO can never replace the pedagogical functions of a real telescope, of course. It can also never substitute for the excitement and challenges of real observing for students at all levels. Possibly, the NVO will excite more interest in astronomy among the public and among students, but the follow-up work - feeding that interest and teaching students how to observe and interpret the real Universe - will continue to place demands on real telescopes and primarily smaller ones to which the public and students can gain access.

Finally, I mention the policy for operating small telescopes within the U.S. ground-based "system" which the AASC has proposed. To quote from their recommendations, "Universities should assume the responsibility for purchasing, instrumenting, and operating *small* (emphasis in original) telescopes needed for their students and faculty." One can read this as a recommendation to the U.S. national observatory not to be concerned about small telescopes, since it is the role of the universities to provide them. Unfortunately, this begs the question of how astronomers at colleges or small universities that do not have telescopes (of any size) at good sites, and can not afford them, are supposed to continue to do their science and teaching. The important function of access to a variety of telescopes, based on merit proposals alone, that the NOAO has traditionally provided to U.S. astronomers is fast disappearing and will soon be gone altogether if the AASC recommendations are interpreted to mean that NOAO should close small and medium-sized telescopes.

A more enlightened approach might be to recognize that important science can be done by exploiting the time domain and that smaller telescopes at the national observatories should be committed to these projects, which can be extremely cost effective, rather than closed or privatized. If a telescope has outlived its scientific usefulness, then it clearly should be shut down. Before that is done, however, a careful study of potential scientific costs and benefits should be made. No one doubts, in fact, that small and medium-sized telescopes are useful and needed at the major research universities. Why should we, therefore, assume that they are not needed by the 30% or more of U.S. astronomers who are not at major research universities but have thought of the national observatory as a place where they can pursue important observational science? It seems that U.S. astronomy is in danger of heading in two orthogonal directions - towards building a wonderful survey instrument to exploit the time domain while, at the same time, diverting resources away from its smaller telescopes which will be needed to capitalize on the full scientific potential of the LSST. Of course, there is a long way to go along these roads and, hopefully, the slightly wrong heading at present, will be corrected (perhaps by the rest of the world, not the U.S.) before it becomes much of a problem.

6. REFERENCES

Ehgamberdiev, S. A., Baijumanov, A. K., Ilyasov, S. P., Sarazin, M., Tillayev, Y. A., Tokovinin, A. A. & Ziad, A. 2000, A&AS 145, 293

Hartmann, L.W. 1998, Accretion Processes in Star Formation, (Cambridge U. Press, New York)

Hartmann, L.W. 2001, AJ 121.1030

Herbig, G. H. 1960, ApJ 131, 632

Herbig, G. H. 1962, Adv. in A&Ap 1, 47

Carpenter, J., Hillenbrand, L. A. & Skrutski, M. F. 2001, AJ 121, 3160

Herbst, W. 1989, AJ 98, 2268

Herbst, W. 2001, in Small Telescope Astronomy on Global Scales, edited by Bohdan Paczynski, Wen-Ping Chen, and Claudia Lemme, ASP Conf. Series, Vol. 246 (ASP, San Francisco)

Herbst, W., Bailer-Jones, C. A. L., Mundt, R., Meisenheimer, K., Wackermann, R., & Wolf, Ch. 2001, A&A, in preparation

Herbst, W., Booth, J. F., Chugainov, P. F., Zajtseva, G. V., Barksdale, W., Covino, E., Terranegra, L., Vittone, A. & Vrba, F. 1986, ApJ 310, L71

Herbst, W., Herbst, D. K., Grossman, E. J. & Weinstein, D. 1994, AJ 108, 1906

Herbst, W., Rhode, K. L., Hillenbrand, L. A. & Curran, G. 2000, AJ 119, 261

Herbst, W. & Shevchenko, V. S. 1999, AJ 118, 1043

Herbst, W. & Wittenmyer, R. 1996 BAAS 189, 4908

Horne, J. H. & Baliunas, S. L. 1986, ApJ 302, 757

Kawaler, S. 2001, in Small Telescope Astronomy on Global Scales, edited by Bohdan Paczynski, Wen-Ping Chen, and Claudia Lemme, ASP Conf. Series, Vol. 246 (ASP, San Francisco)

Mahdavi, A. & Kenyon, S. J. 1998, ApJ 497, 342

Mandel, G. & Herbst, W. 1991, ApJ 383, L75

McKee, C. J. & Taylor, 2001, Astronomy and Astrophysics in the New Millenium, (National Academy Press, Washington, D.C.)

O'dell, C. R. 2001, ARA &A 39, 990

Paczynski, B., Chen, W-P, & Lemme, C. 2001, Small Telescope Astronomy on Global Scales, ASP Conf. Series, Vol. 246 (ASP, San Francisco)

Palla, F., & Stahler, S. W. 1999, ApJ, 525, 772

Rebull, L. M. 2001, AJ 121, 1676

Shu, F. H., Shang, H., Gounelle, M., Glassgold, A. E. & Lee, T. 2001, ApJ 548,1029

Scargle, J. D. 1982, ApJ 263, 835

Stassun, K. G., Mathieu, R. D., Mazeh, T., & Vrba, F. J. 1999, AJ 117, 2941
Vrba, F. J., Herbst, W. & Booth, J. F. 1988 AJ 96, 1032

Chapter 19

Searches for Galactic Halo Remnants

Nigel C. Hambly
University of Edinburgh, Royal Observatory
Edinburgh, Scotland UK

Ben R. Oppenheimer
University of California at Berkeley
Berkeley, California USA

Abstract: There has recently been renewed interest in the possibility that very old, cool white dwarfs may make a significant contribution to the Galactic dark halo mass budget. New results, both theoretical and observational, have shown that such objects have blue optical/NIR colors despite their low atmospheric temperatures. These results have prompted a number of new searches for very cool degenerates, particularly using archival Schmidt plates. Follow-up spectroscopy and imaging are being pursued using telescopes with apertures $\leq$ 4-m. We discuss these results and the likely direction of future searches with emphasis on the role that 'small' telescopes have played and will play in this topical area of research.

Key words: white dwarfs, galactic structure, dark matter

1. INTRODUCTION

White dwarfs (WDs) are the evolutionary end–point of all intermediate/low mass (M < 8M$_\odot$) stars. These stellar 'fossils' provide a record of the past history of the formation and evolution of the Galaxy, given certain assumptions about star formation and stellar evolution.

In the final decades of the last century, sources of field WDs were dominated by a few wide–angle surveys employing 1-m class telescopes. For

T.D. Oswalt (ed.), The Future of Small Telescopes in the New Millennium, Vol. III, 295-302.

hot WDs, the Palomar–Green/Edinburgh–Cape/Kitt–Peak–Downes (Green et al. 1986; Kilkenny et al. 1997; Downes 1984) photographic surveys produced large numbers of hot degenerates, these being identified via blue UB colors and subsequent spectroscopic follow–up. Typical WD numbers (not including composites/CVs) for those surveys were 26% of all blue objects to B ~ 16 (PG) and 15% to B ~ 16.5 (EC, which is complete to redder stars). For intermediate temperature and cool WDs, the major source of candidates has been the monumental Luyten proper motion survey (LHS/NLTT), candidates being initially selected for follow–up via 'reduced proper motion' (e.g. Evans 1992, and references therein). The key point to note for the detection of WD samples is that, with the possible exception of those at the extremes of the temperature scale, photometry alone is not enough to uniquely identify WDs—ideally some other criterion, e.g. reduced proper motion must be available. Furthermore, proper motions allow discrimination between population types (disk/halo) via kinematics; current disk age estimates may well be biased by the inclusion of ancient, cool halo WDs (see later).

Intermediate temperature WDs are currently studied in relation to several topical research areas. For example, recent studies of these stars (Maxted & Marsh 1998; Iben et al. 1997) have revealed that about 10% are binaries that have interacted during their evolution with either white dwarf or low mass main–sequence star companions. The value of this fraction is relevant to testing the possibility that merging double white dwarfs are the progenitors of Type Ia supernovae which will impact the latter's use as a cosmological probe (Maxted & Marsh 1998; Perlmutter et al. 1999). Hot WDs are interesting as severe test cases for theoretical model atmospheres under extremes of temperature and pressure and also as direct descendants of the central stars of planetary nebulae. Rare and/or unusual examples of WDs (e.g. very massive WDs close to the Chandrasekar limit; highly magnetic objects) are of course interesting objects in themselves.

For cool WDs, the key science result that has emerged from the surveys is the Galactic disk luminosity function (hereafter LF) and its application to disk age estimates. This in turn provides a means of establishing a hard lower limit to the age of the Galaxy and ultimately the Universe (e.g. Oswalt et al. 1996; Knox et al. 1999 and references therein). It is the low–luminosity end of the WDLF that is most sensitive to model age differences, and hence cool WD samples carry the most weight when determining the disk age. Presently, cool WD samples are limited to of order a few objects per luminosity bin (e.g. Oswalt et al. 1996; Knox et al. 1999) and sampling statistics alone yield a random error of ~ 1 Gyr in age determinations (Wood 1997). Systematic uncertainties from evolutionary models are slightly larger than this (as judged by comparison of different model LFs—e.g. Knox et al. 1999). However, it is clear that advances in interior and atmospheric physics for WD models will reduce these errors over the next few years, probably to the extent that observational sampling errors dominate. It is calculated (Wood 1997) that an increase in factor ~ 10 in sample size would reduce such statistical errors to ~ 0.25 Gyr. An order of magnitude increase in the field cool WD sample size would enable measurement of the disk age to a precision of ~ 2% (it is currently uncertain at the > 10% level Oswalt et al. 1996; Knox et al. 1999).

The above introduction gives a flavor of the many active research areas concerning WDs. Others include ZZ Ceti stars (Wu & Goldreich 2001 and references therein) and the ever–growing field of cataclysmic variables (Marsh 2000). For a comprehensive review of WD properties and their application to cosmochronometry see Fontaine et al. (2001). However, the current 'hot topic' in WD research is probably that of the halo LF in relation to the nature of the dark lensing bodies (MACHOs) in the Galactic halo. The rest of this article will concentrate on that particular problem.

2. GALACTIC HALO WDS AND DARK MATTER

There is currently major interest in the possibility that a significant fraction of the total mass of the Galactic halo may consist of cool WDs (Hambly et al. 1999; Chabrier 1999; Ibata et al. 1999; Oppenheimer et al. 2001a). The most recent results from the MACHO project indicate a population of dark compact objects having $m \sim 0.5 M_{\odot}$, these objects contributing between 8% and ~ 50% of the total mass of the dynamically inferred halo dark matter (Alcock et al. 2000). It is interesting to note that the coolest degenerates show rather bizarre optical to IR SEDs (Hansen 1998; Hodgkin et al. 2000). This could be used as a detection criterion; however we note that the assumption of a given set of optical/IR colors as a selection tool is potentially dangerous given the differing degrees to which the known coolest WDs show these peculiar SEDs (Oppenheimer et al. 2001b). Moreover, we re-emphasize that proper motions provide the means for population discrimination. The addition of kinematic information to a cool WD sample of sufficient size will allow the construction of separate LFs for the disk, thick disk and halo populations. Our ideal science goal for halo cool WDs is therefore to *measure* the local space density of such objects in order to confirm them (or otherwise) as the source of the microlensing signature, and to measure the age of the halo (a sufficiently large sample of halo WDs would of course enable age dating of that Population).

The idea that very cool white dwarfs may exist in sufficient numbers to contribute significantly to the halo missing mass is a controversial one—some of the problems and their possible solutions are discussed, for example, by Richer (1999). One of the chief objections to the hypothesis that the halo contains large numbers of ancient, cool WDs is that the high proper motion star catalogs of Luyten do not contain large numbers of such stars (Flynn et al. 2001). However, arguments continue (e.g. Monet et al. 2000}) as to the completeness of the LHS catalog at faint magnitudes and high proper motions—e.g. the proto-typical cool halo object WD0346+246 (Hambly et al. 1997; Hambly et al. 1999; Hodgkin et al. 2000) is *not* in the LHS catalog. Clearly, wide–angle, deep proper motion surveys with automated measurement and processing are needed to resolve this issue; moreover, since it has become clear that the coolest H-atmosphere WDs can have rather strange SEDs, it is important that any new search must not be prejudiced as to the colors of the targets sought.

We have reported (Oppenheimer et al. 2001a) on the results of such an automated search using UK/ESO/Palomar Schmidt plate material measured on the fast, precision microdensitometer SuperCOSMOS at Edinburgh (see Hambly et al. 2001; data and details are available online at http://www-wfau.roe.ac.uk/sss). The first release of data consisted of several 1000s of square degrees around the South Galactic Cap (hereafter SGC). We applied an automated process to produce a catalog of high proper motion stars with $\mu >$ 0.33 arcsec yr^{-1} in the region, using accurate positional work and multiple–pass pairing. This survey has an advantage over previous studies in that we have three independent epochs available (two R and one B) to detect high–μ stars with minimal spurious contamination; we additionally have I band data for many fields. In this way, we have generated a high–μ catalog down to R= 19.75 over >4000 square degrees of the SGC. Comprehensive spectroscopic follow–up of all subluminous objects in the reduced proper motion diagram revealed a population of high space velocity WDs with distinctly non disk–like kinematics. We interpreted these results as indicative of the presence of a hitherto unrecognized population of very cool, halo WDs having a local space density at least ~ 2% of the 'standard' dark matter halo, and therefore a plausible directly detected counterpart to the dark microlensing bodies. Currently, parallax observations are being pursued for the sample using the Danish 1.5-m telescope at ESO. If the results are confirmed, and these cool WDs are truly part of the dark matter halo, then this project surely falls into the category of 'big science with little telescopes'.

However, the dark matter interpretation has been hotly disputed. For example, Reid et al. (2001) claim that the observations are merely consistent with the thick disk population assuming the latter to have a somewhat larger local normalization than previously thought. Moreover, it is possible that the high velocity tail of the WD kinematic distribution is the result of the 'runaway' star phenomenon (Koopmans & Blandford 2001; Davies et al. 2001). In such a scenario, high velocity objects are produced when one high mass component of a multiple star system goes supernova, leaving its companion to fly off tangentially with a large amount of orbital kinetic energy. While it seems fairly clear that cool WDs cannot make up *all* of the dark halo (e.g. our own results; Monet et al. 2000; Flynn et al. 2001) a consensus is emerging that some kind of biased IMF is required to explain the observed numbers of cool WDs whatever their population of origin. In any case, at the very least these new discoveries are interesting in their own right.

It is not the purpose of this article to discuss this controversy. What we would like to do is to speculate as to how observations over the next few years may be able to resolve the problem. Model halo WDLFs predict a peak in the LF at $17 < M_V < 18.5$ (Chabrier 1999). Clearly, probing distances of more than one thick–disk scale length ($d \sim 1$ kpc) is not feasible with small aperture facilities; however we would argue that pencil–beam surveys with 8m/HST/NGST are *not* the best way to go. The original two–epoch HDF results (Ibata et al. 1999) which yielded several Galactic halo WD candidates have not been confirmed by third epoch observations; the unfeasibility of spectroscopic confirmation for the WDs tentatively identified via proper motions and two–color photometry was always a problem. In any case, a large increase in WD samples is the key to smaller statistical errors and more definitive statements concerning the relative contributions of the thin–disk/thick–disk/halo populations.

3. SURVEY REQUIREMENTS

The ‘standard’ Galactic dark matter halo has a local mass density near the Sun of $\sim 10^{-2}$ M⊙ pc^{-3}. Current estimates of the local density of remnants from the known stellar (Pop. II) halo are $\sim 10^{-5}$ M⊙pc^{-3} (e.g. Gould et al. 1998). We calculated (Oppenheimer et al. 2001a) that the local mass density of WDs exhibiting halo–like kinematics is at least $\sim 10^{-4}$ M⊙pc^{-3}, i.e. $\sim 1\%$ of the dark halo and $\sim$ 10x more than expected from the known stellar halo alone. In order to sample this population with good statistics, dispel the controversy mentioned above, and verify the local MACHO counterparts as cool WDs, we would ideally like to go more than an order of magnitude better (in terms of sampled volume) than our current survey. This requires a sample volume of $\sim 10^7$ pc^3. However it is important to note that for distinct kinematics we must restrict our WD survey to those directions where proper motions (i.e. tangential velocities) provide a good indication of kinematic population, since cool WD radial velocities are impossible to measure. Such directions are essentially $\sim \pi$ str (1/4 of the whole sky) in the direction of the Galactic poles and the Galactic anticenter (e.g. Koopmans & Blandford 2001). For a sampled volume of $\sim 10^7$ pc^3 over 1/4 of the sky, a distance limit of $d \sim 134$ pc is required, at which our cool halo WDs ($M_V \sim 18$) will have $V \sim 23$.

Since we identify targets via reduced proper motion (RPM), our ideal survey has an additional requirement that the relative proper motion must be accurate enough to distinguish targets via RPM. In the definition

$H = m + 5\log\mu + 5$, propagation of errors shows that $\sigma_H \approx 2.2\sigma\mu/\mu$ assuming astrometric errors dominate. For $\sigma_H < 0.5$ mag (to yield a relatively clean RPM diagram), $\mu > 5\sigma_\mu$, i.e. we require proper motion detections at 5σ or better. Typical centroiding errors for well–sampled images in a 2d array are discussed by King (1983) and Irwin (1985). They found that, in units of the scale size of the image, the centroid error in one coordinate is, to a good approximation, equal to the photometric error—i.e. for a typical imager where the image scale size is approximately equal to one pixel, then a 10σ photometric detection yields an image center accurate to 1/10th of a pixel in either co-ordinate. For two epochs and a total proper motion, the total proper motion error increases by two factors of $\sqrt{2}$ and inversely with the epoch difference.

In order to achieve surveys of the type required here, facilities with apertures *a* in the range $1.0 < a < 4.0$-m will be required. Of course, it is possible to use a larger telescope to make such a survey. For example, Suprime–Cam (Miyazaki et al. 1998) on the Subaru 8-m telescope could be used. However, hundreds of hours of exposure time would be needed, and even if such a survey was considered competitive with all the other projects calling on 8-m time, the resulting candidates at $V \sim 26$ would be difficult to follow up spectroscopically and for parallaxes. The best option is to use the useful sky area available, and survey to the distance required to get the volume required. It is straightforward to show that for a telescope with aperture *a*, field–of–view (FOV) Ω and for an integration time *t*, survey volume *V* goes as $V \propto \Omega\, t^{3/4}\, a^{3/2}$ in the background–limited case. So, all other things being equal, a Schmidt with 1/10th the aperture of a 10-m telescope, but with between 10^2x and 10^3x the FOV, is a more efficient volume sampler; moreover, small facilities can be dedicated to surveys, unlike the 10-m class instruments.

One possibility could be to 'tile' the focal surface of one of the 1.2-m Schmidts with CCDs to yield and unprecedented FOV instrument with moderate sensitivity. The centroiding precision of such a system (with seeing typically ~ 2 arcsec, i.e. image scale ~ 1 arcsec) at $V \sim 23$ should not be great; hence a large epoch difference would be required to compensate and yield stellar proper motions of requisite accuracy. It is fair to say that one of the reasons why the photographic surveys have been so useful in this respect is due to the time baseline (tens of years) between their various epochs of observation.

Perhaps the best opportunity in the future for making a survey of the type described here is with the 'Visible and Infrared Survey Telescope for Astronomy' (VISTA) facility (http://www.vista.ac.uk/). This 4-m telescope will have an optical FOV of ~ 2 square degrees and will reach nominal point source sensitivities of 5σ at $V \sim 26$. At $V \sim 23$, VISTA will deliver astrometric accuracy of order ~tens of milliarcsec. So, a five year baseline will enable proper motion measurement to the required accuracy for a large sample of cool WDs to V ~ 23 over ~ π str—exactly our ideal survey requirement.

Intermediate between the old (1.2-m Schmidts) and the future (4-m VISTA and similar facilities, e.g. CFHT–MegaCam and MMT–Megacam) are the 2.6-m ESO VLT Survey Telescope (VST) and the SDSS 2.5-m at Apache Point. Unfortunately, it appears that currently the SDSS will not be revisited for a second epoch after the current surveys are finished (although much of the survey is being reobserved on timescales of ~ 1 year) and the relative astrometric precision of drift–scan data can be subject to increased scatter with respect to similar observations taken in a conventional stare mode. Also, at the time of writing there is no information on the scope of the VST surveys.

4. CONCLUSION

To date, advances in WD research (and particularly cool WD research) have to a greater extent benefited from the investment of survey time on the 1.2-m Palomar, ESO and UK Schmidt telescopes. For progress to be made over the next 5 to 10 years, it is *vital* that survey facilities in the 2-m to 4-m class are exploited to produce multi–epoch surveys. Survey facilities must not be used once and then closed down; survey areas must be revisited to produce the multi–epoch data that are needed for the detection of stellar proper motions. Time baselines between 1st, 2nd and any subsequent epoch observations must be kept at a constant level of several years to optimize large-scale proper motion measurement.

5. REFERENCES

Alcock C.A. et al, 2000, ApJ, 542, 281
Chabrier G., 1999, ApJ, 513, L103
Davies M.B., King A.R., Ritter H., 2001, MNRAS, submitted (astro–ph/0107456)
Downes R., 1984, PhD Thesis, University of California, Los Angeles
Evans D.W., 1992, MNRAS, 255, 521
Flynn C., Sommer–Larsen J., Fuchs B., Graff D.S., Salim S., 2001, MNRAS, 322, 553
Fontaine G., Brassard P., Bergeron P., 2001, PASP, 113, 409
Green R.F., Schmidt M., Liebert J., 1986, ApJSS, 61, 305
Gould A., Flynn C., Bahcall J.N., 1998 ApJ, 503, 798
Hambly N.C., Smartt S.J., Hodgkin S.T., 1997, ApJ, 489, L157
Hambly N.C., Smartt S.J., Hodgkin S.T., Jameson R.F., Kemp S.N., Rolleston W.R.J., Steele I.A., 1999, MNRAS, 309, L33
Hambly N.C. et al, 2001, MNRAS, in press (astro–ph/0108286)
Hansen B., 1998, Nature, 394, 860

Hodgkin S.T., Oppenheimer B.R., Hambly N.C., Jameson R.F., Smartt S.J., Steele I.A., 2000, Nature, 403, 57

Ibata R.A., Richer H.B., Gilliland R.L., Scott D., 1999, ApJ, 524, L95

Iben I., Tutukov A.V., Yungelson L.R., 1997, ApJ, 475, 291

Irwin M.J., 1985, MNRAS, 214, 575

Isern J., Hernanz J., Garcia–Berro E., Itoh N., Mochkovitch R., 1997, In: White Dwarfs: Proceedings of the Tenth European Workshop on White Dwarfs, Eds. J. Isern, M. Hernanz & E. Garci–Berro (Kluwer: Dordrecht), 113

Kilkenny D., O'Donoghue D., Koen C., Stobie R.S., Chen A., 1997, MNRAS, 287, 867

King I.R., 1983, PASP, 95, 163

Knox R.A., Hawkins M.R.S., Hambly N.C., 1999, MNRAS, 306, 736

Koopmans L.V.E., Blandford R.D., 2001, MNRAS, submitted (astro–ph/0107358)

Marsh T.R., 2000, New Astron. Rev., Vol. 44, Iss. 1-2, p. 119-124

Maxted P., Marsh T.R., 1998, MNRAS, 307, 122

Miyazaki S., Sekiguchi M., Imi K., Okada N., Nakata F., Komiyama Y., 1998, In: Proc. SPIE 3355, Optical Astronomical Instrumentation, ed. S D'Odorico, 363

Monet D.G., Fisher M.D., Liebert J., Canzian B., Harris H.C., Reid I.N., 2000, AJ, 120, 1541

Oppenheimer B.R., Hambly N.C., Digby A.P., Hodgkin S.T., Saumon D., 2001a, Science, 292, 698

Oppenheimer B.R., Saumon D., Hodgkin S.T., Jameson R.F., Hambly N.C., Chabrier G., Fillipenko A.V., Coil A.L., Brown M.E., 2001b, ApJ, 550, 448

Oswalt T.D., Smith J.A., Wood M.A., Hintzen P., 1996, Nature, 382, 692

Perlmutter S. et al, 1999, ApJ, 517, 565

Reid I.N., Sahu K.C., Hawley S.L., 2001, ApJ, submitted (astro–ph/0104110)

Richer H.B., 1999, astro–ph/9906424

Rodriguez E., López–González M.J., López de Coca P., 2000, A&AS, 144, 469

Wood M.A., 1997, In: White Dwarfs: Proceedings of the Tenth European Workshop on White Dwarfs, Eds. J. Isern, M. Hernanz & E. Garci–Berro (Kluwer: Dordrecht), 105

Wu Y., Goldreich P., 2001, ApJ, 546, 469

Chapter 20

Gravitational Microlensing

Bohdan Paczynski
Princeton University
Princeton, New Jersey USA

Abstract: Microlensing has been a useful tool in astrophysics and will continue to be in the future. Many discoveries have been made through the use of microlensing. Microlensing searches made it possible to monitor the apparent brightness of tens of millions of stars every clear night, to process all data in near real time, and to detect the rare microlensing events while they unfold.

Key words: microlensing, optical depth, dark matter

1. EARLY HISTORY, A PERSONAL VIEW

The nature of dark matter is one of the major unsolved puzzles of modern astrophysics. Its existence is inferred from its gravitational pull on stars and galaxies, but we do not know what it is. Speculations range from sub-atomic particles, often called WIMPS, to massive compact objects called MACHOs. There is a precedent for this logic: the planet Neptune was discovered because of its effect on the orbit of Uranus. However, according to some speculations there is no dark matter, rather the laws of gravity or dynamics have to be modified. We have a solar system precedent for this possibility as well: the Mercury's orbit required not another planet, Vulcan, but a refinement of the theory of gravitation.

Ever since Einstein published his theory of general relativity there were many suggestions that the effect of gravitational lensing should be observed in variety of forms. The effect of lensing by our sun was first detected during solar eclipse in 1919: the images of stars visible near the disk of the eclipsed sun were shifted by almost 2 seconds of arc, exactly as predicted by Einstein. It appears that gravitational field curves the trajectories of light rays, and the sources appear to be displaced.

T.D. Oswalt (ed.), The Future of Small Telescopes in the New Millennium, Vol. III, 303-313.

If the lensing mass is compact enough, or far enough, so that its angular size is smaller than the apparent displacement of images of the more distant source, then we should be able to see two separate images of a distant source on the opposite sides of the lens. The first case of a double image of a distant quasar was discovered by Walsh, Carswell & Waymann (1979): the images of 0957+561A and 0957+561B were separated by almost 6 seconds of arc, and a massive galaxy was acting as a gravitational lens. Several dozen double and multiple images of distant quasars and galaxies have been discovered since, and a lot of images and general information can be found on the WWW at: http://cfa-www.harvard.edu/castles/

Can gravitational lensing by ordinary stars be detected? The following theoretical discovery was made independently every few years, by prominent physicists and astrophysicists as well as by a little known waiter from Brooklyn: if the observer and any two stars in our own galaxy are well aligned then two images of the more distant star will be formed by the gravitational field of the closer star acting as a lens. One can make an estimate that the alignment should be better than one part in one hundred million, i.e. the angular separation between the two stars should be less than a milliarcsecond. There is a problem: there are no telescopes capable of resolving such small angles. Another problem: such a precise alignment is very improbable. A lot of technical information as well as the history of gravitational lensing may be found in the excellent book written by Schneider, Ehlers, and Falco (1992).

Fifteen years ago I made pretty much the same discovery (Paczynski 1986). Two independent referees pointed out correctly that there was nothing original in my paper; it was just a compilation of fragments of various ideas. Somehow I managed to persuade the Editor to accept the paper. As it turned out we were all wrong. There was one very new and important ingredient in my paper, the magic words: ‘dark matter’. Without realizing it I bridged the inter-disciplinary gap. A physicist from the nearby Jadwin Hall at Princeton, Dave Bennett, became a frequent visitor to my office at Peyton Hall. He was very seriously interested in the problem of dark matter. He was one of the many who were eager to solve the mystery. My paper proposed a specific way of conducting the search: just monitor the brightness of several million stars for a few years. Whenever any object more massive than planet Earth moves in front of one of those stars, a double image forms. Even though the two images would be seen as one, their total brightness would vary in a predictable way. For the lens as massive as the sun the characteristic time scale for the so-called microlensing event is a month or two. Other things being equal the time scale is proportional to the square root of mass of the lensing object. Several million stars are needed because the expected rate of events is very small.

The principle was fine, but it was considered a science fiction by me as well as all other astronomers, with a possible exception of Ken Freeman from Australia. Dave Bennett was not an astronomer and he did not know that there are all those variable stars that would form a background of 'noise' for the very rare microlensing events. He did not know that the atmospheric seeing is blurring stellar images in a different way every night. He did not know many other things. The bottom line was: he did not know the project could not be done, and so he tried to persuade me that the project was feasible. Well, I knew better, and so he failed to convince me.

But Dave Bennett was a stubborn fellow. He went off to California and tried to persuade Charles Alcock, who was a physicist as well as an astronomer. And the interdisciplinary gap was bridged for the second time. Charles Alcock gave a colloquium in Berkeley, outlining the project: the search for dark matter by means of monitoring the brightness of several million stars in the nearby galaxy, the Large Magellanic Cloud (LMC). The response was enthusiastic. So much so that Jim Rich, visiting from Sacley, France, immediately called his collaborators and advised them to start their own experiment. And so the first two microlens searches were founded: the American-Australian collaboration of 18 participants, called MACHO (MAssive Compact Halo Objects), and the French collaboration of 28 participants, called EROS (Experience de Recherche d'Objects Sombres). Both teams had some astronomers among them, but they were dominated by particle physicists.

I did not know about all these developments at the time. I wanted to have a personal 1-meter telescope, with no TAC (time allocation committee). A friend pointed out to me that there was a truly outstanding instrumentalist at the Warsaw University Observatory, Andrzej Udalski. Being well connected to Warsaw (I lived there till 1981) I contacted Andrzej Udalski, and pretty soon we were dreaming together. Now all we needed were the funds, and naturally a good observing site. I can no longer recall how the Las Campanas Observatory, operated by the Carnegie Institution of Washington, came about. It was something like this: one day George Preston called me. He was in Pasadena, California, the headquarters of the Carnegie Observatories, with telescopes at the excellent site in Chile, at Las Campanas. George proposed some form of collaboration between Carnegie, Warsaw and Princeton. The idea was initially vague, but pretty soon two somewhat independent, somewhat related projects took off.

One project was long term and fairly ambitious: the Warsaw University Observatory was to build their own 1-m-class telescope at the Las Campanas site in Chile. The telescope was to be dedicated to long term massive photometric searches of variable stars of all kinds, among them the microlensing events. Carnegie would contribute the site, Princeton the operating expenses. The telescope would be managed from Warsaw. Marcin Kubiak, the director of the Warsaw University Observatory begun the fund raising effort, and he was successful. But the telescope could be operating no sooner than 1995.

In the mean time another, more modest project took shape. A bunch of 8 astronomers from Warsaw, Pasadena and Princeton applied for about 60 nights in the 1992 season on the Swope 1-m telescope, the oldest one at Las Campanas. We were to monitor about 1 million stars in the galactic bulge, close to the galactic center. The idea was to check if we could recognize microlensing events where they were guaranteed to be present, as there are many ordinary stars in the galactic disk, between us and the million stars located in the galactic bulge. As a by product we were going to find thousands of variable stars, a real treasure for us, as our team was all made of professional variable stars observers. It is worth pointing out that the same variable stars were, at least initially, just a noise from the point of view of particle physicists in their quest for dark matter. We were lucky: we got between 70 and 80 nights on the telescope for each of the four observing seasons: 1992, 93, 94, 95.

What we needed now was the name—so we could be recognizable as a team. Our first clumsy attempts stimulated George Preston to create a list of about 10 acronyms, all of them great. We e-mailed the list to Warsaw, so Polish astronomers could decide which should be our name. And so the OGLE, the Optical Gravitational Lensing Experiment came to be.

The next major event was the early morning phone call from Charles Alcock. It was September 1993, and I was in Warsaw for a few days. I was told that MACHO had their first microlensing event, or the candidate event as it was cautiously named. That was fascinating! I was very excited. It was good, as my 1986 idea was sound. It was not so good as OGLE had no event yet. It turned out that MACHO had had their event for some weeks, but they were waiting for more data. However, they were notified by Jim Rich that EROS had two events to be announced at a meeting of underground physics in Italy. And so MACHO decided to make their announcement at the same meeting. Unfortunately, OGLE had nothing to announce, and even if we had, we were in no way connected to the underground physics.

Within a week of my return to Princeton I got a phone call from Marcin Kubiak in Warsaw: the first OGLE event has been found! By that time the media were full of excitement about the discovery of dark matter. Unfortunately we were one week too late, and we missed the boat. We also knew that MACHO and EROS submitted their announcement papers to Nature which is well known for a very short publication cycle. There was no way we could make it for the same issue. On top of that I had very bad personal relations with Nature at the time, and this would not help with a speedy publication. Fortunately, Marcin Kubiak happened to be the editor of Acta Astronomica, a quarterly journal. Normally, to wait 3 months for a publication would be of no use in this high speed month of October 1993. By some incredible twist of luck the most recent issue had been just sent to the publisher. Kubiak called and asked: 'can you hold printing for two days?'. The publisher agreed, and the discovery paper was written and recycled several times between the authors scattered on two or three continents, all in 48 hours - the marvels of computer networking. As it turned out the issue of Acta Astronomica with the OGLE paper (Udalski et al. 1993) arrived to some libraries a few days ahead of Nature with the papers of the two competing groups (Alcock et al. 1993, Aubourg et al. 1993).

For several months there were no new official reports from MACHO and EROS teams, but there was a general feeling that the rate of events towards the LMC was too low to account for all the dark matter. Perhaps those few events were caused by ordinary stars in our galaxy? In the spring of 1994, back in Princeton, I received two papers on the same day. One was published in the Astrophysical Journal, and its title was: 'Microlensing Events: Thin Disk, Thick Disk, or Halo?' (Gould et al. 1994). These were three different sites within our galaxy in which the lensing events could reside. The paper demonstrated that there were too few stars in our Galaxy at all those sites to account for the MACHO and EROS events observed towards the LMC. The second was a paper from Nature, for me to referee. An unknown post-doc Kailash Sahu (1994) pointed out that the lenses towards the LMC might be just ordinary stars in the LMC itself. I was struck with the simplicity of the idea. While there was no way to prove that Sahu was right, I recommended publication. The paper is still controversial, but it is very often referenced, as there is still no fully convincing way to explain the event rate to the LMC. However, a consensus is gradually emerging that Sahu was right.

The personal account of the early development of modern microlensing searches as presented in this section is based on my story which was written in 1994 for the Internet electronic publication: 'Matters of Gravity' (http://xxx.lanl.gov/abs/gr-qc/9409004). It has been only slightly updated and edited for this publication. It may not be accurate, and the others may have different recollections.

2. GRAVITATIONAL MICROLENSING: 1992 – 2001

The following is a summary of results obtained during the last decade. Again, this is my personal selection out of hundreds of papers.

The dark matter is still a mystery. With one or two dozen microlensing events detected towards the Magellanic Clouds it is more or less agreed that

MACHOs do not account for more than 20% of dark matter, perhaps their contribution is even smaller (Milsztajn et al. 2000, and references therein).

While microlensing events are not related to dark matter, they became a useful tool in astrophysics. Almost one thousand events were detected towards the Galactic Bulge. Several nice complications were predicted theoretically and confirmed observationally. Most stars are in binary or in multiple systems. If the binary separation is comparable to the cross section for gravitational microlensing (the so-called Einstein ring) then the lens becomes 'astigmatic' and creates optical caustics which lead to spectacular brightness variations of the lensed star. Several dozen such events were discovered. While the star crosses the caustic the finite extent of the lensed star leads to a modification of the observed light curve: the distribution of light over the stellar disk, the limb darkening, can be measured. Several such determinations were made. If the lensing event has a long duration then the relative motion of the source, the lens, and the observer (located on Earth) can no longer be treated as linear. The effect of Earth's motion around the sun generates a measurable modulation of the microlensing light curve, the so called 'parallax effect'. Almost ten such events were detected.

The description of numerous scientific results obtained till 1995 can be found in the review: Paczynski (1996). A more recent overview is provided at the conference: 'Microlensing 2000' (Menzies 2000) and by Gould (2001).

What are the most important specific results related to microlensing itself? The several binary events detected towards the Magellanic Clouds are definitely due to double lenses located in the Clouds. The statistics of hundreds of lensing events towards the Galactic Bulge indicates that the number of binary lensing events is 10 to 20 times smaller than the number of single events. Therefore, given several binary events towards the Magellanic Clouds it is reasonable to expect that all, or almost all single lenses are also in the Clouds, and not in the Galactic halo. However, we have no direct evidence for the location of any single lens in that direction. A recent review of the subject was given by Graff (2000). It points to the Magellanic Clouds location of the lenses, as originally proposed by Sahu (1994).

The best determination of the optical depth towards the Galactic Bulge was obtained by Popowski et al. (2000), who restricted the analysis to red clump giants. The optical depth turned out to be about 2 parts in a million, i.e. somewhat smaller than the early OGLE and MACHO estimates, but still uncomfortably large from the point of view of current galactic models (Binney 2000). This offers a new twist to the relevance of microlensing towards our understanding of dark matter: microlensing provides additional support to the claim that the dark matter distribution does not have a cusp at the Galactic center (Binney and Evans 2001).

While there were various claims that a planet was detected by means of microlensing, no case was conclusive, and there is a broad consensus that no planet has been found so far with this technique (Alcock et al. 1998, Bond 2001b, Albrow 2001). There is a tantalizing possibility that stellar mass black holes were responsible for several very long events (Bennett et al. 1999, 2001), with the longest event (Mao et al. 2001) having the time scale of almost 2 years on OGLE scale, or almost 4 years on MACHO scale. The evidence is so far statistical in nature, i.e. it is not definite yet.

A definite discovery of single stellar mass black holes would be of considerable interest (Agol & Kamionkowski, 2001). A major progress in image processing, the so called Difference Image Analysis (DIA), has been developed with OGLE data by Alard and Lupton (1998) and Alard (2000), and implemented to 0.5 terabytes of OGLE data by Wozniak (2000). This 'image subtraction' method is gaining broad acceptance as the best way to detect and to measure variable objects in crowded fields, and in general in fields with a complicated background.

The observational landscape has changed over the last few years. The MACHO team suffered the ultimate Y2K problem: it was terminated on December 31, 1999 (as planned). The EROS team upgraded its instrumentation a few years ago, but it appears to have no plans to operate past the end of 2002. The OGLE team had four years of operation (1997-2000) with its own 1.3-m telescope at Las Campanas, equipped with a 2K x 2K CCD camera (Udalski et al. 1997). Over 500 microlensing events were discovered, mostly towards the Galactic Bulge (Udalski et al. 2000, Wozniak et al. 2001). OGLE has undergone a major upgrade in June 2001. Its new mosaic CCD camera has 8K x 8K pixels, and hopes for another major expansion in the next 5 years, or so. A new team MOA, a collaboration between New Zealand and Japan, is delivering dozens of microlensing events per year in real time Bond et al. 2001a). Several teams, in particular PLANET, carry out follow-up photometric observations searching for signatures of planets (Albrow 2001).

The up to date information about status of OGLE, with links to all major microlensing projects can be found on the Web at:

http://www.astrouw.edu.pl/~ogle/
http://bulge.princeton.edu/~ogle/

3. THE FUTURE OF GRAVITATIONAL MICROLENSING

In my view the main accomplishment of the past microlensing searches was the demonstration that it is possible to monitor the apparent brightness of tens of millions of stars every clear night, to process all data in near real time, and to detect the rare microlensing events while they unfold. The real time detection and recognition of all kinds of transients requires the same technology, and has a broad range of applications. If a search is conducted using standard photometric bands, like OGLE, the massive photometric data can be directly related to the rest of astrophysics. This way the distance to the Large Magellanic Cloud was determined with a higher accuracy than ever before (Udalski 2001, and references therein). The color magnitude diagrams and the light curves of thousands of variables can be used to study Galactic structure, and to test stellar evolution. The more sky is covered, the more often the observations are made, and the deeper they reach, the better it is for science. Let me concentrate on microlensing events. Perhaps the most exciting future discoveries will be planets, with masses as low as that of the Earth, and the determination of the mass function of stars in the galaxy, all the way from brown dwarfs to stellar mass black holes. A major problem is that for most lensing events only one useful parameter is measured: the event time scale. For some events the photometric parallax effect provides a second observable (Refsdal 1966, Gould 1994). A third observable is needed to measure the lens mass directly. Fortunately, it is likely that within the next few years the Very Large Telescope Interferometer (VLTI) at ESO will be capable of resolving double images formed by a number of microlensing events (Delplancke et al. 2001), thereby providing a complete solution: the mass, the distance and the transverse velocity of those lenses for which the parallax effect will also be measured. This will allow the determination of masses and the distribution of stellar mass black holes in our Galaxy.

The single most important parameter which determines the effectiveness of a search is the data rate, as microlensing events are very rare. For a given data rate the sensitivity to various event time scales depends on the frequency of observations. At least 5 good data points covering an event are needed to be reasonably confident that the event is real. Hence, the past searches, with nightly sampling, had low detection efficiency for events shorter than 10 days, and a negligible efficiency for 1-day events. As a result the combined MACHO and EROS effort was only able to put an upper limit to the optical depth to planetary mass lenses (Alcock et al. 1998), and several years of follow-up studies also established only an upper limit (Albrow et al. 2001, Bond et al. (2001b).

With the new 8K x 8K CCD camera OGLE-III will increase the data rate by an order of magnitude, it will increase the number of monitored stars to about 300 million, and it will detect about 1000 microlensing events per year. For the first time a microlensing project should be capable of detecting several microlensing events by Jupiter mass planets.The reasoning is very simple: geometric cross-section for gravitational microlensing is proportional to the lens mass, hence a Jupiter, with a mass of 0.003 solar has a cross-section 100 times smaller than a typical star of 0.3 solar masses. The increase of the data rate implies the ability to detect signal from the less common events. However, the time scale of a microlensing event is proportional to a square root of the lens mass, i.e. a typical Jupiter induced event will have a duration about 10 times shorter than a typical stellar event, i.e. it will be about one or several days long. In order to increase sensitivity to such short events the observing procedure of OGLE-III will be different than adopted in past searches. Approximately half the time the telescope will monitor some fields several times every clear night, and the other half will be used to monitor a much larger number of fields once every several nights. About once a month the fields will be 'rotated', so that after several years all fields will be covered sparsely most of the time, and frequently for one month. This schedule will dramatically increase the sensitivity to short duration planetary microlensing events, and it will provide a very long time baseline to establish that the relevant stars are stable. The past experience demonstrated that the best criterion for a microlensing nature of light variation is a combination of two elements: a long-term light stability of a target star, and a good fit of the light curve to a microlensing model.

Obviously, the exact split of telescope time between the two observing modes: the number of observations per night in the high data sampling mode, and the time interval for the low data sampling mode, will be established with a series of experiments, as there are too many unknowns to fix all numbers ahead of time.

In addition to the search for Jovian planets through their gravitational microlensing OGLE-III will also search for transits, like those discovered for HD 209458 (Brown et al. 2001, and references therein). The best targets for planetary transits are old stars near the main sequence turn-off point, as they are relatively bright, numerous, and compact (cf. Gaudi 2000). However, the galactic bulge stars are too faint for a 1.3-m telescope. Fortunately, there are very many galactic disk stars which are much brighter, and these will be perfect targets for the search for planetary transits.

While ground based observations of transits may reveal planets with sizes comparable to Jupiter or Saturn, the photometric errors are likely to make it impossible to detect Earth size planets this way. However, even less massive planets can be detected through microlensing (Paczynski 1996, and references therein). The only technical problem is the need for the data rate even larger than OGLE-III. This may be difficult today, but it is likely to be relatively easy in just five years, with a steady reduction in the cost of CCD detectors and computers.

It is interesting that microlensing events were detected not only towards the Galactic Bulge, but also in other directions in the Milky Way (Derne et al. 2001, Udalski & Szymanski 1999). While planetary systems in the Galactic Bulge may be different than in the disk, those which are in the disk some 1 or 2 kiloparsecs away from us should have the same statistical properties as planetary systems at several parsecs from us. It is important to establish what fraction of nearby stars is likely to have Earth mass planet prior to spending several billion dollars on a direct search for nearby systems.

Even OGLE-III will cover only about 100 square degrees in the sky, i.e. only 0.25% of the whole sky. The same technology: frequent photometric and astrometric measurements of as many stars as feasible, as often as feasible is a major goal for the proposed Large Synoptic Survey Telescope, also known as the Dark Matter Telescope (Tyson et al. 2000). For some projects it may be best to use a single telescope, which is as large as possible. For some other projects, like frequent imaging of all sky in a search for rapid transients, it is better to have a large number of smaller telescopes. As far as I can tell there is a sound science case to be made for the OGLE-like project to grow through many more factors of 10 in the data rate. As there is a learning curve to every undertaking, it is better to proceed in a number of steps, each no more than a factor of 10 larger than the previous one, rather than to proceed in mega-steps of several orders of magnitude at once (Paczynski 2001).

A major problem that has to be solved in order to fully utilize scientific potential of major searches is the issue of data archive and access. This is a generic problem for all present and future surveys, and we may expect that the major technical problems will be solved not by every team separately, but that a general solution will be worked out as the necessary step to develop any Virtual Observatory (Brunner et al. 2001).

It is very encouraging that nature offers us new possibilities, and human ingenuity formulates and solves new attractive problems. When I was checking this article for mistakes two very interesting papers appeared on the electronic preprint server: one theoretical (An & Gould 2001), another observational (An et al. 2001). For the first time the masses of the lensing objects, the two components of a binary star, were measured using gravitational microlensing. This strengthens my conviction that the most interesting results from the future massive variability searches, initiated by the searches for microlensing events, will be unexpected, true discoveries.

Note added at proof: OGLE monitored several million stars on 32 nights in 2001 every 12 minutes, searching for planetary transits. Periodic transits were discovered for 59 stars (Udalski et al. 2002a,b). Several groups made spectroscopic follow-up observations on large telescopes, searching for low amplitude reflex motion of the stars. The first definite detection of such a motion was reported by Kanacki et al. (2003) for OGLE-TR-56, indicating a "hot Jupiter" with an average density of 0.3 gm cm^{-3}. The orbital period determined by the OGLE is only 1.2 days, the shortest for any known planet.

4. REFERENCES

Agol, E., & Kamionkowski, M. 2001, astro-ph/0109539
Alard, C. 2000, A&A Suppl., 144, 363
Alard, C., and Lupton, R. H. 1998, ApJ, 503, 325
Albrow, M. D. et al. (PLANET team) 2001, ApJ, 556, L113
Alcock, C. et al. (MACHO team) 1993, Nature, 365, 621.
Alcock, C. et al. (EROS and MACHO teams) 1998, ApJ, 499, L9
An, Jin H., & Gould, A. 2001, astro-ph/0110095
An, Jin H. et al. (PLANET team) 2001, astro-ph/0110068
Aubourg, E. et al. (MACHO team) 1993, Nature, 365, 623.
Bennett, D. P. et al. (MACHO team) 1999, BAAS, 195, 3707
Bennett, D. P. et al. (MACHO team) 2001, astro-ph/0109467
Binney, J. 2000, astro-ph/0004362 (cf. Menzies 2000)
Binney, J. and Evans, N. W. 2001, astro-ph/0108505
Bond, I. A. et al. (MOA team), 2001a, astro-ph/0102181
Bond, I. A. et al. (MOA team), 2001b, astro-ph/0102184
Brown, T. M. et al. 2001, ApJ, 552, 699
Brunner, J. B. et al. 2001, astro-ph/0108381, in press in ASP Conf.
Delplancke, F., Gorski, K. M., & Richichi, A. 2001, A&A, 375, 701
Derne, F. et al. (EROS team), 2001, A&A, 373, 126
Gaudi, B. S. 2000, ApJ, 539, L59
Gould, A. P. 1994, ApJ, 421, L71
Gould, A. P., Miralda-Escude, J., Bahcall, J. N. 1994, ApJ, 423, L105
Gould, A. P. 2001, PASP, 113, 903
Graff, D. S. 2000, astro-ph/0005521 (cf. Menzies 2000)
Konacki, M. et al. 2003, astro-ph/0301052 (in press, Nature)
Mao, S. et al. (OGLE team), 2001, astro-ph/0108312
Menzies, J. W. (Editor) 2000, ASP Conf. Proc. Ser. 239: 'Microlensing 2000, A New Era of Microlensing Astrophysics'
Milsztajn, A. et al. (EROS team), 2000, A&A, 355, L39
Paczynski, B. 1986, ApJ, 304, 1.
Paczynski, B. 1996, ARA&A, 34, 419
Paczynski, B. 2001, astro-ph/0108112, in press in ASP Conf. 246
Popowski, P. et al. (MACHO team) 2000, astro-ph/0005466 (cf. Menzies 2000)
Refsdal, S. 1966, MNRAS, 134, 315
Sahu, K. 1994, Nature, 370, 275.
Schneider, P. Ehlers, J., & Falco, E. E. 1992, 'Gravitational Lensing', Berlin: Springer-Verlag
Udalski, A. 2000, AcA, 50, 279
Udalski, A., et al. (OGLE team) 1993, AcA, 43, 289
Udalski, A., et al. 2002a, AcA, 52, 1
Udalski, A., et al. 2002b, AcA, 52, 115
Udalski, A., Kubiak, M. and Szymanski, M. (OGLE team) 1997, AcA, 47, 319
Udalski, A. & Szymanski, M. (OGLE team), 1999, ATel, 39, 1
Udalski, A. et al. (OGLE team) 2000, AcA, 50, 1
Tyson, J. A., Wittman, D., & Angel, J. R. P. 2000, astro-ph/0005381
Walsh, D., Carswell, R. F., & Waymann, R. J. 1979, Nature, 279, 381
Wozniak, P. R. 2000, AcA, 50, 421
Wozniak, P. R. et al. (OGLE team) 2001, astro-ph/0106474

Chapter 21

Dynamics, Star Formation and Chemical Evolution in the Nearby Galaxies from Studies of Their Stellar Systems

Mary Kontizas
University of Athens
Athens, GREECE

Evangelos Kontizas
National Observatory of Athens
Athens, GREECE

Abstract: The 2- to 4-m class telescopes have to be dedicated to telescope time-consuming projects for which our present knowledge is based on small statistical samples. Spectroscopy (high resolution, 2-D, multi-object) and narrow band photometry provide the appropriate tools for these telescopes in order to face important astrophysical problems in our Local Group of galaxies. The Magellanic Clouds and the other nearby galaxies offer excellent targets for a) Astrophysical questions such as dynamics of star clusters and galaxies, star formation, evolution of galaxies and b) serve as templates for the remote galaxies studies. We therefore propose three main areas of observing projects: 1) *Dynamical studies* of star clusters and associations in galaxies (isolated and interactive environments) have been of great interest lately due to the fast development of theory and the numerical simulations by huge dedicated computers. Observations of a large sample of clusters can provide information on i) the stellar content of their cores and particularly the binary population in them ii) search for coronae of light stars iii) tidal tails due to the gravitational field of the parent galaxy iv) mechanisms of cluster formation. 2) *Kinematical studies* of large stellar structures in galaxies such as carbon stars, planetary nebulae, stellar complexes, giant shells, spiral structures and kinematics of the nearby dwarf galaxies in order to understand star formation mechanisms in various ages and environments. 3) *The Chemical evolution of galaxies*, based on good metallicity and abundance determinations, is so far rather poorly investigated because of the huge amount of work needed for both obtaining the data and reducing them. Now the new sophisticated instruments and the

T.D. Oswalt (ed.), The Future of Small Telescopes in the New Millennium, Vol. III, 315-329.

reduction tools enable us to plan a pilot mega-project towards the understanding of galaxy evolution.

Key words: kinematics, abundances, star formation, magellanic clouds, local group

1. INTRODUCTION

The galaxies are the building blocks of the luminous Universe. Their morphological, chemical and dynamical properties provide the link between our present epoch and the far distant early Universe. The nearby galaxies can be the targets to investigate with high spatial resolution the stellar populations, the dynamical structure and the star formation history in general. On the other hand theories of stellar evolution are now including more sophisticated ingredients, with a variety of metallicities and dust properties allowing a more precise comparison with observations.

The dwarf galaxies are a very important class of galaxies. Our Milky Way is surrounded by a large number of dwarf galaxies with the most known ones the Magellanic Clouds, our nearest neighbor galaxies after the Sagittarius dSph, which is actually merging with our galaxy and is under tidal disruption. They are very important objects cosmologically since they represent the most faint and small galaxies in the Universe. If "bottom-up" is the common mechanism of forming giant galaxies, then the investigation of those objects within the nearest members of our Local Group is becoming the necessary observing plan of the future in order to understand the properties of the very first galaxies ever formed. If on the other hand we also witness the formation of dwarf galaxies now then the investigation is even more urgent and has to be planned carefully on a global scale. We have to define and search what is primordial and what is recent. Have we found or do there exist such dominant characteristics allowing this separation? How do such characteristics (if defined) manifest in their integrated properties in order to allow us to classify the remote galaxies? We therefore propose the study of the stellar populations in these objects in order to understand:

- *Dynamical investigations* of large subcomponents in the nearby galaxies (such as carbon star systems, star cluster systems, PNe, supercomplexes, supershells and their relation to the morphology and formation of bars, disks, spirals.
- *Chemical evolution of galaxies* (age-metallicity relations, metallicity gradients, abundances of specific important elements) and combine both groups of studies in order to
- *understand star formation*, its primordial mechanisms and find the dominant parameters in such different environments.

The latter is true in our Local Group where the morphological variety spans all or almost all Hubble types; the metallicities are up to one order of magnitude different. We have "isolated" and "interacting" galaxies, a really important laboratory.

The observations needed for all these investigations are survey type and of course detailed investigations of selected specific targets. We need to observe thoroughly a large number of objects in all categories mentioned and this is a work that only dedicated small telescopes to each project can do.

Among the observational tools are:

- spectroscopic studies (multi-object and high resolution)
- narrow band photometry (Stromgren, Hα, Hβ, specific lines)
- multiwavelength studies (near IR, optical)

All these tools can (or are already on the way) be improved with the new generation detectors, which must equip the telescopes of this class in order to transform them in instruments of very high efficiency. They can definitely provide first class data which will answer some of the basic astrophysical questions of our epoch and will definitely open new ones. Their observations will serve as follow up targets of the existing large facilities and of the very large telescopes of the next decade. We know very well that in science both statistical methods and data from extended surveys and very analytical work go hand by hand to face important problems.

2. DYNAMICS – KINEMATICS OF LARGE STELLAR SYSTEMS

Radial velocities offer the principal information for studying the dynamics of the different stellar systems and population substructures in galaxies. The nearest galaxies are very important targets for surveying these systems and have the opportunity to understand their complex morphology. Since observational evidence continues to build our confidence of the importance of interactions among galaxies in the development of their various subpopulations, such as star clusters, carbon stars, planetary nebulae, stellar complexes. Very often none of these systems can be described in terms of a single population and appear to comprise subcomponents.

The Magellanic Clouds with their richness in all kinds of stellar systems and their proximity are primary targets for large surveys. However this is also true for our Galaxy (Norris 1997) and much more work has to be done in the future to tackle this problem. "Top-down" or "bottom-up" is the primary galaxy formation mechanism, how much can these stellar subsystems tell us?

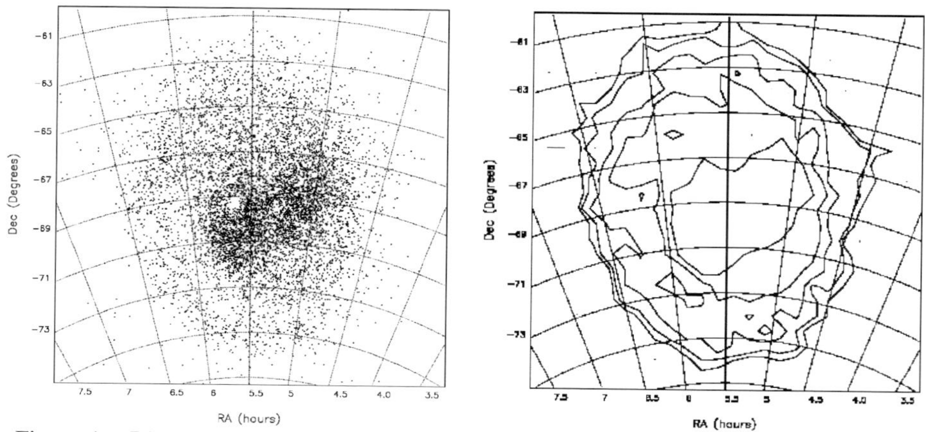

Figure 1. Distribution of carbon stars in the LMC (left) Isodensity contour map for the carbon stars in the LMC using a pixel element 40×40 arcmin^2. The contours are 1, 2, 5, 10, and 20 carbon stars per pixel (right), from Kontizas et al. 2001, A&A.

2.1 STELLAR COMPONENTS IN NEARBY GALAXIES

2.1.1 The Carbon Stars

The carbon stars have been extensively studied in the LMC and SMC. Azzopardi (1999) gives a comprehensive review of all surveys; and their importance from Schmidt telescopes and from radial velocity studies (Morgan & Hadzidimitriou 1995, Kontizas et al. 2001, Kunkel et al. 1997b, Hadzidimitriou et al. 1997, Kunkel et al. 1997b). Kunkel & Demers (2000) have used these investigations to explain the LMC/SMC interaction. Their distribution and isocontours for the LMC are illustrated in Figure 1.

A complex structure was revealed and many new questions have arisen, for the SMC kinematics and the LMC similarly. Are the outermost carbon stars in LMC debris left along the SMC orbit? Why there is no angular velocity or very weak in the SMC? How certain is this? It is clear that we need to answer these questions in order to understand the interaction of MW with the two Clouds and the one between them (Gardiner 1996), 0.4-0.2 *Gyr* ago. Radial velocities and spectroscopically determined abundances are needed to help with this problem.

2.1.2 The Planetary Nebulae

The Planetary Nebulae are the stage of stellar evolution at the last steps just before the Asymptotic Giant Branch (AGB). For our Galaxy their study is more complicated due to distance variation and reddening. In the MCs, the distance is almost the same and the field reddening law (Morgan 1997; Dopita 1999). Ground-based spectrophotometry, absolute flux, expansion velocity and density information permits us fully self-consistent diameters, ages, masses and abundances to be derived for the nebulae, so their place on the HR can be reliably determined. A large number of PNe in the MCs have been studied (Dopita 1999; Richer et al. 1999) which revealed an age-metallicity relation similar to that determined by the star clusters (Girardi et al. 1995; Kontizas et al. 2001). A typical distribution of PNe in the LMC is shown in Figure 2. Several of the PN are associated with actual clusters and it is necessary to prove this with extra spectroscopic kinematical studies. Such survey work will prove the connection of PN with clusters in the MCs, enhance the sample of PN and improve the accuracy for metal abundance determination (Pagel & Tautvaisiene 1999).

2.1.3 The Stellar Complexes

Stellar complexes are defined as groups of OB associations, clusters and field stars with young and massive stellar components first recognized by Shapley in the LMC as "small irregular star clouds." He stressed that nearly all of them appear to be distinct physical organizations. Later on Lucke & Hodge (1970), Martin et al. (1976) and Efremov (1979) have suggested

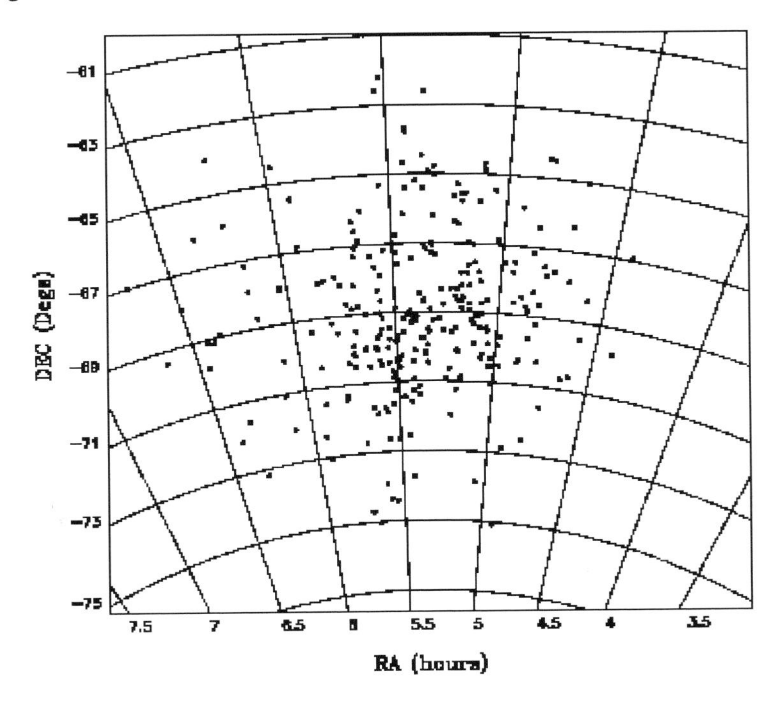

Figure 2. The distribution of Pne in the LMC (private communication, D. H. Morgan).

discriminating between stellar associations, aggregates and complexes. Feitzinger & Braunsfurth (1984) found in the LMC three levels of hierarchical clustering of OB associations and HII regions with characteristic scales of 100, 400 and 1500 pc. Baade (1963) was first to stress that star formation in the LMC occurs in two scales in associations with dimensions from 10-100 pc (Kontizas et al. 1994; Gouliermis et al. 2000) and huge areas of ~500 pc (Maragoudaki et al. 1998). The even larger scale is that of superassociation, the first example of which is the 30 Dor. region in the LMC (Baade 1963; Ambartusmian 1964).

A stellar complex is similar to a *superassociation* but with much fewer associations, members, and dimensions (Efremov 1997; Kontizas et al. 1996, Maragoudaki et al. 1998, Maragoudaki et al. 2001). Data for the Galaxy, the LMC, SMC, M31 and M33 show that complexes are at least 100 times more numerous than superassociations, whereas the ratio of the oldest stars in them is only 10:1. Therefore only 1 out of 10 complexes passed through the stage of superassociation and was totally associated to the burst of star formation. They consequently offer the signature of those that are relics **of** *superassociations* (Efremov 1997). The LMC and SMC are hosting such objects, which appear extremely important in studying the star formation mechanisms. The radial velocities for these regions are so limited that we could say nothing complete exists. We need kinematical observations to understand whether these objects are bound in the LMC. Efremov (1997) has given a few examples where complexes with similar ages have different stellar components. The principal difference is that one class contains stellar associations and young clusters whereas the others contain a very dense stellar component with no clustering at all. What is responsible for a star-forming region reaching the same age with and without clustering? Metallicity is a natural thought, but we need to check this with spectroscopic abundance and narrow band photometry. This investigation leads us to find evidence of star formation is different metallicity environments. We expect that in lower metallicity regions the formation of bound stellar groupings is much higher.

Results in populous LMC clusters do not always agree with these theoretical expectations, but our observations are still too poor to lead to any accurate conclusions (Efremov 1997).

A detailed study in the IR and X-rays shows that the complexes are associated with SNRs and high IR emission (Maragoudaki et al. 2001). They show characteristics of "starburst" galaxies. Are these regions prototypes of intense star forming activity in galaxies?

Extended metallicity studies and internal kinematics are needed to understand the physics involved. Prevot et al (1989) have measured velocities of ~200 supergiants. Their internal velocity is 6.6 ± 1.3 km/sec. These 200 stars reveal 18 very dense stellar complexes. This sample is very small and need to be systematically increased and extended to SMC.

2.1.4 Supergiant Shells and Filamentary Structures

The Interstellar Medium (ISM) in spiral and irregular galaxies reveal a complex variety of filaments, loops and shells manifesting a violent ISM (McGray & Snow 1979). It has been suggested that they are formed by the collective action of fast stellar winds and supernova explosions of massive stars in multiple OB associations (Points et al. 1999). The LMC is an excellent site to study these gigantic shells and smaller ones. Because of its proximity it is possible to study the physical structure of supergiant shells and resolve the associated stellar content to determine their formation mechanism.

The kinematic structure of these shells can be studied by radio, X-rays. In the optical range, CCD images in emission lines enable us to investigate the structure of the warm ($\sim 10^4$ K) ionized gas while high dispersion spectra provide their kinematics of high angular resolution. The largest structures identified by Goudis & Meaburn (1978) are nine with diameters 600–1400 pc and contain from 25 to 400 blue stars in associations (Meaburn 1980; Kennicutt et al. 1995; Staveley-Smith et al. 1999). Smaller shells represent a different activity of star formation and are equally important to study in detail for their kinematical properties. We must investigate their association with the stellar complexes in a systematic and complete survey, since both structures lead to a comprehensive study of star formation activity and its manifestation at "star burst", superluminous, AGN stage in galaxies.

2.1.5 Star Cluster Systems

The role of star clusters as tracers of stellar and dynamical evolution in a galaxy is well known. A particularly interesting and exhausting investigation is the study of NGC 1886 galactic cluster by Nordstroem & Andersen (1997). They observed systematically a large number of clusters spectroscopically to derive accurate memberships, to search for binaries, mass segregation effects and Mass Function.

The LMC cluster system has been detected exhaustively (Kontizas et al 1988; Kontizas et al. 1990; Zaritsky et al. 2000) and the typical distribution of the cluster systems in both Clouds is given by Irwin (1991) in Figure 3. It is very important to see in all three diagrams how well the LMC subpopulations follow the morphology (Harris & Zaritsky 1999; Holtzman et al. 1999; Zepf 1999) of the parent galaxy.

If a survey with this goal is planned for the next 5-8 years in order to study a large number of clusters and associations in the MW and MCs, then stellar evolution theories (with chronology as one of the principal achievements) and dynamical evolution in stellar systems will be checked with very important implication in the study of dynamical evolution and star formation in galaxies (Massey et al. 2001; Meibom et al. 2001; Efremov & Elmegreen 1999).

Dynamical evolution of star clusters was theoretically being developed and still is, with special reference to the effects due to the various locations in a galaxy and the role of binaries in its evolution. The most important questions arising from star cluster investigations can be summarized:

- how efficient is star cluster formation in different environments
- what is the role of star clusters in the early evolution of galaxies
- how the clusters' dynamical evolution is associated to star formation mechanism
- how the star clusters relate to the other large stellar components in galaxies and its subcomponents
- what is the role of binaries in clusters. The variable star populations such as Cepheids, RR Lyrae.

2.1.6 Young Clusters and Star Formation – Mass Segregation

The "embedded" clusters in the MW and the very young globulars first found in the MCs are the best targets to be studied in order to trace the dominant physical parameters and mechanisms of star formation. In young star clusters and associations, mass segregation (although at different dynamical stages) reflects the loci where massive stars form in a protocloud.

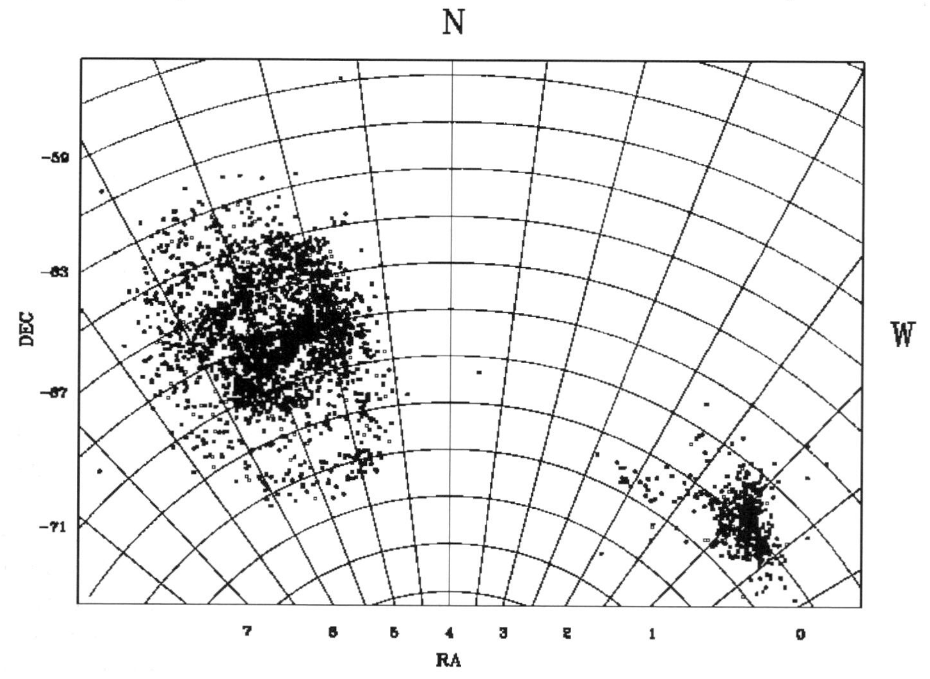

Figure 3. The cluster systems in LMC & SMC as given by Irwin (1991).

Theoretically there are conditions where both central and peripheral massive star formation occurs (Murray & Lin 1996, Kontizas et al 1998; Keller et al. 2000, Bonnell 1999, Bonnell & Davis 1998). It is therefore urgently needed to plan a pilot project for a large number of clusters and associations in the MW and in the Magellanic Clouds.

- The role of binaries and/or Be stars in clusters has become a very hot issue lately with modern facilities in moderate-size ground based telescopes (Bonnell 1999, Keller et al. 2000) and their connection to blue strugglers.
- An issue coming from mass segregation is the distribution of faint stars during its dynamical evolution. If the faint stars escape to the outer regions due to energy equipartition, they must be found mainly in the outskirts and possibly form a corona around the clusters' periphery (Kontizas et al. 1997).
- How is the IMF varying from the center to the periphery.

These characteristics can tell us whether some are primordial or not, implying that they are associated to the gas dynamics and star formation conditions (Bonnell et al. 1998).

A number of less crowded clusters must be selected and particularly the numerous open clusters in the Magellanic Clouds and the associations, at different locations in the parent galaxy. These clusters should be observed with deep CCDs at narrow band filters and at extended areas around each cluster (Johnson et al. 1999).

The broad band and narrow band filters will be used for Mass Function determination, ages, detection of Be stars. Narrow band observations will be used to derive metallicities (See next section) and their relation with age.

Therefore mass segregation, MF and metallicity in young stellar systems describe some of the characteristics which will enable us to understand their formation in different galaxy environments.

2.1.7 Old Clusters and Dynamical Segregation

We know that a star cluster during its dynamical evolution suffers energy redistribution and after a time scale called mean relaxation time equipartition is achieved. During this particular timescale, massive star segregation is occurring and central depletion of light ones. This time depends strongly on the total mass (number of cluster star members) of the cluster and its central density. These effects must be investigated together with the previous mentioned young stellar systems. We must be able to observe dynamically old clusters and from a very large sample to achieve a wide range of all dynamical stages since the MCs are currently forming clusters, whereas the MW has only old globular clusters. Again a large survey of many different clusters should be programmed for the next decade, exactly to match theory with observations for clusters (Porras et al. 2000; Fregeau et al. 2000; Howell et al. 2000)

- at different locations (i.e. metallicities, field characteristics)
- different ages

2.1.8 Star Clusters and Tidal Tails

If the gravitational field is strong where the cluster is located and/or during the passages of the cluster from perigalacticon of its orbit around the center of the parent galaxy, stars escape faster and often produce tidal tails (Leon et al. 2000; Wilhelm 2001). These tails are very important if we investigate their structure for a large number of clusters in a galaxy, since they trace the effect of the galaxy's gravitational field on massive objects (such as star clusters) and therefore its dynamical characteristics.

This work can be a survey for all old clusters of our nearest galaxies and the Milky Way. We need to investigate large areas around clusters for both:

- Radial velocities in order to separate spectroscopically cluster from field stars
- Deep photometry in order to produce the density distribution around the center of each cluster. Tidal tails will appear as asymmetric density enhancements in outermost cluster regions (Wilhelm 2001).

2.1.9 Binary and Multiple Star Clusters

Binary clusters are found to be quite often in the MCs reaching about 10% of the cluster population. Statistically fewer than 50% are expected to be chance superpositions (Hatzidimitriou 1990; Kontizas et al 1993b). Several investigations have revealed that they are mainly young objects with obvious traces of their dynamical interaction (Kontizas et al. 1995). Their ages give evidence of large populous star cluster formation (Makino 1991).

It is therefore very important again to undertake a large survey of studying the binary clusters in the MCs and understand the necessary conditions where they are formed and how they further evolve.

Photometry and spectroscopy will provide ages, spatial distribution for searching tidal tails between them, and metallicity to check its importance in forming multi-clusters.

2.1.10 Spirals

From our Milky Way it is known that a lot has been learned from kinematical studies of its subcomponents such as spiral arms, discs, halo, bulge (Norris 1997). The nearest galaxies M31 and M33 present also very good targets for radial velocities of the supergiants in the arms of their spirals. They should demonstrate the characteristic deviations from the circular velocities, connected with the spiral density waves which are observed in HI and CO (Efremov 1997). Selected groups of stars should be observed to investigate the behavior of these groups.

2.1.11 Dwarf Spheroidal Galaxies

The dSph orbiting the Milky Way are also very important targets for radial velocity measurements. They are not just ellipsoidal systems with very simple dynamical history. We know that things are more complicated because of the existence of the dark matter and the interaction with the MW (Mateo 1998; Van den Bergh 2000).

Very good investigation in the nearby dSphs have shown that tidal effects are obvious in their density profile and appear to be extended far beyond the tidal radius of the King model. The velocity dispersion is also much higher than expected from a virial equilibrium (2-3 km/sec) giving evidence of the tidal effects from the MW (Hatzidimitriou 1997).

- Systematic multi-wavelength surveys of nearby dwarf galaxies at large areas to search for the density fluctuations due to tidal effects.
- Radial velocities of intermediate accuracy (~7 km/sec) of a very large number of stars, to define membership particularly in the outermost areas.
- Accurate velocities (~3 km/sec) for a more specific kinematical study of the tidal structures.

The majority of the existing telescopes with diameter ≤4-m either high-resolution spectroscopy or multislit can provide the radial velocities and a big pilot project should be planned.

3. CHEMICAL EVOLUTION – ABUNDANCES

The Galaxies of the Local Group serve as our laboratories for understanding star formation and stellar evolution in differing environments. In the Local Group the galaxies currently active in star-formation cover a factor of 10 in metallicity and span a range of Hubble-types from dwarf spheroidal to irr, Sb and Sc.

Uniform surveys in UBVRI, Hα, [SII] and [OIII] have been started for almost all members of our Local Group (Massey et al. 2001). Such surveys may include narrow band photometry so that Mass Functions (MF) and metallicities will enable us to investigate the age-metallicity relations and the IMF in these galaxies. The later two are principal parameters for determining the chemical evolution in a galaxy together with the star formation rate which can be determined by the color magnitude diagrams (CMDs) of various stellar populations at different age. Star clusters are good targets for such a project (Larsen et al. 2001).

The MCs are known to show some gaps in the age-metallicity relation from the existing observations (Kontizas et al 1993a; Pagel & Tautvaisiene 1999; Da Costa 1999; Da Costa & Hatzidimitriou 1998; Dirsch et al. 2000 Kontizas et al. 2001)

The main problems are **a** gap at 2-7 Gyr and the change of slope in the age-metallicity relation at ~0.2-0.4 Gyr. These times seem to coincide with the close encounters of both Clouds and their proximity to MW. Therefore we need to know accurately the exact behavior of these relations. We need to know the triggering of star cluster formation in the triple system of our nearest interacting galaxies. How important are the signatures of this interaction to star formation.

Another issue regarding chemical evolution is the existence of lithium in clusters. This is a very important issue and a survey is already underway but it has to be completed with a large sample of clusters (Deliyiannis & Steinhauor 2001, Boesgaard et al. 2001). *A priori* the observed Li abundance in the atmosphere of normal A stars is representative of the Li content of the cloud from which the protostar is formed (Burkhart & Coupry 1997). Then studying clusters of different ages could, in principle, give some information concerning the galactic evolution of Li abundances.

Finally the question of age-metallicity effects in very old clusters can provide an answer to the critical question. How long star formation lasts in a stellar system? How efficient is it and what is the right proportion of massive stars that regulate the depletion of gas in a stellar system (Wallensteein & Hughes 1999)? Narrow band Strömgren photometry is a good tool to carry out these investigations and of course high-resolution spectroscopy.

4. STAR AND CLUSTER FORMATION

The suggested investigations are mostly aiming to provide very important information on some very open issues of the star formation, its mechanisms and dominant physical parameters. Star formation is one of the main goals for astrophysical research in the 21st century. Despite the great success of the theory of stellar evolution in the past century there is a gaping hole. It neither predicts nor explains how stars form. Such knowledge is critical for understanding how stars form and their planetary systems but also how systems of stars and galaxies evolve. The theory of star formation in clusters is similarly primitive because of the complicated interaction of the cores and protostellar winds in these regions (Murray & Linn 1996, Murray et al. 1999).

Considering that star formation is one of the four major goals for all planned new large telescopes in ground and space, it is absolutely necessary to use now and for the next decade all available small telescope and updated instrumentation for extended surveys which will operate towards the end of the decade. More specifically the proposed studies give information and new challenges towards the following astrophysical questions.

1. From young clusters
   - Where do massive stars form: at the center or the periphery? Under what conditions does one or the other occur.
   - Are binary clusters the predecessors of merging mechanism for the formation of very populous globular clusters?
   - what is the role of binaries
2. From studying large stellar components and structures
   - star formation sites in galaxies and their connection to its morphology and evolution
   - star burst regions (vs) interactions between galaxies
3. From metallicities and abundances
   - Age-metallicity relations and their connection with galaxy interactions
   - the role of Li abundance and its connection to galaxies' chemical evolution.

5. REFERENCES

Ambartsumian, V.A. 1964, IAU Symp. No 20, p. 122

Azzopardi, M. 1999, New Views of the Magellanic Clouds, IAU Symp. 190, eds. Y.-H. Chu, N. Suntzeff, J. Hesser, & D. Bohlender, p. 324

Baade, W. 1963, JRASC 57, 232

Boesgaard, A. M., Deliyannis, C. P., King, J. R., Stephens, A. 2000, ApJ 553, 754

Bonnell, I. A., Davies, Melvyn B. 1998, MNRAS 295, 691

Bonnell, I. A. 1999, The Origin of Stars & Planetary Systems. eds. C.J. Lada & N.D. Kylafis. Kluwer Academic Publishers, p. 479

Burkhart, C. & Coupry, M. F. 1997, A&A 318, 870

Cioni, M.-R. L., Habing, H. J., Israel, F.P. 2000, A&A 358, L9

Da Costa, G. S. & Hatzidimitriou, D. 1998, AJ 191, 1934

Da Costa G. S. 1999, New Views of the Magellanic Clouds, IAU Symp. 190, eds. Y.-H. Chu, p. 397

Deliyannis, C. P. & Steinhauer, A. 2001, AAS 198, 4203

Dirsch, B., Richtler, T., Gieren, W. P., Hilker, M. 2000, A&A 360, 133

Dopita M. A. 1999, Active Galactic Nuclei & Related Phenomena, IAU Symposium 194, eds. Y. Terzian, E. Khachikian, & D. Weedman, p. 199
Efremov, Yu. N. 1979, SvAL 5, 12
Efremov, Y. N. & Elmegreen, B. G. 1999, New Views of the Magellanic Clouds, IAU Symp. 190, ed. Y.-H. Chu, p. 422
Efremov, Y. 1997, in Wide-Field Spectroscopy, E. Kontizas, M. Kontizas, D. H. Morgan, & G. P. Vettolani, eds. Kluwer Academic Pub., p. 145
Feitzinger, J. V. & Braunsfurth, E. 1984, A&A 139, 120
Fregeau, J. M., Joshi, K. J., Portegies Zwart, S., Rasio, F.A. 2000, AAS 196, 4111
Gardiner, L. T. & Noguchi, M. 1996, MNRAS 278, 191
Girardi, L. Chiosi, C. Bertelli, G., Bressan, A. 1995, A&A 298, 87
Goudis, C. & Meaburn, J. 1978, A&A 68, 189
Gouliermis D., Kontizas M., Korakitis, R. Morgan, D. H. Kontizas, E., Dapergolas, A. 2000, AJ 119, 1737
Harris, J. & Zaritsky, D. 1999, ApJ 117, 2831
Hatzidimitriou, D., Bhatia, R. K. 1990, A&AS 84, 527
Hatzidimitriou, D. 1997, in Wide-Field Spectroscopy E. Kontizas, M. Kontizas, D. H. Morgan, & G. P. Vettolani, eds. Kluwer Academic Pub., p. 153
Hatzidimitriou, D., Croke, B. F., Morgan, D. H., Cannon, R. D. 1997, A&AS 122, 502
Holtzman, J. A., Gallagher, J. S., Cole, A. A., Mould, J. R., Grillmair, C. J., Ballester, G. E., Burrows, Christopher J., Clarke, John T., Crisp, David, Evans, R. W. Griffiths, R. E., Hester, J. J., Hoessel, J. G., Scowen, P. A., Stapelfeldt, K. R., Trauger, J. T., Watson, A. M. 1999, AJ 118, 2262
Howell, J. H., Guhathakurta, P., Sarajedini, A., Gilliland, R. L., Albrow, M. D., Brown, T. M., Charbonneau, D., Burrows, A. S., Cochran, W. D., Baliber, N., Edmonds, P. D., Frandsen, S., Bruntt, H., Lin, D. N. C., Vogt, S. S., Choi, P., Marcy, G. W., Mayor, M., Naef, D., Milone, E. F., Stagg, C. R., Williams, M. D., Sigurdsson, S., VandenBerg, D. A. 2000, AAS 196, 4108
Irwin, M. J. 1991, IAU Symp. 148, 453
Johnson, J. A., Bolte M., Stetson P. B., Hesser J. E., Somerville R. S. 1999, AAS 194, 4033
Keller, Stefan C., Bessell, M. S., Da Costa, G. S. 2000, AJ 119, 1748
Keller, S., Kontizas, M., Gouliermis, D., Bellas-Velidis, I., Bessell, M. S., Kontizas, E., da Costa, G. S. 2000, Astronomische Gesellschaft Meeting, Heidelberg, in press
Kennicut, R. C. 1980, MNRAS 192, 365
Kontizas E., Dapergolas A., Morgan D. H., & Kontizas, M. 2001, A&A 369, 932
Kontizas, M., Hadjidimitriou, D. & Metaxa, M. 1987, IAU Symp. No. 126, Reidel Publ., p. 538
Kontizas M., Morgan D. H., Hadzidimitriou D. & Kontizas 1990, A&AS 84, 527
Kontizas E., Kontizas M., Sedmak G., Smareglia R. & Dapergolas A. 1990, AJ 100, 425
Kontizas E., Metaxa M., & Kontizas M. 1988, AJ 96, 1625
Kontizas E., Kontizas M., Michalitsianos, A. G., 1933a, A&A 269, 107
Kontizas, M., E. Kontizas & A. G. Michalitsianos, 1993b, A&A 267, 59
Kontizas, M., E. Kontizas, A. Dapergolas, S. Argyropoulos & Y. Bellas-Velidis, 1994, A&A 107, 1
Kontizas, M., Hatzidimitriou, D., Bellas-Velidis, I., Gouliermis, D., Kontizas, E., Cannon, R. D., 1998, A&A 336, 503
Kontizas, M., D. H. Morgan, E. Kontizas & A. Dapergolas, 1996, A&A 307, 359
Kunkel, William E., Demers, S., Irwin, M. J., Albert, Loic, 1997a, ApJ 488, L129
Kunkel, W. E., Irwin, M. J., Demers, S. 1997b, A&AS 122, 463
Kunkel, W. E., Demers, S., Irwin, M. J. 2000, AJ 119, 2789
Larsen, S. S., Clausen, J. V., Storm, J. 2001, A&A 364, 455L

Leon, S., Meylan, G., Combes, F. 2000, A&A 359, 1096
Lucke, P. B. & Hodge, P. W. 1970, AJ 75, 171
Makino J., Akiyama K., Sugimoto 1991, Ap&SS 185, 63
Maragoudaki, F., Kontizas, M., Kontizas, E., Dapergolas, A., Morgan, D. H. 1998, A&A 338, L29
Maragoudaki et al. 2001, A&A 379, 864
Martin, N., Prevot, L., Rebeirot, E., Rousseau, J. 1975, A&A 51, 31
Massey, P., Hodge, P. W., Jacoby, G. H., King, N. L., Olson, K. A. G., Saha, A., Smith, 2001 CAAS 197, 113.02
Mateo, M. 1998, ArA&Ap 36,435
McCray, R. & Snow, T. P., Jr. 1979, ARA&Ap p. 213-240
Meaburn, J. 1980, MNRAS 192, 365
Meibom, S. C. D., Barnes, S. A., Dolan, C. J., Mathieu, R. D. 2000, AAS 197, 4105
Morgan, D. H., Hatzidimitriou, D. 1995, A&AS 113, 539
Morgan, D. H., 1997, in Wide-Field Spectroscopy, E. Kontizas, M. Kontizas, D. H. Morgan, & G. P. Vettolani, eds. Kluwer Academic Pub., p. 161
Murray, S. D., Lin, D. N. C. 1996, ApJ 467, 728
Murray, S. D., Raymondson, D. A., Urbanski, R. A, 1999, ApJ 517, 829
Nordstoem, B., Andersen, J., Andersen, M. I. 1997, A&A 322, 460
Norris J. E. 1997, in Wide-Field Spectroscopy, E. Kontizas, M. Kontizas, D. H. Morgan, & G. P. Vettolani, eds. Kluwer Academic Pub., p. 133
Pagel, B. E. J. & Tautvaisiene, G. 1999, Ap&SS 265,461
Points, S. D., Chu, Y. H., Kim, S., Smith, R. C., Snowden, S. L., Brandner, W., Gruendl, R. A. 1999, ApJ 518, 298.
Points, A., Cruz-Gonzalez, I. Salas, L. 2000, AAS 196, 4204
Prevot, L., Martin, N., Rousseau, J. 1989, A&A 225, 303
Richer, M. G., Stasinska, G., McCall, M. L. 1999, A&AS 135, 203
Staveley-Smith, L., Kim, S., Stanimirovic, S. 1999, IAU Symposium No. 190, eds. Y.-H. Chu, N. B. Suntzeff, J. E. Hesser & D.A. Bohlender, p. 37
van den Bergh S. 2000, in The galaxies of the Local Group, by Sidney Van den Bergh. Cambridge Univ. Press, 2000, Cambridge Astrophysics Series, vol. 35.
Wallerstein, G., Hughes, J. 1999, from Globular Cluster to Field Stars, Proc. of the 35th Liege International Astrophysics Colloquium, eds. A. Noels, P. Magain, D. Caro, E. Jehin, G., p535
Wilhelm, R. 2001, AAS 197, 4112
Zaritsky, D., Harris, J., Grebel, E. K., Thompson, I. B. 2000, ApJ 534, L53
Zepf, S. 1999, AAS 194, 4015Z

Chapter 22

The Livermore Optical Transient Imaging System

Hye-Sook Park
Lawrence Livermore Natioanl Laboratory
Livermore, California USA

Dieter H. Hartmann
Clemson University
Clemson, South Carolina USA

G. Grant Williams
Steward Observatory
Tucson, Arizona USA

Abstract: The 1997 discovery by BeppoSAX of X-ray afterglow emission from Gamma-Ray Bursts allowed rapid optical follow-up observations with small ground-based telescopes. These observations led to the detection of optical afterglow emission, which established sub-arcsecond positions. Subsequent observations with large-aperture telescopes on the ground and the Hubble Space Telescope established redshifts and host galaxies for a significant number of bursts. Thus Gamma-Ray Bursts (GRBs) were recognised as the most energetic explosions in the universe. The release of a large amount of energy in a small volume then leads to a fire "ball" (or a jet), which reaches relativistic expansion velocities, whose interaction with the surrounding medium leads to afterglow emission.

These observations greatly advanced our understanding of this phenomenon by establishing the energy scale, testing the relativistic physics of the expanding fireball, and probing the geometry of the emitting plasma. Several theoretical models were developed, which commonly involve formation of an accreting black hole and production of a geometrically focussed MHD jet as the central engine. To test this model, we search for *prompt* optical transients (POTs) with the automated LOTIS telescope located at Livermore. To date, in addition to the first POT detection of GRB990123 by the ROTSE experiment, two more GRB events (GRB021004 and GRB021211) were detected with POT by RIKEN, KAIT and Super-LOTIS systems. All these early optical signals show that the early time light-curves are different from the late time

T.D. Oswalt (ed.), The Future of Small Telescopes in the New Millennium, Vol. III, 331-342.

afterglow curves. More POTs need to be observed in order to understand the GRB progenitor environment.

We also describe Super-LOTIS experiment, a robotic 0.6-m telescope located at the Kitt Peak National Observatory, which is dedicated to rapid searches for POT emission. Super-LOTIS follows the GRB light curve to V ~ 21, allowing coverage from seconds to hours of the afterglow phase. The impact of this and similar small robotic telescopes on the study of GRB physics is significant, but periods when no burst afterglow is observable can be utilised for independent investigations. We describe programs on variable stars, supernova light curves, and a search for novae in M31.

Key words: gamma-ray bursts, afterglows, optical transients, robotic telescopes

1. INTRODUCTION

1.1 Gamma-Ray Bursts in a nutshell

Cosmic Gamma-Ray Bursts (GRBs) are short flashes of high energy X- and γ-radiation occurring with an isotropic distribution on the sky and apparently without recurrence. The duration distribution is bimodal, with peaks at 0.5-s and 30-s. Bursts can last from less than 0.1 to over 1,000 seconds and exhibit fluctuations on millisecond time scales. The GRB power spectrum typically peaks in the 100-keV to 1-MeV range, and spectral hardness (flux ratio in two bands, one at "high" energies and the other at "low" energies) correlates with duration. Short GRBs are on average harder. It is common to speak of short/hard and long/soft burst classes. GRB observations have recently been reviewed by Fishman and Meegan (1995) and Klose (2000).

Their brightness distribution and angular isotropy led to an intense debate on their distance scale, reminiscent of the famous Curtis-Shapley debate at the National Academy in April 1920 on the nature of spiral nebulae. The GRB debate is described in Volume 107 of the Publications of the Astronomical Society of the Pacific. After February 28, 1997 the solution became obvious.

GRB970228 became the first success story, when the Italian-Dutch X-ray satellite BeppoSAX (www.asdc.asi.it/bepposax) discovered a fading X-ray afterglow that allowed a rapid (8 hours after the burst) and accurate (arcmin) localization. This in turn led to the detection of fading optical emission (with afterglow flux following a power law $F_{opt} \propto t^{-1.5}$), and eventually a faint (V ~ 26) host galaxy at z = 0.695. Since February 1997, such success stories have unfolded in similar fashion many times, and established their cosmological distance scale and the fact that all GRBs are associated with a host galaxy. Details of many multi-wavelengths afterglow observations can be found at www.aip.de/People/JGreiner/ and the review by van Paradijs,

Kouveliotou, & Wijers (2000). Figure 1 shows the afterglow of GRB010222 as an example.

Afterglows are explained with a "fireball model" in which a relativistic flow of ejecta interacts with the surrounding medium. The observations revealed breaks in the afterglow evolution of some bursts, which are interpreted as the manifestation of a jet in which beamed flows make a transition from fast (i.e. relativistic) to slow (non-relativistic) motion. Including solid angle beaming corrections, the observed fluxes imply energy releases of ~ $5x10^{50}$ ergs (Frail et al. 2001), a value that is very similar to the famous value 1 foe = 10^{51} ergs often quoted for supernovae. However, we note that the binding energy of a neutron star is in excess of 10^{53} ergs, that 10^{51} ergs is the kinetic energy of the expanding remnant, and that only 10^{49} ergs generate the UV-Opt-IR light curve that render supernovae so prominent. The term "hypernova" is often used to describe bursts and their afterglows, but this word is frequently misinterpreted as shorthand for "the GRB model", of which there are several.

The formation of a fireball is inevitable when such a large amount of energy is injected on a short time scale into a small volume, as indicated by the time profiles. The nature of the underlying GRB engine is an open question.

What causes the burst phenomenon is not yet fully understood, but leading models commonly invoke the formation of a rapidly accreting black hole. There are many routes to such a system, and two examples are the merger of a compact double neutron star system (the merger model, see e.g. Ruffert & Janka 1998) and core collapse of a massive, rapidly rotating star (collapsar model, see e.g. MacFadyen & Woosley 1999). Some numerical simulations suggest that these two distinct models correspond to the

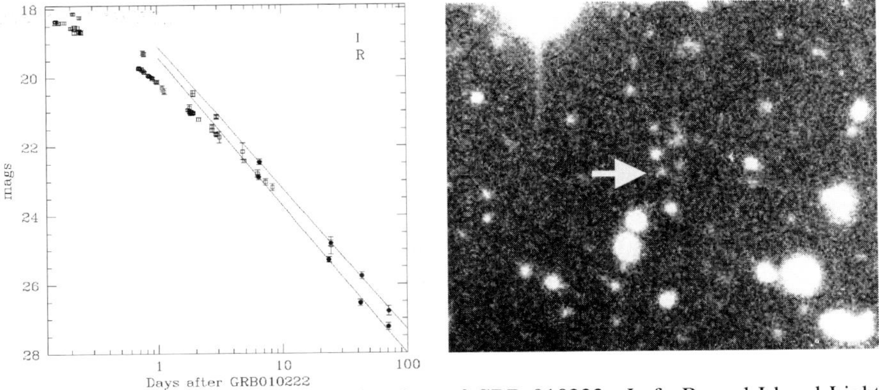

Figure 1. The multi-wavelengths afterglow of GRB 010222. Left: R- and I-band Light curves as posted on J. Greiner's website, and the R-band image obtained with the 90" Bok Telescope (KPNO, Steward Observatory) ten days after the GRB. The brightness of the OT at that time was R ~ 23.5 (Henden et al. 2001). The break in the curve is due to the jet nature of the GRB.

observed duration bimodality, but this has not yet been firmly established. If true, one expects long bursts (collapsars) to be associated closely with star formation activity (short life of massive stars). The shorter bursts from mergers should be less tightly linked to regions of star formation (because there is a substantial amount of time between binary formation and the final merger event, and because these systems have large space velocities that can carry them far from their birth sites). These predicted trends have yet to be established. The current set of bursts with afterglows is limited, for technical reasons, to the class of long-soft events. The presence of a host galaxy in all cases supports the idea that these GRBs trace the cosmic star formation history. Whether or not a similar link can be established for short-hard GRBs remains an open question. The currently operational HETE-2 mission (space.mit.edu/HETE/) is sensitive to short GRBs, but the information provided by HETE has not yet led to a significant number of afterglow observations for this GRB class.

2. THE LIVERMORE OPTICAL TRANSIENT IMAGING SYSTEM

2.1 LOTIS

During the CGRO era BATSE produced rapid (seconds), but rather large (degrees) burst locations. Instruments designed to detect prompt optical emission based on BATSE positions were thus forced to cover a substantial fraction of the sky. While all sky monitoring is a feasible option, as demonstrated by CONCAM (e.g., Nemiroff & Rafert 1999), the flux limit is generally not sufficient to significantly constrain the prompt afterglow. An automatic slew with a small telescope allows a fast response to magnitudes of V ~ 10, or even fainter, with integration times comparable to gamma ray burst durations (~10-s). The **L**ivermore **O**ptical **T**ransient **I**maging (LOTIS) system represents this approach (see Figure 2).

Figure 2. The LOTIS experiment at LLNL, Livermore, CA

LOTIS is based on four Cannon EF 200-mm f/1.8L lenses with an aperture of 110-mm each. The total field of view covered by LOTIS is 17.4° x 17.4°, sufficient to cover most BATSE locations. The delay between receiving a trigger and pointing at the burst location is typically 10-s, so that many burst positions were imaged while gamma-ray emission was still in progress. No prompt optical transients (POTs) were recorded in the period from October 1996 to July 1997 (e.g., Park, et al. 1997), during which five bursts were imaged simultaneously. The magnitude limit for a 10-s exposure (no filter) is mag ~ 14.5, with a CCD thermoelectrically cooled to T = -10 °C. The 15 μm pixels of the Loral 442A 2kX2k CCD correspond to 15" pixels on the sky.

Until the end of Compton Observatory LOTIS had responded to 145 GRB triggers, and for more than a dozen events upper limits on prompt optical afterglow emission were obtained (Williams et al. 1999; Williams 2000). Similar constraints for seven bursts were obtained with ROTSE (e.g., Kehoe et al. 2000). While these limits provide useful constraints on burst models, detection of POT emission is of course more valuable. Such detection has so far only been accomplished once (GRB990123), with ROTSE (Akerlof et al. 1999). The peak brightness of the POT, V ~ 9, is astounding for a source at redshift z = 1.6 (Kelson et al. 1999).

The demise of Compton Observatory (2000) discontinued BATSE locations, and the search for prompt emission focussed on the more accurate positions from the HETE satellite (space.mit.edu/HETE/), launched in October 2000. The four LOTIS telescopes were aligned to cover one common FoV, and R and V filters were added. LOTIS thus covers a FoV much larger than HETE error boxes, and provides multi-color photometry to V ~ 15. The most recent LOTIS sky patrol observations of GRB010921 (52 minutes after the burst) limit the afterglow flux to V = 14.2 and mag = 15.3 without filter (see GCN 1114). To increase the probability to detect prompt OTs much fainter than this level, we developed Super-LOTIS with a smaller FoV (appropriate for response to HETE and other space-based burst detectors in planning stage or under development, such as Swift) and much higher sensitivity.

2.2 Super-LOTIS

The Super-LOTIS telescope is a Boller & Chivens 0.6-m reflector of f/3.5, with a $(0.84°)^2$ FoV. The telescope (Figure **3**) was originally constructed for Lick Observatory, but recently decommissioned. We refurbished it with computer control devices, a coma corrector, and an electronic CCD imager (see hubcap.clemson.edu/~gwilli/LOTIS). Super-LOTIS saw first light on Feb. 25, 1999 in Livermore, and was then moved to Kitt Peak National Observatory in April 2000, where it began routine operation in Oct. 2000. The system is fully robotic and can slew to a new GRB location within ~ 20-s (depending on the actual angular offset between burst and telescope at the time of the trigger). The CCD system is identical to that used in LOTIS, but cooled to T = -30 °C. Integrating for 30 seconds results in a magnitude limit of R ~ 18.5. The 15-μm CCD pixel corresponds to an angular resolution of 1.5" on the sky.

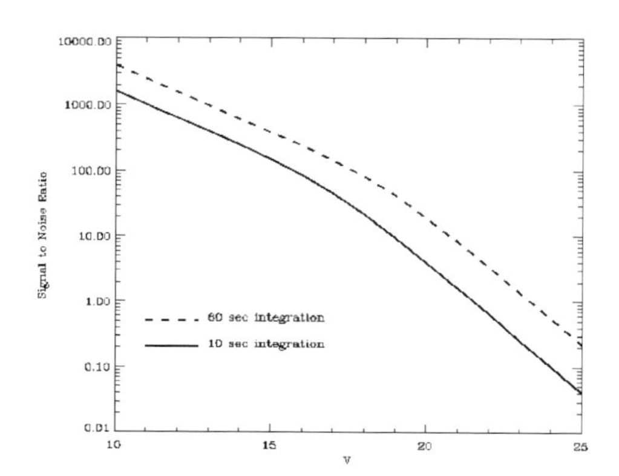

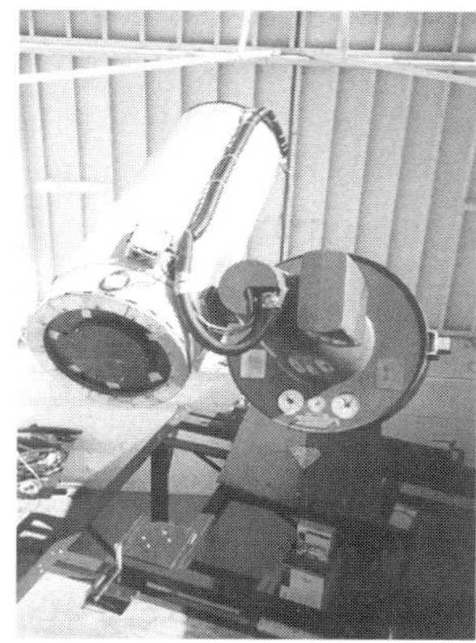

Figure 3. Super-LOTIS at KPNO. Depending on the desired Signal-to-Noise ratio, POT detection down to V ~ 20 requires an integration of about one minute. If prompt optical emission is brighter than V ~ 17, an integration of time of only ten seconds is sufficient.

Super-LOTIS saw first light on Feb. 25, 1999, and has operated at KPNO since then. During engineering testing in March, GRB990308 was imaged starting 1700-s after the GRB (Park et al. 2000). No fading or flaring source brighter than V = 15 was detected. The bright limit was caused by weather conditions at the Livermore site, which prevented a deeper search.

3. LOTIS & SUPER-LOTIS NON-GRB PROGRAMS

3.1 Nova Rates and Distributions

When there is no ongoing afterglow bright enough to be detectable, Super-LOTIS carries out a pre-programmed patrol program targeted at a variety of objects. The large field of view of Super-LOTIS is well suited for a large-scale search for novae in M31. The apparent size of M31 on the sky is 92' X 197' (e.g., Holmberg 1958). This extremely large extent requires a mosaic of images, but Super-LOTIS does not need more than 6 images to cover the full field. It is advantageous to cover M31 with an even larger set

Figure 4. M31 observed with the large FoV of LOTIS. Super-LOTIS covers this field with a mosaic of either 3x3 or 3x4 exposures. The bulge receives a reduced exposure time.

of exposures, to allow nova detection well into the halo and also the outskirts of the disk (see Figure 4).

The nova rate in M31, as well as the bulge-to-disk ratio, is not well known. In the Milky Way the situation is even worse, because of the disadvantaged solar location in the disk. The Galactic nova rate is $R_N(MW) \sim 35 \pm 11\ yr^{-1}$ according to Shafter (1997), or $R_N(MW) \sim 41 \pm 20\ yr^{-1}$ according to Hatano et al. (1997). The rough similarity of M31 and the Milky Way suggests that the rate of M31 is comparable.

Novae are cataclysmic variables involving close binary systems consisting of a late-type star filling its Roche lobe, which causes accretion onto the white dwarf companion. Stellar population synthesis studies of binary systems by Yungelson, Livio, & Tutukov (1997) imply that the nova rate of a galaxy is determined by its current star formation rate. In other words, the nova rate is a tracer of a young stellar population, and not of - as it is often stated - an old population. One thus expects that the observed nova originate predominantly from the disk of a galaxy, and to a lesser degree from the bulge. This appears to be in disagreement with Galactic nova observations (e.g., Patterson 1984), but significant dust extinction in the ISM severely hampers the interpretation of these data. On the other hand, observations of novae in nearby galaxies (M31, M51, M87, M101; Shafter et al 2000) do not support the theoretical predictions from Yungelson et al. (1997). The nova disk-to-bulge ratio is thus an unsettled issue.

Observations of novae in M31 have a long tradition, reaching back to the early part of the century (Hubble 1929). Recent observations by Ciardullo et al. (1987) and Capaccioli et al. (1989) concluded that the novae in M31 were dominated by the bulge component. However, Monte Carlo simulations by Hatano et al. (1997) suggest that projection effects caused some disk novae to be misidentified as bulge novae. These authors conclude that the nova rate in M31 is disk dominated. Obviously, new dedicated nova surveys of M31 are needed, especially in the outer disk. The global nova rate in M31 is about 29 $\pm$ 4 yr^{-1} (e.g., Shafter, Ciardullo, & Pritchet 2000), so that a survey with Super-LOTIS should detect about one event every two weeks, while M31 is visible on the sky. The nova peak brightness is $M_v \sim -9$, so that a distance modulus of μ(M31) ~ 24.4 (e.g. Freedman & Madore 1990) implies peak magnitudes V ~ 15.4, which can be obtained by Super-LOTIS with a very short exposure (see Figure 4).

3.2 Variable Stars

LOTIS requires little supervision for data taking. It starts every night, performs data acquisition, responds to weather changes, and archives data automatically. We have been archiving LOTIS all-sky data for many years. To illustrate the quality of archived data, we will discuss here the results from three months of an imaging campaign on SS Cygni.

SS Cygni was discovered in 1896 by Louisa Wells of the Harvard College Observatory. From photographic plates taken over a period of 40 days, its intensity was observed to vary from magnitude 7.2 to 11.2. Since then SS Cygni is now known as a cataclysmic variable and is one of the most observed objects amongst all variable stars. It is believed to be a binary system consisting of a red dwarf star and a white dwarf. The strong gravity of the dense white dwarf pulls a stream of gas off of the companion star. This gas collects in an accretion disk around the white dwarf. A large amount of energy is released when the disk material falls onto the surface of the white dwarf, causing the intensity of the system to brighten across the spectrum.

We analyzed images taken from June to September 2000. During this period, LOTIS acquired 60-80 images/day of SS Cygni. The images were taken over a wide range of weather and sky conditions including three full-moon cycles, foggy nights, rainy nights, as well as many clear nights. The challenging task in applying sophisticated photometry algorithms to extract quantitative light curves from the images is to try to separate the intensity variations that are due to the local conditions from the variability of the stellar objects. This is done by looking at the intensities of nearby non-variable stars. Figure 5 shows SS Cygni when it was in the quiescent state (June 5, 2000 image) and when it was in an outburst state (Sept. 11, 2000 image). The brightening of SS Cygni is clearly visible in these images. The

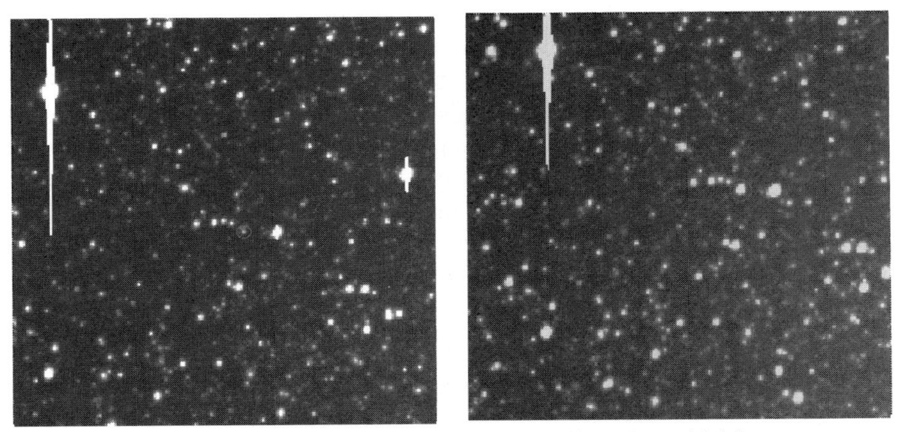

Figure 5. SS Cygni in quiescence (left), and in outburst (right).

September 2000 outburst was also viewed by Chandra and EUVE, studying the nature of energy flow from the companion red giant to the white dwarf. See articles at xrtpub.harvard.edu/chronicle/0300/aavso.html and details on www.space.com/scienceastronomy/astronomy/amateurs_sscyg_001010.htm.

Basic analysis steps are 1) dark subtraction, 2) astrometric corrections, 3) isolation of all stellar objects, 4) normalization of the intensity to reference stars, and 4) entering the results into a database of stellar magnitudes vs. time. All these algorithms are automatic. The resulting light curve of SS Cygni is shown in Figure 6. From a comparison standard star data we

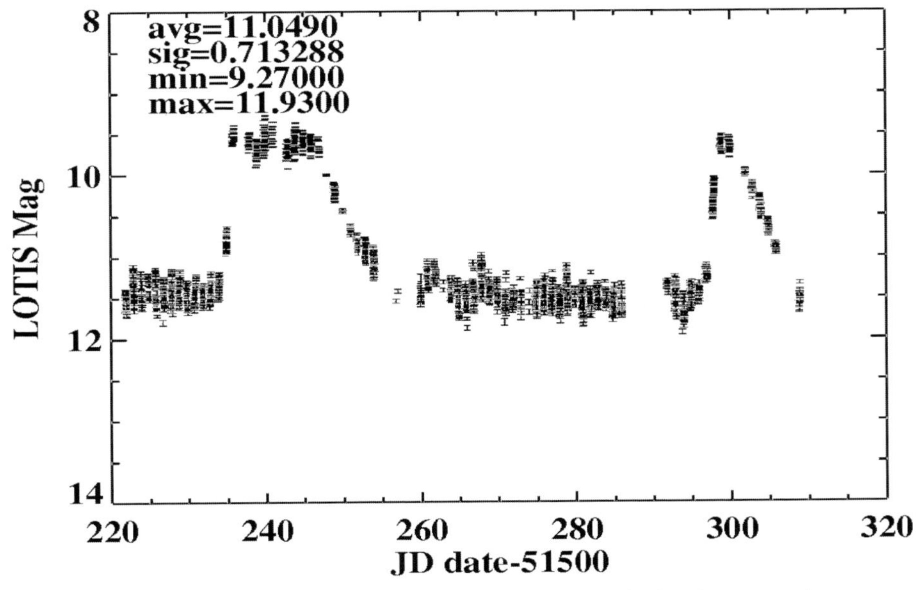

Figure 6. Light curves of the cataclysmic variable SS Cygni, showing two outbursts.

deduce an accuracy of the relative photometry of $\delta V \sim 0.03$ for a $V \sim 11.5$ star. We observed two outbursts of this variable and since we have continuous coverage, we are able to measure the rising phase of the outbursts. We are now extending these algorithms to cover a larger field of view to produce a database containing many more objects. Accurate and systematic monitoring of the variability of many stellar objects is of interests to many astronomers and we show that small telescopes can produce high quality measurements.

3.3 Type Ia Supernovae

In the past decade Type Ia supernovae have been used to establish the accelerated expansion of the Universe (e.g., Bahcall et al. 1999, Leibundgut 2001), and studies of the CMB fluctuation spectrum (e.g., de Bernardis et al. 2000) provide strong evidence for a spatially flat universe (predicted by most inflation models). Together, the SNIa and CMB observations suggest that 70% of the energy density in the universe is dark energy and only 30% is matter (of which only ~1/6 is baryonic). To further support this emerging cosmological framework, one needs observations of SNIa at high z, and major efforts along these lines (e.g., SNAP) are under development. NGST will extend such studies to Type II supernovae. It must be pointed out, however, that the success of these studies rests on a solid understanding of SNIa theory (e.g., Hillebrandt & Niemeyer 2000) and an extensive set of calibrators in the local universe. Here even small aperture telescopes can make a contribution, and we plan an observational campaign to establish multi-color photometry for nearby supernovae. SNIa are often considered standard candles, but it is known that their light curves vary considerably (corresponding to a wide range of synthesized ^{56}Ni, from 0.1 to 1.0 solar masses). However the observational Phillips relation between luminosity and decline rate allows multi-color template fitting (e.g., Phillips et al. 1999) of the light curves, which makes SNIa useful "standard candles". For statistical studies it is important to assemble a large set of well sampled lightcurves, and we plan to use Super-LOTIS for automated observations of SNIa in the local universe.

3.4 The Next Galactic Supernova

The gravitational binding energy released in a core collapse (Type II) super-nova, $E_{bind} \sim 3 \times 10^{53}$ ergs, is mostly (>99%) carried away by neutrinos. Only one foe (10^{51} ergs) emerges as kinetic energy of the ejecta, which eventually powers the light from a long-lived supernova remnant. The bright optical SN light show that can outshine a whole galaxy for days to weeks adds up to the tiny amount of $E_{opt} \sim 10^{49}$ ergs, but provides essential information on several key parameters of the exploding star. The first light to emerge from a SNII is the UV and X-ray dominated flash resulting from shock breakout (e.g., Shigeyama et al. 1987; Woosley 1988; Ensman & Burrows 1992). This occurs about 30 minutes to 10 hours after core bounce, depending on the envelope properties of the star. This short-lived, super-bright phase of the supernova "afterglow" has never been observed. Afterwards, the shock and energy deposited by fresh radioactivity in the ejecta powers the late-time light-curve, which is commonly observed for weeks to months, or even years in some cases.

To observe the emission from shock breakout, and the transition to the late-time evolution, an early warning system is required. Such a system could be based on the detection of neutrinos generated by the core collapse and the subsequent cooling phase of the new-born neutron star. This phase lasts for approximately 10 seconds (e.g., Burrows 2000), so that there is plenty of time (in principle) to react with robotic and non-robotic telescopes, as long as the location accuracy is sufficient. Such a Super Nova Early Warning System (SNEWS) is in fact operational (e.g., Habig, 1999), and is capable to alert the community within less than a minute of a Galactic supernova.

To derive a source position one can either use the directionality of the scattering reaction $\nu + e^- \rightarrow \nu + e^-$ (neutrinos of all three flavors, and their anti-particles participate) in a single detector. Alternatively, arrival time information from a global network of detectors can be used to triangulate the event, similar to the IPN method employed for GRBs. Detailed simulations (Beacom & Vogl 1999) suggest that the scattering method is superior, and that error box sizes of 5° (Super-Kamiokande) to 20° (SNO) can be derived. This uncertainty is large, but LOTIS, and even Super-LOTIS, has a large enough FoV to image positions provided by SNEWS. A successful super-nova campaign of course requires that telescopes maintain a high duty cycle. A global response network of small aperture telescopes could respond to an alert from SNEWS, to obtain the position of the supernova at the earliest possible moment and rapidly direct large telescopes for follow-up studies.

4. REFERENCES

Akerlof, C. et al. 1999, Nature 398, 400

Bahcall, N. A., Ostriker, J. P., Perlmutter, S., & Steinhard, P. J. 1999, Science 284, 1481 (see also astro-ph/9906436 for an update)

Barthelmy, S. D., et al. 2000, in "Gamma-Ray Bursts", AIP 526, eds. R. M. Kippen et al., p. 731

Beacom, J. F., & Vogl, P. 1999, Phys Rev D60, 033007, astroph/9811350

Burrows, A. 2000, Nature 403, 727
Capaccioli, M., et al. 1989, AJ 97, 1622
Ciardullo, R., et al. 1987, ApJ 318, 520
de Bernadis, P., et al. 2000, Nature 404, 955
Ensman, L., & Burrows, A. 1992, ApJ 393, 742
Fishman, G. J. and Meegan, C. A. 1995, ARAA 33, 415
Frail, D. A., et al. 2001, ApJ Letters, in press (astro-ph/0102282)
Freedman, W. L., & Madore, B. F. 1990, ApJ 365, 186
Habig, A. 1999, astro-ph/9912293
Hartmann, D. 1999, Proc. Natl. Acad. Sci., Vol. 96, p. 4752
Hartmann, D. H., MacFadyen, A., & Woosley, S. E. 2000, in "Gamma-Ray Bursts", AIP 526, eds. R. M. Kippen et al., p. 653
Hatano, K., Branch, D., Fisher, A. & Starrfield, S. 1997, MNRAS 290, 113
Hillebrandt, W. & Niemeyer, J. C. 2000, ARAA 38, 191
Henden, A. A., et al. 2001, ApJ, submitted: "Afterglow of GRB010222"
Holmberg, E. 1958, Lund Medd. Ser.II, No. 136
Hubble, E. 1929, ApJ 69, 103
Kehoe, R., et al. 2001, in "Supernovae and Gamma-Ray Bursts", eds. M. Livio, N. Panagia, & K. Sahu (Cambridge Univ. Press), p. 47
Kelson, D., et al. 1999, IAU Circ. 7096
Klose, S. 2000, Reviews in Modern Astronomy 13, 129
MacFadyen, A. I., & Woosley, S. E. 1999, ApJ 524, 262
Park, H.-S., et al. 1997, ApJ 490, L21
Patterson J. 1984, ApJS 54, 443
Phillips, M. M., et al. 1999, AJ 118, 1766
Park, H.-S., et al. 2000, in "Gamma-Ray Bursts", AIP 526, eds. R. M. Kippen et al., p. 736
Ruffert, M., & Janka, H.-T. 1998, A&A 344, 573
Shafter, A. W. 1997, ApJ 487, 226
Shafter, A. W., Ciardullo, R., & Pritchet, C. J. 2000, ApJ 530, 193
Shigeyama, T., Nomoto, K., Hashimoto, M., & Sugimoto, D. 1987, Nature 328, 320
van Paradijs, J., Kouveliotou, C., & Wijers, R. 2000, ARAA 38, 379
Vrba, F. J., Hartmann, D. H., & Jennings, M. C. 1995, ApJ 446, 115
Williams, G. G., et al. 1999, ApJ 519, L25
Williams, G. G. 2000, Ph.D. thesis, Clemson University, Aug. 2000, available at compton.as.arizona.edu/LOTIS/pictures - phdthesisNEW3
Williams, G. G., et al. 2000, in "Gamma-Ray Bursts", AIP 526, eds. R. M. Kippen et al., p. 250
Woosley, S. E. 1988, ApJ 330, 218
Yungelson, L., Livio, M., & Tutukov, A. 1997, ApJ 481, 127

Chapter 23

The Importance of Small Telescopes to Cosmological Research

John Huchra
Harvard-Smithsonian Center for Astrophysics
Cambridge, Massachusetts USA

Abstract: Over the last several decades, research with small telescopes has contributed significantly to cosmology. We tell the story of one continuing endeavor, the study of the large-scale structure of the Universe. In the 20th century, LSS studies started with 2D surveys done on patrol cameras, moved to large Schmidt telescopes and then on to 3D redshift surveys on telescopes of all sizes. Modern LSS programs such as the 2dF, Sloan and 2MASS redshift surveys are still based on catalogs made on "small" telescopes, 1.2- to 2.5-m in diameter. The definition of small is changing, however, but 1- to 2-m class telescopes still fill a niche for innovative and important projects, even in cosmology, and thus have a place in the national infrastructure for astronomy.

Key words: galaxies, surveys, maps, redshifts, omega, mass, luminosity function

1. INTRODUCTION: MAKING MAPS

Extragalactic astronomy as we know it started after the first quarter of the last century with Hubble's conclusive proof that galaxies were "extragalactic" (1926), followed by his discovery of the universal expansion law (Hubble 1929). Although these discoveries were produced by the behemoths of the day, primarily the 100-inch (2.5-m) Hooker reflector at Mt. Wilson, Shapley, once convinced to give up the local hypothesis, threw himself into the task of more accurately mapping the nearby Universe with the Harvard Patrol Cameras in Massachusetts, Peru and South Africa. The first uniform (in this case, magnitude limited) 2-D map of galaxies was the Shapley-Ames catalog of the mid-1930's (Shapley & Ames 1932). The Shapley-Ames catalog contains ~1300 galaxies with types and magnitudes, and although smaller than the previous century's catalogs of fuzzy objects (e.g. the NGC and IC of Dreyer), attempted to stick to galaxies and provide accurate magnitudes.

T.D. Oswalt (ed.), The Future of Small Telescopes in the New Millennium, Vol. III, 343-354.

Maps of the galaxy distribution improved significantly only about 30 years later with the advent of surveys on a new generation of extremely wide-field small telescopes, the Schmidts. In the early 20th century, Bernhardt Schmidt developed a design for a telescope with a spherical primary, a glass correcting plate and a curved focal plane that provided coma, astigmatism and sphericial aberration free wide-field imaging. Fritz Zwicky and his collaborators installed first an 18-inch and finally a 48-inch Schmidt telescope at Mt. Palomar. A survey of the accessible sky was funded by the National Geographic Society. Zwicky and colleagues produced the first modern large catalog of galaxies and galaxy clusters from this survey in the late 1960's (Zwicky et al. 1961-68), and at the same time the Russian group headed by Vorontsov-Velyaminov produced a similar catalog (Vorontsov-Velyaminov et al. 1962-74).

When similar wide-field Schmidt telescopes became available in the southern hemisphere, similar catalogs were produced (c.f. West and Lauberts; Lauberts 1982). Deeper maps were actually made on smaller telescopes — the most famous perhaps the Shane-Wirtanen maps made with the dual 10-inch astrograph at Lick Observatory (Shane & Wirtanen 1967). The Princeton group, led by Jim Peebles (c.f. Peebles & Hauser 1974; Groth and Peebles 1977), pioneered the development of the statistical approach to galaxy clustering, albeit only 2-D statistics; the world began to cry out for real 3-D data. The Zwicky catalog showed significant clustering on the sky (c.f. Turner & Gott 1976) and the Shane-Wirtanen showed what appeared to be a giant filament covering half the northern sky. The only way the real topology of the structures was going to be determined was with redshift data.

With the advent of accurate Cosmic Microwave Background dipole measurements in the 1970's, followed by very clear determinations of the motions of nearby galaxies (Aaronson et al. 1982; Lynden-Bell et al. 1988) attempts to produce all-sky maps were made by several groups (Sharp 1986; Lynden-Bell & Lahav 1988). However, the need for 3-D maps only became more acute.

Here it is worth noting that almost all the optical/IR data discussed above was taken with small and dedicated telescopes, often of survey nature. This includes the photometry that went into the IR Tully-Fisher and the D_n-σ work on the large-scale motions of galaxies. The only exception to that was Hubble's pioneering work on the distance scale, which used the 100-inch telescope, the largest of its time, and the velocity dispersion studies of Lynden-Bell et al. for their work on flows. The workhorse mappers were the astrographs and mid-sized Schmidt telescopes and the twin 36-inch telescopes at Kitt Peak and Cerro Tololo. The IR Tully-Fisher relation was discovered using the KPNO 36-inch (Aaronson, Mould & Huchra 1979).

2. THE FIRST 3-D SURVEYS

Although there were a fairly large number of galaxy velocities known by the mid-1970's, the largest actual complete and magnitude-limited sample covering any wide area only contained a few hundred galaxies (c.f. Schechter 1974). A small number of narrow field surveys were being undertaken (c.f. Chincarini & Rood 1971; Tifft & Gregory 1976), mostly centered on galaxy clusters. Only large telescopes had modern detectors and spectrographs and there was not sufficient access by any one group to do major programs. The only exception to that was Allan Sandage and his work on the radial velocities for the Revised Shapley-Ames Catalog (Sandage & Tammann 1981).

A breakthrough came in 1976 when Steve Shectman built a new and inexpensive type of digital scanner for the 52-inch telescope at the then McGraw Hill Observatory (Shectman & Hiltner 1976). By then a number of Peebles' students and colleagues, including Marc Davis at CfA, were vitally interested in getting large number of redshifts. I, too, had gotten into the act about that time as well in collaboration with some of my Caltech colleagues, and we began measuring redshifts with the image tube spectrographs on the Palomar 1.5-m and the Kitt Peak 0.9-m. At CfA we had the interest and we also had available a moderately poor (spherical mirror) 1.5-m telescope that was much underutilized (Huchra 2001; See Figure 1).

Figure 1. A cluster of small telescopes at the F. L. Whipple Observatory used for "Cosmological" observations. On the left is the dome of the mighty Tillinghast 1.5-m that has been used for the CfA redshift surveys and is now in use for the 2MASS redshift survey. On the right is the dome of the northern 1.3-m 2MASS telescope, twinned at CTIO, which has produced a superb near-IR imaging survey of the whole sky. At the center is the dome of the 1.2-m telescope used for a variety of programs including photometry of galaxies for the HST H0 Key Project and for photometry of bright SN Ia for the High-Z SN team.

That was the key: modern instrumentation on a dedicated small telescope. The first CfA redshift survey was born in 1977, we took our first data in 1978 and we completed the survey of 2399 galaxies and 1 star (Zwicky had made a mistake) brighter than 14.5 magnitude in June of 1981. The maps were somewhat startling, but not overwhelming. However we did have the first significantly large and complete galaxy sample to play with to study the statistics of the galaxy luminosity function and 3-D galaxy clustering. In that initial survey we had the first fairly good data pointing towards a low-density universe (Davis & Huchra 1982).

3. THE CFA SURVEY

The second CfA survey was started almost four years after the first had been completed. Again, it was enabled by the combination of a good instrument on a small and available telescope and by the previous surveys done on the Palomar Schmidt telescope. Without the Zwicky catalog, we probably would not have started the project and we certainly would have adopted a different morphological approach. CfA2 was a ten-year survey of 18,000 galaxies, but done in strips so that intermediate results could be analyses. The initial results showed conclusively that the distribution of galaxies was considerably more complex than had been previously thought (de Lapparent, Geller and Huchra 1985) and those results were followed up by the discovery of the Great Wall, the (still) largest apparently coherent structure in the Universe (Geller & Huchra 1989). Figure 2 shows the resulting distribution.

This was followed by the most detailed studies of the luminosity function as a function of morphological type (Marzke et al 1994a; 1994b; Figure 3), and the use of the survey system to probe extragalactic structures at low galactic latitude. Along the way, we also completed various subsets of the IRAS galaxy surveys and the Optical redshift Survey (c. f. Fisher et al 1995). The complete dataset has been available on the web since 1995 at the URL: (http://www.cfa-www.harvard.edu/~huchra/zcat). Again the cosmological implications of this work, essentially done completely with small telescopes, has been profound.

4. 2DF, SDSS, AND BEYOND

The next generation of redshift surveys are more complex. Their goals are primarily similar to those of the early CfA surveys: study the statistics of galaxy clustering both nearby and at high redshift. The two main low redshift galaxy surveys are the 2dF (two degree field) survey, which is being done at

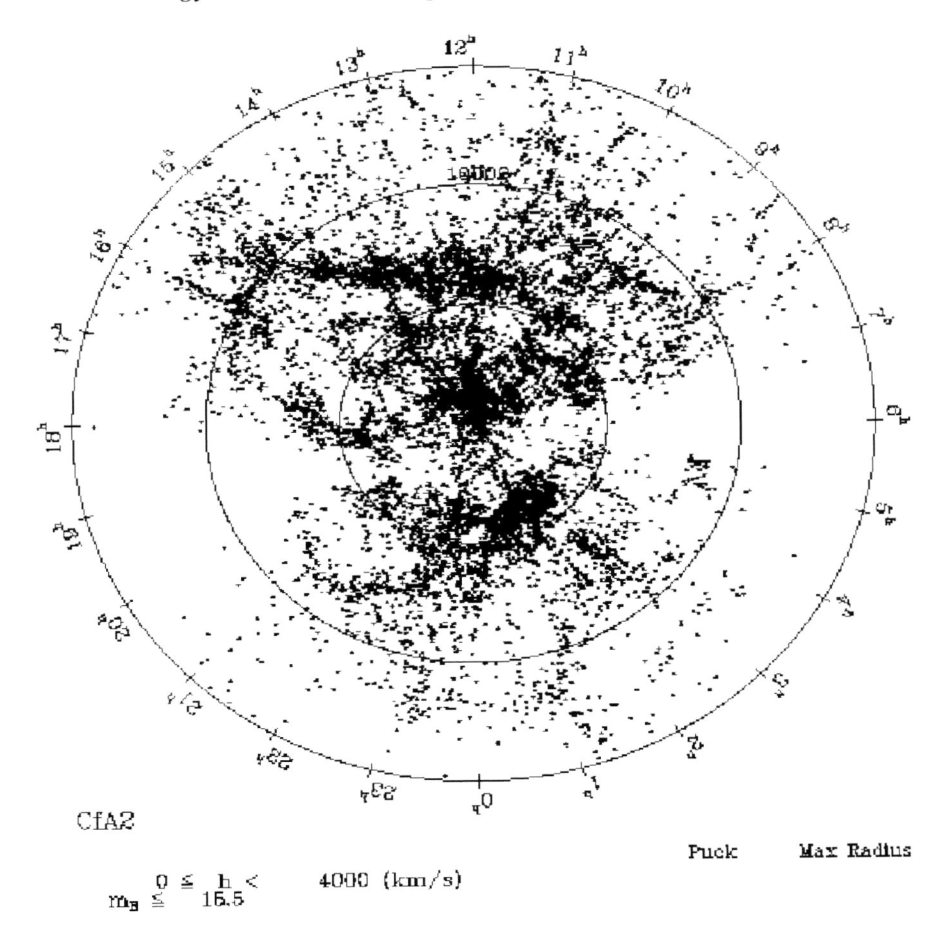

Figure 2. Plot of the redshift distribution of the CfA2 survey in cylindrical coordinates. We are at the center. The height/thickness of the cylinder above the equatorial plane is 4000 km/s and its radius is 15,000 km/s. The Great Wall, a.k.a. the A1367-Coma-Hercules supercluster, Local Supercluster and Pisces-Perseus supercluster are all easily visible.

the 4-m AAT and the Sloan Digital Sky Survey (SDSS) which utilizes a dedicated 2.5-m telescope in New Mexico. The 2dF survey is still based on photographic plates taken with the 1.2-m SRC Schmidt telescope in Australia, but the SDSS survey is providing its own, deep, digital catalog from a giant CCD mosaic camera on the parent telescope. This dedicated 2.5-m is revolutionizing the field with not only the survey five color digital and photometric imaging but also follow-up multi-fiber spectroscopy of galaxies and quasars. Unfortunately, the spectroscopic survey will only be complete to a limit that is 4-5 magnitudes brighter than the imaging survey, a limitation of spectroscopy with small telescopes.

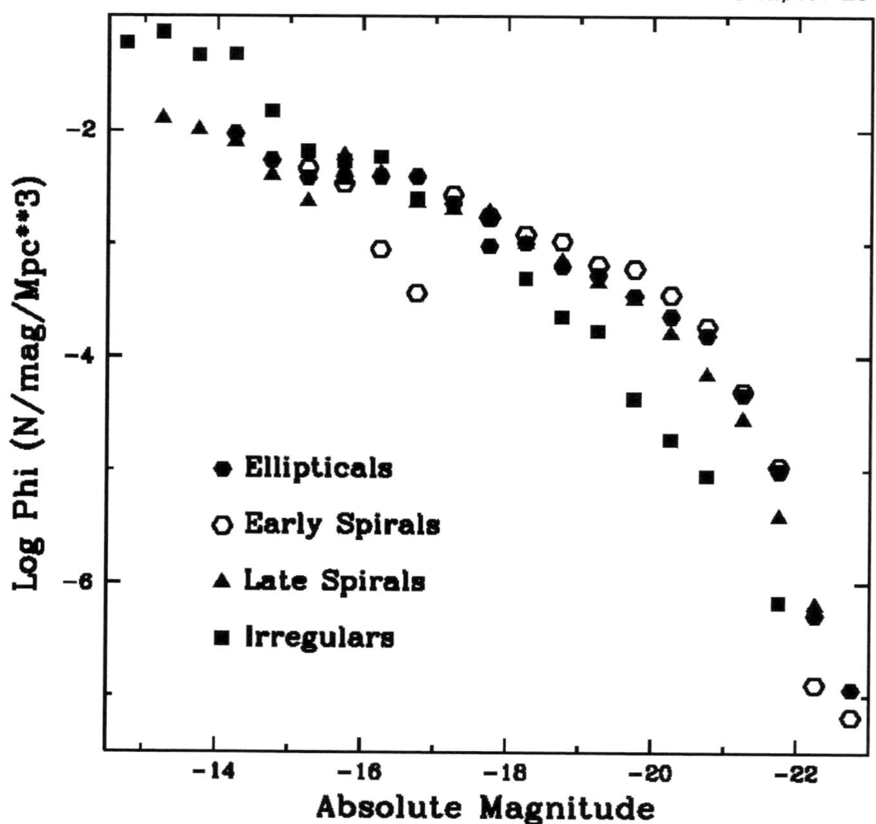

Figure 3. The galaxy luminosity function as a function of morphological type (adapted from Marzke et al. 1995). Note the steeply sloped LF for the very latest types. The normalization is uncertain.

Nonetheless, the plan is to obtain spectra with this "small" telescope for a million extragalactic objects. Already the SDSS has discovered the half-dozen highest redshift quasars known and it appears as if the survey of these objects has found the redshift at which the Universe reionized after its initial recombination (into neutral H and He) after a few hundred thousand years (as seen in the CMB; Becker et al. 2002; Djorgovski et al. 2002). This is yet another spectacular cosmological discovery essentially driven by work with a small (2.5-m) telescope.

5. 2MASS

Last but not least on my list of small telescope projects currently making great cosmological contributions is the soon to be released Two Micron All-Sky Survey (2MASS). 2MASS was conceived over a decade ago as a small explorer satellite (NIRAS, Fazio et al 1988) to make an all-sky near infrared map to a limit 10,000 times fainter than the previous Two Micron Sky Survey of Neugebauer and Leighton (1969). After review by NASA, it was decided to do this project from the ground to save money. The first ground-based proposal was submitted in 1991 and the project was finally funded and approved in 1993 (Kleinmann et al 1994). Two matched 1.3-m telescopes at CTIO and FLWO and two matched 3-channel IR cameras (Figure 4) have now been used to make an essentially complete three band map of the sky containing over 3 million extended sources (mostly galaxies) and 200 million stars (see Figure 5).

The 2MASS survey is extremely uniform. Our goal was to produce a survey with a photometric zero point uniform at the 1% level around the whole sky. This is absolutely necessary to provide the accuracy required to measure the galaxy density field and dipole and the galaxy correlation function to the requisite accuracy. 2MASS is much less affected by extinction than optical surveys: $A_K < 0.1\ A_B$, and the spectral energy distribution of light from old stars, the best tracer of the baryonic masses of galaxies, peaks around 1.6 microns.

2MASS is thus the only survey which can be used to construct a uniform, all-sky, three-dimensional map of the local universe. We are now engaged in

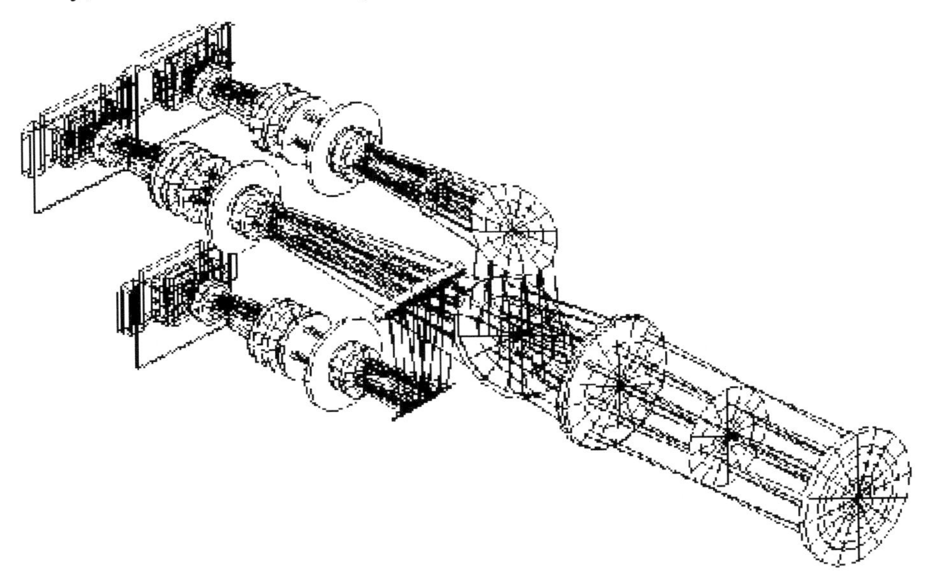

Figure 4. Ray trace of the 3-channel JHK_S 2MASS camera designed and built by Mike Skrutskie and Telec Optics. The design uses dichroics to enable simultaneous observation of all three colors, with the direct (minimum reflective loss) path reserved for the K band.

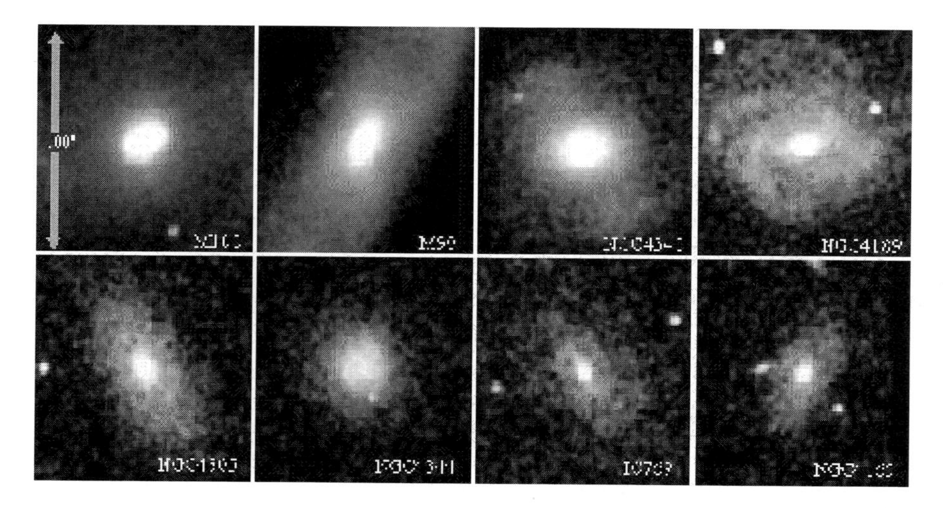

Figure 5. Some bright galaxies captured by 2MASS.

a progressive redshift survey (working to fainter and fainter magnitudes) to study dark matter and large-scale structure through the local density field and its effects on the peculiar velocities of galaxies. In particular, matching the local velocity and density fields measures a combination $\beta = \Omega_M{}^{0.6} / b$ of the matter density Ω_M and the bias factor b. Determinations based on the density field derived from the IRAS (60μ) galaxies (Dekel et al. 1993, Schmoldt et al. 1999) tend to produce high values of β, while those based on optically-selected galaxy samples (Santiago et al. 1996) tend to give low values. At some level the problem can be attributed to the different bias factors for IRAS and optically selected galaxies (Baker et al. 1998). The IRAS samples and the newer PSCz sample provide poor sampling on scales smaller than 1000 km s^{-1} and are strongly biased against dense regions.

Optically selected catalogs have severe problems with extinction and the non-uniformities of the photographic surveys from which they are derived. Even the really local density field is not in good shape. In the Tonry et al. (2000) analysis of the Local Supercluster there are many missing elements to the density model, including possible local sources for the observed quadrupole field and the "Local Anomaly", despite the high quality of their data on the local flow velocities. Because of the problems in determining the true shape and overdensity of the Local Supercluster, Tonry et al. (2000) could not determine Ω_M. Any attempt to reduce systematic uncertainties must begin with a much cleaner determination of the local density field of galaxies.

The 2MASS redshift survey is now underway. Information on the survey can be found at http://cfa-www.harvard.edu/~huchra/2mass/. In the north, we are once again using the FLWO 1.5-m and are supplementing those

Figure 6. The 6dF fiber plate holder for the 1.2-m SRC Schmidt telescope. This instrument has 100+ fibers over a 30 square degree field-of-view and is one of the most efficient means available for doing wide-area redshift surveys.

observations with observations at the McDonald Observatory, Arecibo, Nancay and Green Bank. In the south, we are using a panoply of small telescopes: the 1.5-m at CTIO, the 1.5-m at ESO and the 1.2-m SRC Schmidt in Australia. The key redshift program in the south is again being done on a revitalized small telescope. 2MASS is the object catalog for the 6 Degree Field Survey (6dF). This project utilizes a CCD spectrograph and the multi-fiber system that places ~120 fibers in the 30 square-degree field of the Schmidt (Figure 6). Typical exposure times for K=12.5 magnitude galaxies are less than 2 hours per field, so on good nights it is possible to obtain well over 500 galaxy redshifts.

As of this writing, redshifts are available for over 90% of the galaxies in 2MASS brighter than K_S=11.25 over the whole sky and above galactic latitude 5° (Figure 7). We expect to complete the 11.25 magnitude sample of about 25,000 galaxies by the end of 2002, and the 12.2 magnitude sample of 125,000 galaxies by 2005. The full million will take me past retirement! We have a preliminary measurement of the K-band galaxy luminosity function (Kochanek et al. 2001), and expect to have the local density field analysis of the 11.25 mag sample by the end of this year. Once again a triumph for small telescopes.

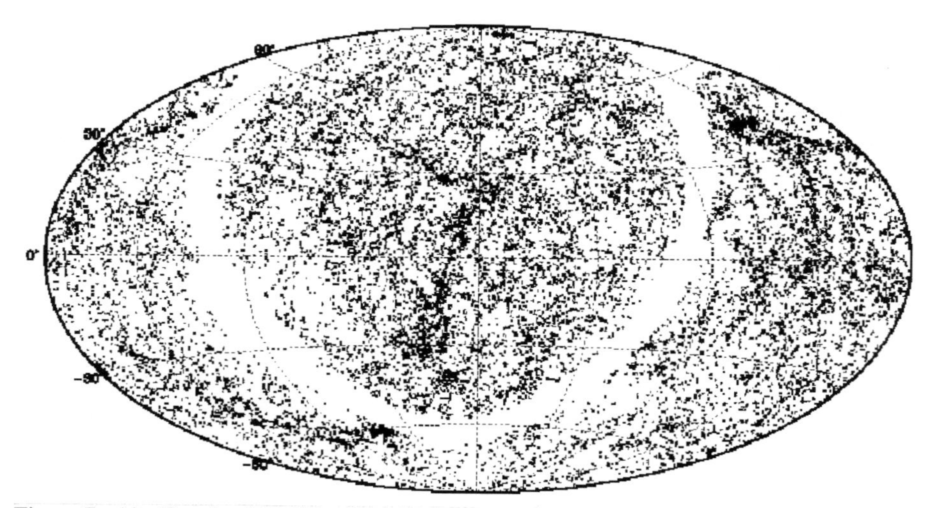

Figure 7. Aitoff projection of the 2MASS galaxies brighter than $K_S=11.25$ with measured redshifts, as of March 2002. There are about 20,000 objects in this plot. Some old friends are readily apparent including the Local Supercluster near the center of the plot, and the Pisces-Perseus supercluster in the upper right.

6. MODERN DAY MUSINGS

There is a long and glorious history of important and fundamental research with small telescopes. Even in cosmology, small telescopes have been key in recent work on the Hubble constant and on mapping large-scale structure. The IRTF relation was "discovered" on the KPNO 0.9-m and calibrated on the mighty KPNO 0.1-m; large surveys such as the CfA Redshift Survey were done on small telescopes (the FLWO 1.5-m) and it is not possible to assess the great contribution to astronomy made by the Palomar Sky Survey, which was done on a 1.2-m telescope. Small telescopes are absolutely necessary for key projects, trying out novel ideas, testing instrumentation, teaching and just the general practice of astronomy.

However, there is no doubt that "the times, they are a-changing." There are numerous issues that need to be addressed in assessing the need for small telescopes in general and the need for small telescopes at the National Observatories. A short list includes the need for access, amateur astronomers, federal versus state versus private funding, the synergy of small telescopes feeding large telescopes, and last but not least, budgets— operations, site costs, modern instrumentation, peripherals.

The good news is that, despite all this, many new small telescopes are being built. And, amateurs now run 1-m class telescopes. The bad news is that with fixed budgets, the National Observatories can no longer "easily" provide access to every type of telescope and instrument. There is no right of access to facilities of every kind for every scientist; the NSF exists to enable science in general. A more positive way of thinking about our current problem at NOAO is that it's really an opportunity. A long time ago I once said "the 4-m is the 60-inch of the future." The astronomical community is getting the Gemini telescopes plus new access to 4-m telescopes like WIYN and SOAR. In reality, our national facilities are just getting bigger by a factor of two! We're redefining "small."

If the community wants to push for more or continued support for small telescopes, the issues to address *must* be science based. We must take the high road. Individual key projects like the Sloan DSS, training and education, the development of special purpose consortia (not Federally funded) are all good arguments. It is absolutely clear that small telescopes are needed and indeed are a national treasure. New small telescopes are being built right now, like the 2MASS telescopes, justified by important scientific programs best done on small telescopes. It is also clear that the community will need some creativity in finding funding sources and will likely need to find new ways of operating them.

7. ACKNOWLEDGMENTS

This work was supported in part by NASA for the 2MASS project and by the Smithsonian Institution. I would like to thank my colleagues in this work, Mike Skruskie, Tom Jarrett, Tom Chester, Roc Cutri, Steve Schneider, Jeff Mader and Lucas Macri from the 2MASS galaxy team, and Rae Steining, the fearless 2MASS project manager. As always thanks are due to Marc Davis, Margaret Geller and Michael Strauss for LSS work, to Will Saunders, Quentin Parker, Matthew Colless, Brian Boyle and the rest of the 6dF team, and to the support staffs of all the observatories involved in these cosmological programs.

8. REFERENCES

Aaronson, M., Mould, J. & Huchra, J. 1979, ApJ 229, 1
Aaronson, M., Huchra, J., Mould, J., Schechter, P. & Tully, B. 1982, ApJ 258, 64
Baker, J., Davis, M., Strauss, M., Lahav, O., & Santiago, B. 1998, ApJ 508, 6
Becker, R. et al 2002, AJ in press (astro-ph/0108097)
Chincarini, G. & Rood, H. 1971, ApJ 168, 321
Davis, M. & Huchra, J. 1982, ApJ 254, 437
Dekel, A., Bertschinger, E., Yahil, A., Strauss, M., Davis, M., Huchra, J. 1993, ApJ 412, 1
de Lapparent, V., Geller, M. & Huchra, J. 1986, ApJL 302, L1

Djorgovski, S., Castro, S., Stern, D., Mahabal, A. 2002, ApJL in press (astro-ph/0108069)
Fazio, G. et al. 1988, private communication (NIRAS proposal)
Fisher, K., Huchra, J., Strauss, M., Davis, M., Yahil, A. & Schlegel, D. 1995, ApJS 100, 69
Geller, M. & Huchra, J. 1989, Science 246, 897
Gott, J. R. & Turner, E. 1976, ApJ 209, 1
Groth, E. & Peebles, P. J. E. 1977, ApJ 217, 385
Hubble, E. 1926, ApJ 63, 236
Hubble, E. 1929, PNAS 15, 168
Huchra, J. 2001, Mapmaker, Mapmaker Make me a Map, in Our Universe, A. Stern, ed., (Cambridge: Cambridge)
Kleinmann et al. 1994, in Infrared Astronomy with Arrays: The Next Generation, Ian McLean, ed., (Dordrecht: Kluwer), p. 219
Kochanek, C., Pahre, M., Falco, E., Huchra, J., Mader, J., Jarrett, T., Chester, T., Cutri, R. & Schneider, S. 2001, ApJ 560, 566
Lauberts, A. 1982, The ESO Uppsala Survey of the ESO (B) Atlas, ESO, Garching
Lynden-Bell, D., Faber, S., Burstein, D., Davies, R., Dressler, A., Terlevich, R. & Wegner, G. 1988, ApJ 326, 19
Lynden-Bell, D. & Lahav, O. 1988, in Large Scale Motions in the Universe, Rubin & Coyne, eds. p. 187
Marzke, R., Huchra, J. & Geller, M. 1994a, ApJ 428, 43
Marzke, R., Geller, M., Huchra, J. & Corwin, H. 1994b, AJ 108, 437
Neugebauer, G. & Leighton, R. 1969, The Two-Micron Sky Survey. A Preliminary Catalogue, NASA Special Publication
Peebles, P. J. E. & Hauser, M. 1974, ApJS 28, 19
Sandage, A. & Tammann, G. 1981, A Revised Shapley-Ames Catalog of Bright Galaxies, (Washington: Carnegie)
Santiago, B. et al. 1996, ApJ 461 38
Schechter, P. 1976, ApJ 203, 297
Shane, C. D. & Wirtanen, C. 1967, Pub. Lick Obs. 22, Part I
Shapley, H. & Ames, A. 1932, Annals of the Harvard College Observatory 88, No 2, p 43
Sharp, N. 1986, PASP 98, 740
Shectman, S. & Hiltner, W. 1976, PASP 88, 960
Schmoldt, I. et al. 1999, AJ 118, 1146
Tifft, W, & Gregory, S. 1976, ApJ 205, 696
Tonry, J., Blakeslee, J., Ajhar, E. & Dressler, A. 2000, ApJ 530, 625
Vorontsov-Velyaminov, B. A., Krasnogorska, A. and Arihpova, V. 1962-74, Morphological Catalogue of Galaxies, (Moscow: Moscow State University) (MCG)
Zwicky, F., Herzog, Wild, P., Karpowicz, M., & Kowal, C. 1961-1968 Catalogue of Galaxies and of Clusters of Galaxies, (Pasadena: California Institute of Technology)

Index

- **Volume 270**: **Dayside and Polar Cap Aurora**
  Authors: Per Even Sandholt, Herbert C. Carlson, Alv Egeland
  Hardbound, ISBN 1-4020-0447-8, July 2002
- **Volume 269**:**Mechanics of Turbulence of Multicomponent Gases**
  Authors: Mikhail Ya. Marov, Aleksander V. Kolesnichenko
  Hardbound, ISBN 1-4020-0103-7, December 2001
- **Volume 268**:**Multielement System Design in Astronomy and Radio Science**
  Authors: Lazarus E. Kopilovich, Leonid G. Sodin
  Hardbound, ISBN 1-4020-0069-3, November 2001
- **Volume 267: The Nature of Unidentified Galactic High-Energy Gamma-Ray Sources**
  Editors: Alberto Carramiñana, Olaf Reimer, David J. Thompson
  Hardbound, ISBN 1-4020-0010-3, October 2001
- **Volume 266: Organizations and Strategies in Astronomy II**
  Editor: André Heck
  Hardbound, ISBN 0-7923-7172-0, October 2001
- **Volume 265: Post-AGB Objects as a Phase of Stellar Evolution**
  Editors: R. Szczerba, S.K. Górny
  Hardbound, ISBN 0-7923-7145-3, July 2001
- **Volume 264: The Influence of Binaries on Stellar Population Studies**
  Editor: Dany Vanbeveren
  Hardbound, ISBN 0-7923-7104-6, July 2001
- **Volume 262: Whistler Phenomena**
  Short Impulse Propagation
  Authors: Csaba Ferencz, Orsolya E. Ferencz, Dániel Hamar, János Lichtenberger
  Hardbound, ISBN 0-7923-6995-5, June 2001
- **Volume 261: Collisional Processes in the Solar System**
  Editors: Mikhail Ya. Marov, Hans Rickman
  Hardbound, ISBN 0-7923-6946-7, May 2001
- **Volume 260: Solar Cosmic Rays**
  Author: Leonty I. Miroshnichenko
  Hardbound, ISBN 0-7923-6928-9, May 2001
- **Volume 259: The Dynamic Sun**
  Editors: Arnold Hanslmeier, Mauro Messerotti, Astrid Veronig
  Hardbound, ISBN 0-7923-6915-7, May 2001
- **Volume 258: Electrohydrodynamics in Dusty and Dirty Plasmas**
  Gravito-Electrodynamics and EHD
  Author: Hiroshi Kikuchi
  Hardbound, ISBN 0-7923-6822-3, June 2001
- **Volume 257: Stellar Pulsation - Nonlinear Studies**
  Editors: Mine Takeuti, Dimitar D. Sasselov
  Hardbound, ISBN 0-7923-6818-5, March 2001
- **Volume 256: Organizations and Strategies in Astronomy**
  Editor: André Heck
  Hardbound, ISBN 0-7923-6671-9, November 2000
- **Volume 255: The Evolution of the Milky Way**
  Stars versus Clusters
  Editors: Francesca Matteucci, Franco Giovannelli
  Hardbound, ISBN 0-7923-6679-4, January 2001

- **Volume 254:** **Stellar Astrophysics**
  Editors: K.S. Cheng, Hoi Fung Chau, Kwing Lam Chan, Kam Ching Leung
  Hardbound, ISBN 0-7923-6659-X, November 2000
- **Volume 253:** **The Chemical Evolution of the Galaxy**
  Author: Francesca Matteucci
  Hardbound, ISBN 0-7923-6552-6, May 2001
- **Volume 252:** **Optical Detectors for Astronomy II**
  State-of-the-art at the Turn of the Millennium
  Editors: Paola Amico, James W. Beletic
  Hardbound, ISBN 0-7923-6536-4, December 2000
- **Volume 251:** **Cosmic Plasma Physics**
  Author: Boris V. Somov
  Hardbound, ISBN 0-7923-6512-7, September 2000
- **Volume 250:** **Information Handling in Astronomy**
  Editor: André Heck
  Hardbound, ISBN 0-7923-6494-5, October 2000
- **Volume 249:** **The Neutral Upper Atmosphere**
  Author: S.N. Ghosh
  Hardbound, ISBN 0-7923-6434-1, July 2002
- **Volume 247:** **Large Scale Structure Formation**
  Editors: Reza Mansouri, Robert Brandenberger
  Hardbound, ISBN 0-7923-6411-2, August 2000
- **Volume 246:** **The Legacy of J.C. Kapteyn**
  Studies on Kapteyn and the Development of Modern Astronomy
  Editors: Piet C. van der Kruit, Klaas van Berkel
  Hardbound, ISBN 0-7923-6393-0, August 2000
- **Volume 245:** **Waves in Dusty Space Plasmas**
  Author: Frank Verheest
  Hardbound, ISBN 0-7923-6232-2, April 2000
- **Volume 244:** **The Universe**
  Visions and Perspectives
  Editors: Naresh Dadhich, Ajit Kembhavi
  Hardbound, ISBN 0-7923-6210-1, August 2000
- **Volume 243:** **Solar Polarization**
  Editors: K.N. Nagendra, Jan Olof Stenflo
  Hardbound, ISBN 0-7923-5814-7, July 1999
- **Volume 242:** **Cosmic Perspectives in Space Physics**
  Author: Sukumar Biswas
  Hardbound, ISBN 0-7923-5813-9, June 2000
- **Volume 241:** **Millimeter-Wave Astronomy: Molecular Chemistry & Physics in Space**
  Editors: W.F. Wall, Alberto Carramiñana, Luis Carrasco, P.F. Goldsmith
  Hardbound, ISBN 0-7923-5581-4, May 1999
- **Volume 240:** **Numerical Astrophysics**
  Editors: Shoken M. Miyama, Kohji Tomisaka, Tomoyuki Hanawa
  Hardbound, ISBN 0-7923-5566-0, March 1999
- **Volume 239:** **Motions in the Solar Atmosphere**
  Editors: Arnold Hanslmeier, Mauro Messerotti
  Hardbound, ISBN 0-7923-5507-5, February 1999